Biological Invasions in the South American Anthropocene

Fabián M. Jaksic • Sergio A. Castro

Biological Invasions in the South American Anthropocene

Global Causes and Local Impacts

Springer

Fabián M. Jaksic
Center of Applied Ecology
and Sustainability
Pontificia Universidad Católica de Chile
Santiago, RM - Santiago, Chile

Sergio A. Castro
Universidad de Santiago de Chile
Santiago, RM - Santiago, Chile

Translation from the Spanish language edition: Invasiones biologicas en Chile: causas globales e impactos locales by Fabián M. Jaksic & Sergio A. Castro © Ediciones Universidad Católica de Chile, 2014. Original Publication ISBN 978956141406-8.

ISBN 978-3-030-56381-3 ISBN 978-3-030-56379-0 (eBook)
https://doi.org/10.1007/978-3-030-56379-0

Jointly published with EdicionesUC

This Springer imprint is published by the registered company Springer Nature Switzerland AG
The registered company address is: Gewerbestrasse 11, 6330 Cham, Switzerland

I dedicate this book with love to the Jaksic Gil family.

Fabián M. Jaksic

I thank and dedicate this book to my family, who have always expressed support and interest in the work I do. To Paola, Camilo, Juan Pablo, and Juan.

Sergio A. Castro

Preface

The book *Biological Invasions in the South American Anthropocene: Global Causes and Local Impacts* is organized into two parts. The first comprises eight self-contained chapters, which provide a conceptual and bibliographic review of the field of invasion biology. Under an explicit theoretical framework, tools are provided to understand the complexity of the phenomenon of biological invasion, with emphasis on ecological aspects. This phenomenon is visualized as a process of geographical expansion that occurs in sequential stages, ranging from the initial dispersion or departure from the native range to the arrival and subsequent expansion in the invaded area.

In the second part of the text, we review the ecology of several exotic species that have invaded Chilean and Argentine territory, which in the light of the available information have resulted in useful models as case studies. In these chapters, the reader will find the development of extensive analyses focused on ecology, as well as on the historical, economic, management, and social perspectives of exotic species in new territories. In particular, we analyse the invasion of the European rabbit (*Oryctolagus cuniculus*), the wild boar (*Sus scrofa*), the North American beaver (*Castor canadensis*), the American mink (*Neovison vison*), and the exotic fish (several Salmonid species).

Santiago, Chile

Fabián M. Jaksic
Sergio A. Castro

Acknowledgements

We thank the Center for Applied Ecology and Sustainability (CAPES) for the opportunity to undertake writing of this book and CONICYT PIA/BASAL FB0002 for financing.

About This book

This book provides a conceptually organized framework to understand the phenomenon of biological invasions at the Anthropocene global scale. Most advances toward that aim have been provided from North American and European researchers, with fewer contributions from Australia and South Africa. Here we fill the void from the Neotropics, focusing on the research experience in South American countries, with a strong emphasis on Argentina and Chile.

The text is divided into two parts: The first half comprises self-contained chapters, providing a conceptual, bibliographic, and empirical foundation in the field of invasion biology, from an Anthropocene perspective. The second half reviews the ecology, biogeography, and local impacts in South America of exotic species groups (European rabbit, Eurasian wild boar, Canadian beaver, North American mink, and Holarctic freshwater fishes), which are shown to be useful models for case studies of global relevance.

Contents

Part II Case Studies

About the Authors

Fabián M. Jaksic has a degree in biology from the University of Chile, a PhD in zoology from the University of California (Berkeley), is an American Ecological Society's certified senior ecologist, and a member of the Chilean Academy of Sciences. He is currently a Professor and Director of the Center of Applied Ecology and Sustainability of the Pontificia Universidad Católica de Chile (CAPES-UC), and is the author of over 230 WoS-indexed scientific articles, over 50 book chapters, and of 12 books. Jaksic has also received numerous awards and prizes, including Fullbright and Guggenheim scholarships, Chile's Presidential Chair in Science, and Chile's National Prize in Natural Sciences.

Sergio A. Castro holds a BS degree and a PhD in biological sciences with a major in ecology from the Pontificia Universidad Católica de Chile. He is currently an Associate Professor at the School of Chemistry and Biology of the Universidad de Santiago de Chile. Castro has developed research in the field of plant ecology and biogeography, as well as on plant and vertebrate invasions. He is the author of 50 WoS scientific articles and 6 book chapters and coauthor of 3 books. He is also an Associate Researcher at the Center of Applied Ecology and Sustainability of the Pontificia Universidad Católica de Chile (CAPES-UC).

Part I
Conceptual Framework

Chapter 1
Introduction to the Phenomenon of Biological Invasions

1.1 Biological Invasions

Biological invasions can be defined as phenomena in which a species extends its geographic range, occupying regions in which it was not previously present (Elton 1958; Mack et al. 2000; Kolar and Lodge 2001; Lockwood et al. 2007). Despite the apparent simplicity of this definition, from an ecological point of view, invasions are a complex and, by the way, fascinating type of phenomenon. The complexity of biological invasions lies in the fact that the geographic range, i.e., the geographic area occupied by a certain species (Table 1.1), is a spatial and temporal dynamic attribute, the size of which depends on ecological and evolutionary processes operating under different scales and hierarchies (Brown and Lomolino 1998; Sax et al. 2005; Lockwood et al. 2007).

In the case of species whose populations are in clear population decline or in danger of extinction, for example, the structure of their range has been affected not only by the reduction in size but also by the segmentation of their continuity, which results in fragmented subpopulations in danger of disappearing. Biological invasions are the opposite phenomenon, where the range is growing or expanding (Fig. 1.1). From these considerations, it is possible to establish that both invasions and extinctions are related phenomena as long as they are conceived as different expressions of range dynamics.

Nevertheless, expansions of the geographical range can occur in various ways, depending closely on the mode of dispersal of organisms. On the one hand, the range can grow gradually through short-distance dispersal, where individuals gradually occupy space from the distributional edge (Fig. 1.1a), and thus it grows progressively. On the other hand, if organisms can be dispersed over long distances, the expansion of the range will show a saltational dynamics so that it could now grow from several fronts simultaneously (Fig. 1.1b). Of course, both types of dispersal (short and long distance) can occur in combination, depending on the species or type of organism (Fig. 1.1c). Additionally, the extent of the range not only depends

F. M. Jaksic, S. A. Castro, *Biological Invasions in the South American Anthropocene*, https://doi.org/10.1007/978-3-030-56379-0_1

Table 1.1 Definition of some terms in bioinvasion ecology, in the sense that they will be used in the text

Alien species: An exotic species that inflicts an environmental impact
Archaeobiota: Set of species introduced in pre-Columbian times (approximately before 1500). It can be used relative to flora (i.e., archaeoflora) or fauna (i.e., archaeofauna)
Dispersal: Process of propagation, migration, or spatial displacement of organisms, also called propagules. Primary dispersion: Initial dispersion process that occurs from the original range of distribution to another region. Secondary dispersion: Process of dispersion that involves displacement into the recipient area
Exotic species: Species present in a region to which it has arrived. Synonyms: Non-native species, non-indigenous species
Founding colony: Group of individuals that arrive in a certain area
Introduced species: a species that has been inoculated into a given region, by either voluntary or involuntary human action
Introduction: Inoculation of organisms by human action
Invasive species: This concept will be used under two meanings: (a) exotic species in the process of expansion in a container area; (b) exotic species that inflicts an impact on the environment where it is found
Minimum residence time: The lapse of time from the first registration of an exotic species in a territory to the present time
Native species: Species whose range originally belongs to a certain area. Synonyms: Autochthonous species, indigenous
Naturalized species: Exotic species whose population in the colonized area is able to reproduce spontaneously, without direct human intervention. Synonyms: Established species or feral species
Neobiota: Set of species introduced in post-Columbian times (approximately after 1500). It can be used relative to flora (i.e., neoflora) or fauna (i.e., neofauna)
Propagules: Organisms or biological unit that is capable of dispersal
Range: Geographical distribution of a species. A distinction can be made between primary and secondary range as a reference to the original distribution or in the invaded region, respectively
Spread: Expansion of the geographical range
Unnaturalized species: An exotic species whose population in the colonized area is not capable of spontaneous reproduction

on the organisms and their dispersal capacity. Indeed, the dispersion over short or long distances can be affected by the presence of spatial barriers that could constitute insurmountable pitfalls to the expected expansion. This is how different types of barriers can be differentiated, some of them permeable, others semi-permeable, and still others completely impervious to trespassing. In this way, it can be established that the nature of these barriers and their spatial distribution may also affect range growth.

Another important factor to consider is the time scale at which biological invasions occur. These may involve long time scales (e.g., thousands of years) associated, for example, with changes in geographical or geological conditions that, after removing physical barriers, end up facilitating dispersal. This is the case of the Great American Interchange (Webb 1991), an event that implied the reciprocal invasion of the South American and North American biota after the formation of the

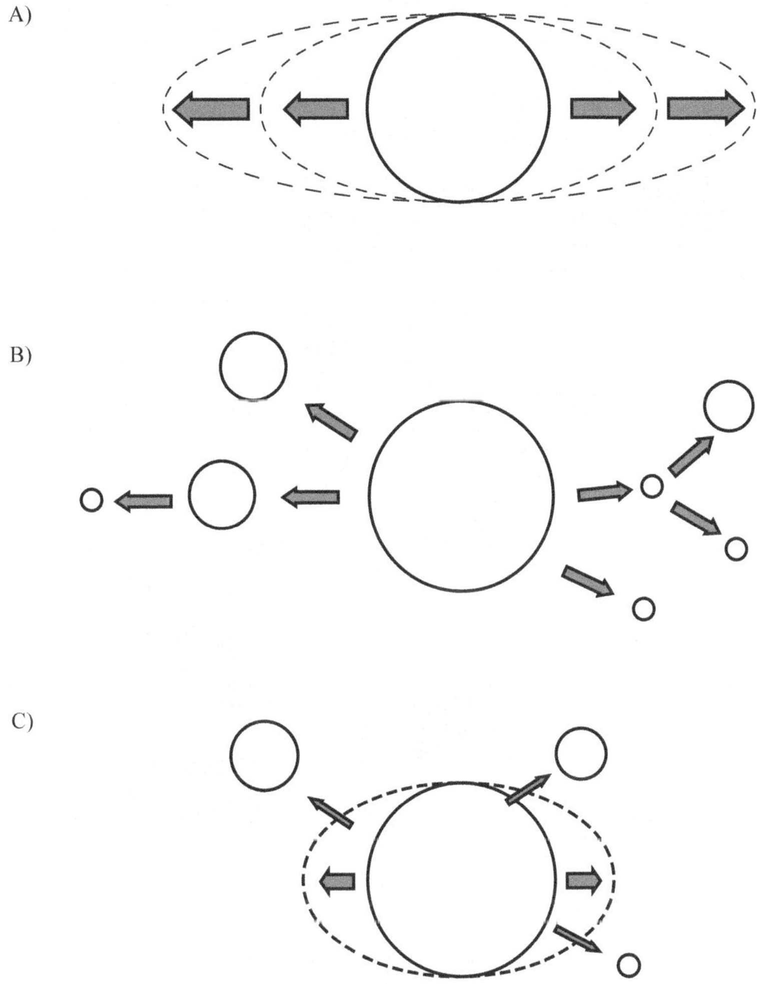

Fig. 1.1 Geographical range expansion modes. (**a**) Gradual spread mediated by short distance dispersion from the periphery of the range; (**b**) saltational spread mediated by long distance dispersion; (**c**) spread combining short and long distance dispersion

Isthmus of Panama (Fig. 1.2). As a result of this exchange, South America currently contains a mixed fauna and flora represented by lineages originated in situ, which coexist with lineages whose evolutionary ancestors originated in the Nearctic or Palearctic regions.

Another well-documented example in Chile is that which describes changes in the range of distribution of flora and vegetation in response to climatic modifications, specifically those changes that occurred during glacial and interglacial periods (Villagrán et al. 1998). It is known that during the glaciations the flora was distributed in non-glaciated sites known today as "refuges" and that as the glaciers

Fig. 1.2 Great American Interchange as representation of processes of reciprocal invasion of species involving South America and North America (Webb 1991)

retracted after the progressive warming of the earth's atmosphere, the species colonized areas free of ice. Unlike the previous example – in which a geological bridge connected two continents – in this second case, it is observed that the modification of the climate is the determining factor of the retreat of the glacial hull and the expansion of the geographical range of the organisms.

Other types of invasions have occurred in much shorter time lapses. In fact, in a few decades numerous species have been able to arrive and settle in remote areas where they were not previously found. Unlike the previous examples, these invasions are fundamentally related to the activities of exchange of goods and services that humans develop on the planet, a historical moment that has been called the Anthropocene (Malhi 2017). To illustrate with an example, it is possible to mention the case of the California poppy (*Eschscholzia californica*) whose original distribution corresponds to the west coast of the United States and part of Mexico (Fig. 1.3a) but which is currently present in several locations in southern Europe, Asia, Australia, and South America (Fig. 1.3b, c; Randall 2002). Its presence in Chile

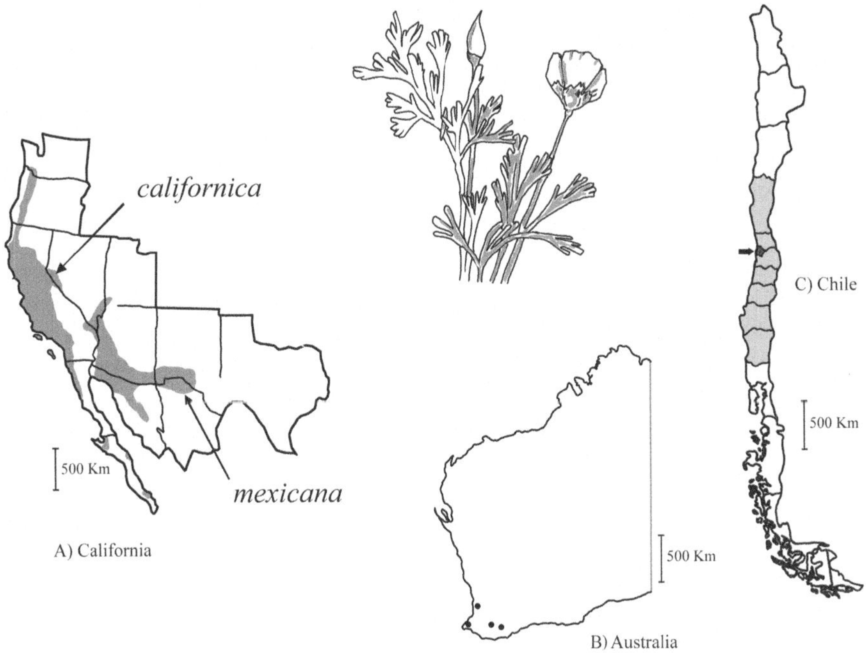

Fig. 1.3 Invasion of *Eschscholzia californica* in Australia and Chile (Randall 2002). (A) Original distribution of *Eschscholzia californica* on the west coast of North America, where two subspecies are recognized, *Eschscholzia californica californica* and *Eschscholzia californica mexicana*; (B) distribution of *Eschscholzia californica* in Australia, indicating the sites where specimens have been collected on the southwest coast of the continent; (C) distribution of *Eschscholzia californica* in Chile, indicating the geographic extension (about 1200 km, approximately) that the species currently occupies; the arrow indicates the locality where it would have been introduced initially

dates back to the first half of the nineteenth century, when it was introduced as an ornamental plant (Johow 1948). However, after its escape this species expanded territorially in the country, occupying at present a latitudinal extent greater than 1200 km and an altitudinal distribution that reaches 2000 m elevation (Arroyo et al. 2000; Peña-Gómez et al. 2014) (Fig. 1.3c).

Another case of an invasion that has taken place in the very short term is exemplified by the presence of the European rabbit (*Oryctolagus cuniculus*) currently present in different regions of the globe where it has arrived after being introduced (Flux and Fullagar 1992). It has been indicated that from the Iberian Peninsula (original range of the species), it was taken to North Africa and that the Phoenicians introduced it into Egypt and Asia Minor. At the same time, the Romans were responsible for taking it to northern Europe, and the Normans would have introduced it to Great Britain. Finally, Spanish, French, and Dutch settlers – between the seventeeth and nineteenth centuries – introduced it to all their colonies and to some 600 oceanic islands (Crosby 2004; Camps-Rabadà 2008). Interestingly, the practice of introducing domestic animals such as the European rabbit was accompanied by the involuntary introduction of other less attractive small mammal species included in the cabotage, such as rats and mice (*Rattus norvegicus*, *Rattus rattus*, and *Mus musculus*).

1.2 Text Focus

This text focuses on the latter type of invasions, i.e., those facilitated directly or indirectly by human beings. At least five reasons support our effort. First, it is now considered that the number and rate at which different species are introduced from one region to another, as well as the regions and geographic distances involved, are unprecedented facts in the history of terrestrial biota (Elton 1958; Williamson 1996). This is what, from an ecological point of view, is highlighted as the Anthropocene (Malhi 2017), characterized by the ubiquity and depth of human domination over the planet's ecosystems (Malhi 2017). In South America, for example, since European colonization (which has been going on for slightly more than 500 years), it has been easier for flora and fauna to arrive from regions as distant as Europe, Asia, Africa, and Australia, a situation that could not have occurred without human assistance. Part of this flora, fauna, and diversity of microorganisms has been integrated into the ecosystems to which they have arrived, establishing interactions of all kinds, a fact that from the biological point of view is not trivial.

Second, recent antecedents point out that a direct result of this relocation of species is that the global distribution of biodiversity is in the process of rearranging or reconfiguration (McNeely 2001), all in a "real-time", current, and contingent framework. Again, it is necessary to emphasize that this is one more manifestation of the Anthropocene (Malhi 2017). The occurrence of invasions has given rise to such a mixture of species, biological groups, and geographic origins that some authors have called this period the Homogecene or the conformation of a New Pangea (Mooney and Cleland 2001; Rosenzweig 2001). Because biological invasions currently involve the exchange of species across the globe, one of the most conspicuous manifestations of global change has been recognized in this process (Drake et al. 1989; Vitousek et al. 1996), a human imprint that will be indelible for the remainder of the planet's history.

Third, invasions represent a default modification of biodiversity, since the immediate result is that the number of species in a region increases (Williamson 1996). Although most exotic species can coexist with native species diversity, a small number of the former can modify the structure and functioning of invaded ecosystems, sometimes with undesirable results (Drake et al. 1989; Williamson 1996). In general terms, it has been indicated that the presence of exotic species can lead to changes in the physical-chemical properties of ecosystems, nutrient cycles, productivity, trophic chains, and compositional balance of communities or alter the dynamics of disturbances and succession, among many other aspects. In some cases, the presence of exotic species can cause the local or global extinction of native species, so they are usually considered a threat to biodiversity conservation (Wilcove et al. 1998), although controversy exists (Hulme et al. 2015; Thomas and Palmer 2015).

Fourth, beyond the field of ecology, exotic species are responsible for important economic effects, both desired and undesired by humans (Pimentel et al. 2005). For the United States, for example, it has been estimated that 98% of its food system rests on exotic species, which brings an economic benefit of around 800 billion dollars (Pimentel et al. 2005), while the damage caused by species that act as weeds or pests accumulate costs of the order of 97 billion dollars (Pimentel et al. 2005). These antecedents have served as tools to promote border control policies aimed at controlling the arrival of new species in certain countries, as well as keeping at bay some of the species that are already introduced. In fact, there are so-called black lists that include species that under no point of view should be accepted for entry into certain countries.

Fifth, and finally, it is possible to recognize that some introduced species may affect human health (Williamson 1996). In fact, the spread of infectious diseases can be conceived as invasion events, where pathogenic organisms disperse (i.e., expand their range) to other hosts in the population (Shigesada and Kawasaki 1997). The harmful effect of exotic species may be derived from their involvement as vectors of human pathogens (e.g., viruses, bacteria) or by the production of substances or structures that impact *Homo sapiens* (e.g., toxins, spores, and allergenic pollens).

1.3 Some Fundamental Considerations in Invasion Ecology

Many ecologists perceive invasions as intrinsically negative events. This fact that was noted early by Rosenzweig (2001) is still noticeable in academic circles. Much of this perception can be seen, for example, in the dissemination of initiatives to protect native flora and fauna, in control and eradication programs for exotic species, as well as in policies to prevent the import or export of species. In their historical analysis, Davis et al. (2001) identified the roots of this preconception. These authors pointed out that it would have been cultivated under the post-World War II psychology, where the term "invasion" was evidently associated with that of "threat". Thus, by the 1950s Charles S. Elton (Fig. 1.4), the recognized promoter of invasive ecology in its modern facet, published his book *The Ecology of Invasions by Animals and Plants*, where he incorporated concepts with warlike nuances, and to a certain extent catastrophic, which have ended up permeating not only the scientific community but also the general public (Davis et al. 2001; Cadotte 2006).

From a biological perspective, it seems prudent to consider biological invasions as phenomena with no prior intrinsic value. Rather, its implications or consequences will depend on the framework or context in which the phenomenon is analyzed. For example, conservation biologists are more likely to view invasions as damaging phenomena, while environmental managers interested in the introduction of a species for exploitation will see this differently.

Fig. 1.4 Charles S. Elton (England; 1900-1991), one of the most influential ecologists of the twentieth century. In 1958, he published *The Ecology of Invasions by Animals and Plants*, a text considered a cornerstone for discipline

Hand in hand with the above, another field of discussion in invasion ecology has been concerned with the use and proliferation of a diversity of terms (i.e., jargon), the use of which also merits critical discussion. Terms such as species "advene", "adventitious", "alloctone", "anthropocore", "casual", "colonizer", "ephemeral", "escaped", "established", "exotic", "feral", "foreign", "imported", "introduced", "non-indigenous", "non-native", "plague", "synanthropic", "transient", "translocated", "weed", and "wild" are often used in the literature to refer to exotic species. These terms are often used without a prior definition (Hodges 2008), so several authors have recognized the need to conceptually unify the discipline (e.g., Richardson et al. 2000; Colautti and MacIsaac 2004).

On the one hand, the same term can be used to describe different states or stages of the invasive process, while on the other hand, different terms can be used to describe the same stage of the invasion (i.e., Davis and Thompson 2001; Colautti and MacIsaac 2004). An example of the first situation is the use of the concept of "exotic species"; in some studies, it appears as a synonym for "introduced" species – i.e., applied to the set of non-native species – while in other studies it is used in a more restricted sense, for example, to refer to the set of "naturalized" species (see Richardson et al. 2000). On the other hand, two different terms referring to the same characteristic are "feral" or "naturalized", as both allude to exotic species whose populations can become self-sustaining without human assistance (Richardson et al. 2000).

Let us look at some examples of how some concepts have been used in the scientific literature. After reviewing 157 articles, Richardson et al. (2000) found that the term "naturalized species" was used as a synonym for "invasive species" in 29% of cases, while it was used as a synonym for "exotic" or "non-native" species in 25% of cases. Other uses highlighted more subtle biological differences, such as the ability to establish self-sustainable populations without human intervention in 23% of cases or the ability to establish self-sustainable populations in natural and semi-natural environments in 8% of cases.

The diversity of terms and concepts in the ecology of biological invasions undoubtedly contributes to slowing the advance of scientific knowledge (Davis and Thompson 2001; Rejmánek et al. 2002; Chew and Laubichler 2003; Colautti and MacIsaac 2004). However, beyond the scientific domain, we believe that this semantic richness can be healthily appreciated in light of considering that the phenomenon of biological invasions constitutes an arena of multiple implications and interests and that at the same time it gathers the attention of diverse actors of contemporary societies, among them, scientists (e.g., see Hodges 2008). Under this prism, it seems reasonable – at least for the moment – to live with this terminological diversity in search of establishing bridges of communication with the rest of society. The task of revising and defining these terms and concepts escapes the purposes set forth in this book. Rather, in our work we will propose a conceptual framework from the ecology of invasions, which will allow us to integrate and give a common thread to the diversity of studies in this area.

1.4 Conceptual Framework of Invasions

An important advance towards conceptual unification in invasion ecology was initially proposed by Richardson et al. (2000). Basically, these authors proposed that biological invasions can be conceived as an eminently population-based process, consisting of the crossing of barriers of diverse nature (e.g., geographic, environmental, reproductive, etc.; Fig. 1.5). This framework has had the merit of simplifying terminological complexity, which significantly reduces the number of concepts needed to describe the phenomenon. In this way, terms such as "exotic," "casual," and "naturalized" can be unequivocally applied to species at different stages of crossing these barriers (Fig. 1.5). Although this conceptual model was originally developed specifically for plants, its general structure may be suitable for other types of organisms (Carlton 1985; Kolar and Lodge 2001; Sakai et al. 2001).

Some years later, Colautti and MacIsaac (2004) recognized some limitations to the previous approach, emphasizing the need to develop a neutral terminology, given the persistent imprecisions, preconceptions, and historical load of some concepts. To this end, they proposed a model based on seven states, identified as 0, I, II, III, IVa, IVb, and V (see Fig. 1.6). Each of these states constitutes a stage that describes the population situation in which there is a species. Thus, each exotic species present (or potentially present) in a given region could be ascribed to one of these categories, ranging from stages prior to primary dispersal (State 0) to expansion of the secondary range in the host community (State V). Despite the

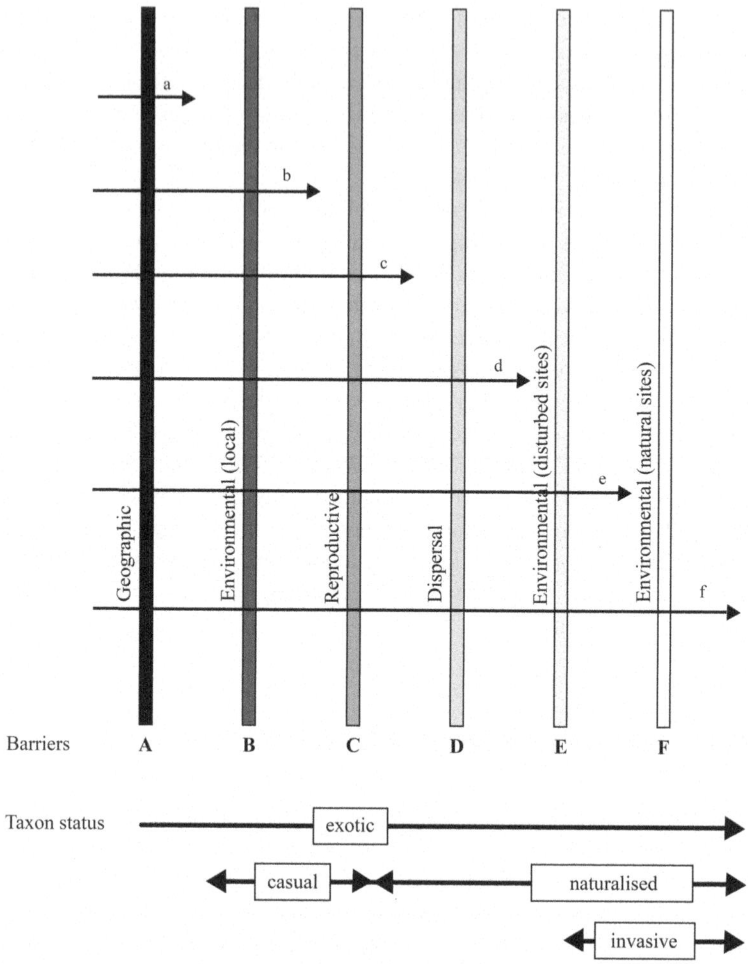

Fig. 1.5 Major demographic barriers that restrict plant expansion and the status they achieve after transposing crossing such barriers (de Richardson et al. 2000). (**a**) Geographic barrier; (**b**) environmental barriers at the point of introduction; (**c**) reproductive barriers; (**d**) local and regional dispersion barriers; (**e**) environmental barriers at disturbed sites; and (**f**) environmental barriers at natural sites

attractiveness and potential usefulness of this proposal, it has not been adopted by the scientific community more than a decade after its nomination.

From the previous discussion, we rescued the conception of the transfer of stages of type A → B, where A and B correspond to two population states mediated by a process of transit between them, represented by the arrow (→). These transit processes allow the crossing of barriers of different nature, such as those already distinguished by Richardson et al. (2000), geographic, population, and reproductive, among others (Fig. 1.5).

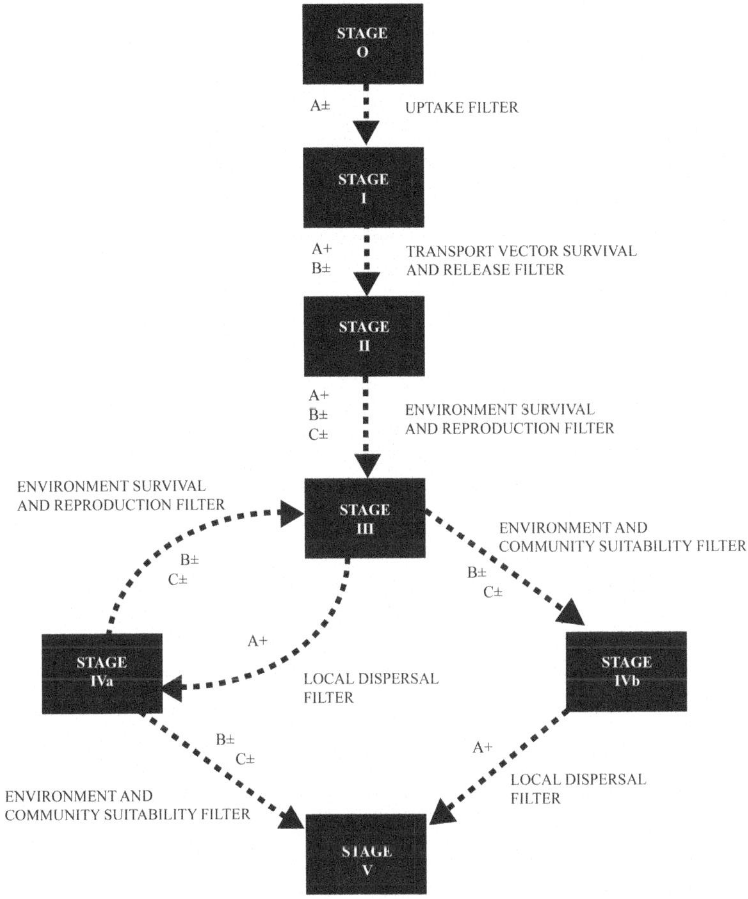

Fig. 1.6 Conceptual framework proposed by Colautti and MacIsaac (2004) to define terms in invasion ecology. It begins with propagules residing in a region (potential invaders in State 0). They must pass through a series of filters that prevent the transition to later phases. States III to V are subdivided according to the abundance and distribution of non-native species. Within this framework, a non-native species may have a localized distribution and be sparsely abundant (State III), widespread but sparsely abundant distribution (State IVa), localized but abundant distribution (State IVb), or generalized and abundant distribution (State V). Three factors can affect the likelihood of transition for each state: (**a**) propagule pressure, (**b**) environmental requirements (physical-chemical), and (**c**) biotic interactions with the community

Initially, our scheme distinguishes native species in their original range of distribution, which may eventually be dispersed (i.e., primary dispersal; see Table 1.1) beyond this range (Fig. 1.7). This distinction is conceptually relevant, as it recognizes that the study of invasions can be approached from the region of origin, where native species that are exotic in other regions reside, as well as native species that could potentially be dispersed to other regions. For the transition from one stage to the next, dispersal will be required to be directly or indirectly facilitated by humans.

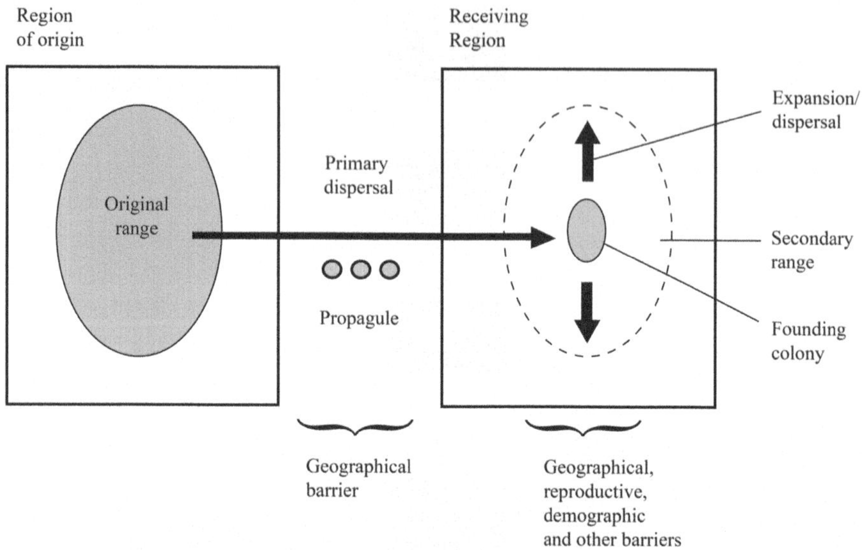

Fig. 1.7 Conceptual model of the invasion process, which distinguishes the transit of barriers of diverse nature, which must be transposed by the propagules that disperse into a recipient region, and then expand by deploying the secondary range. The founding colony is the initial group of organisms that have managed to reach the recipient community and may eventually expand its secondary range

This fact constitutes a central characteristic to define our phenomenon of interest, because – as will be seen later – in no other process will human assistance be a necessary condition.

As has traditionally been recognized, primary dispersal (Table 1.1) may be intentionally facilitated by humans, as is the case with species introduced for crop exploitation, or unintentionally, as is the case with many accidentally introduced species (Heywood 1989). Regardless of this mechanism, only a small fraction of the organisms that initiated the primary dispersal (i.e., propagules, Fig. 1.7) will survive and successfully reach the destination region (Kolar and Lodge 2001; Sakai et al. 2001). According to what we currently know of this stage, most of them will not be able to overcome this first hurdle (Lockwood et al. 2005). To illustrate this fact, Williamson (1996) suggested the so-called rule of 10, stating that about one tenth of the species that disperse finally arrive at a destination area and that, in turn, only 10% of them will be able to establish themselves as a population in their new area. Although this rule is difficult to confirm, the message to be communicated is how unlikely it will be to successfully cross this first barrier (Lockwood et al. 2005).

From this perspective, each species that has successfully arrived in a new area can be considered exotic or non-native (see Table 1.1), since its origin is in a different region (Richardson et al. 2000). At this stage, a new population or founding colony is formed (see Table 1.1), destined to extinction or establishment (Fig. 1.7). In some cases, alien species can only persist as a result of human assistance, as is the case with some intentionally propagated species. In other cases, the colony may survive and reproduce independently of human action. This distinction has made it possible to distinguish the condition of feral or naturalized species (see Table 1.1), terms that are frequently used in the literature. However, even so, these terms do not have a clear definition, because many naturalized organisms in practice are established only in anthropized environments, where they find the habitat conditions conducive to their establishment; in these cases, established populations depend indirectly on human intervention on the habitat and are not capable of establishing themselves in less anthropized environments. It is important to point out that those naturalized species that are in the process of expansion can also be recognized by the name of invasive species, alluding to the fact that they are taxa in the process of expanding their range (Table 1.1).

Finally, the persistence of exotic species over time, regardless of the mechanisms involved, may be associated with various effects, both desired and undesired. Here it is important to consider that these effects can be (though not always) proportional to the abundance and level of expansion of the secondary range. At one extreme is the archetype of an exotic species that achieves spatial domination, thus exerting profound modifications to the structure and functioning of ecological systems. At the other extreme are those species that reach low population sizes and have minimal effects on the recipient region (Williamson 1996). Thus, the concept of invasive species offers a second meaning, because it may allude to those exotic species (naturalized or not) that exert some type of impact in the secondary range (Table 1.1).

1.5 Objectives and Scope of the Text

Based on the multiple implications of biological invasions, we have tried to develop in this text a synthetic, organized, and coherent framework for the study of biological invasions, addressing conceptual problems such as those linked to the field. We have spun our story into numerous publications, some of them seminal in the discipline, others influential, which have inspired our work. Based on this, and as a central focus, we intend to organize and highlight the experience of numerous researchers in the ecology of invasions, an experience that is rescued in South America and that should permeate our society. In this way, we hope to cover the need for a text on the topic of invasion ecology from a local and regional perspective, as well as delineate the strengths and weaknesses of the research carried out to date in our South American continent, with emphasis on Chile and Argentina. In this we are encouraged by the conviction of being faced with a phenomenon of great biological interest, which in our opinion has been insufficiently addressed in this part of the globe.

As can be seen from the references we use and the scope of our work, this exercise has included a large number of articles and titles, an important part of which have been published in journals, texts, or reports with restricted circulation. At the same time, we note that some of these studies were not originally conducted under the conceptual framework presented here, so in these cases, we have adjusted them to draw our own conclusions. Studies of biological invasions are currently growing in South America (Quiroz et al. 2009), but at a lower rate than in other regions of the globe (Pyšek et al. 2008). For this reason, we hope that this publication will contribute to strengthening and stimulating research in the region.

References

Arroyo MTK, Marticorena C, Matthei O, Cavieres L (2000) Plant invasions in Chile: present patterns and future predictions. In: Mooney HA, Hobbs RJ (eds) Invasive species in a changing world. Island Press, Washington, DC, pp 385–421

Brown JH, Lomolino MV (1998) Biogeography. Sinauer, Sunderland

Cadotte MW (2006) Darwin to Elton: early ecology and the problem of invasive species. In: Cadotte MW, McMahon SM, Fukami T (eds) Conceptual ecology and invasion biology: reciprocal approaches to nature. Springer, The Netherlands, pp 15–33

Camps-Rabadà J (2008) Los primeros conejos llegaron a América iniciado noviembre de 1493. Información Veterinaria 9:21–24

Carlton JT (1985) Transoceanic and interoceanic dispersal of coastal marine organims: the biology of ballast water. Oceanogr Mar Biol Annu Rev 23:313–371

Chew MK, Laubichler MD (2003) Natural enemies: metaphor or misconception? Science 301:52–54

Colautti RI, MacIsaac HJ (2004) A neutral terminology to define invasive species. Divers Distrib 10:135–141

Crosby AW (2004) Ecological imperialism: the biological expansion of Europe, 900–1900. Cambridge University Press, Cambridge

Davis MA, Thompson K (2001) Invasion terminology: should ecologists define their terms differently than others? No, not if we want to be of any help. Ecol Soc Am Bull 82:206

Davis MA, Thompson K, Grime JP (2001) Charles S. Elton and the dissociation of invasion ecology from the rest of ecology. Divers Distrib 7:97–102

Drake JR, Mooney HA, di Castri F, Groves RH, Kruger FJ, Rejmánek M, Williamson M (1989) Biological invasions: a global perspective. In: Scientific committee on problems of the environment (SCOPE). Wiley, Chichester

Elton CS (1958) The ecology of invasions by animals and plants. University of Chicago Press, Chicago

Flux JEC, Fullagar PJ (1992) World distribution of the rabbit, *Oryctolagus cuniculus*, on islands. Mammal Rev 22:151–205

Heywood VH (1989) Patterns, extents and modes of invasions by terrestrial plants. In: Drake JA, Mooney HA, di Castri F, Groves RH, Kruger FJ, Rejmánek M, Williamson M (eds) Biological invasions: a global perspective. Wiley, Chichester, pp 31–60

Hodges KE (2008) Defining the problem: terminology and progress in ecology. Front Ecol Environ 6:35–42

Hulme PE, Pauchard A, Pyšek P, Vilà M, Alba C, Blackburn TM, Bullock JM, Chytrý M, Dawson W, Dunn AM, Essl F, Genovesi P, Maskell LC, Meyerson LA, Nuñez MA, Pergl J, Pescott OL, Pocock MJO, Richardson DM, Roy HE, Smart SM, Štajerová K, Stohlgren T, van Kleunen M, Winter M (2015) Challenging the view that invasive non-native plants are not a significant threat to the floristic diversity of Great Britain. Proc Natl Acad Sci U S A 112:2988–2989

Johow F (1948) Flora de las plantas vasculares de Zapallar. Rev Chil Hist Nat 49:8–566

Kolar CS, Lodge DM (2001) Progress in invasion biology: predicting invaders. Trends Ecol Evol 16:199–204

Lockwood JL, Cassey P, Blackburn T (2005) The role of propagule pressure in explaining species invasions. Trends Ecol Evol 20:223–228
Lockwood JL, Hoopes M, Marchetti M (2007) Invasion ecology. Blackwell Publishing, Massachusetts
Mack RN, Simberloff D, Lonsdale WM, Evans H, Clout M, Bazzaz FA (2000) Biotic invasions, causes, epidemiology, global consequences, and control. Ecol Appl 10:689–710
Malhi Y (2017) The concept of the Anthropocene. Annu Rev Environ Resour 42:25.1–25.28
McNeely JA (2001) The great reshuffling: human dimensions of invasive alien species. IUCN, Gland/Cambridge
Mooney HA, Cleland EE (2001) The evolutionary impact of invasive species. Proc Natl Acad Sci U S A 98:5446–5451
Peña-Gómez FT, Guerrero PC, Bizama G, Duarte M, Bustamante RO (2014) Climatic niche conservatism and biogeographical non-equilibrium in *Eschscholzia californica* (Papaveraceae), an invasive plant in the Chilean Mediterranean region. PLoS One 9:e105025
Pimentel D, Zúñiga R, Morrison D (2005) Update on the environmental and economic costs associated with alien-invasive species in the United States. Ecol Econ 52:273–288
Pyšek P, Richardson DM, Pergl J, Jarošik V, Sixtová Z, Weber E (2008) Geographical and taxonomic biases in invasion ecology. Trends Ecol Evol 23:237–244
Quiroz C, Pauchard A, Cavieres LA, Anderson CB (2009) Análisis cuantitativo de la investigación en invasiones biológicas en Chile: tendencias y desafíos. Rev Chil Hist Nat 82:497–505
Randall RP (2002) A global compendium of weeds. RG & FJ Richardson, Melbourne
Rejmánek M, Richardson DM, Barbour MG, Crawley MJ, Hrusa GF, Moyle PB, Randall JM, Simberloff D, Williamson M (2002) Biological invasions: politics and the discontinuity of ecological terminology. Ecol Soc Am Bull 83:131–133
Richardson DM, Pyšek P, Rejmánek M, Barbour MG, Panetta FD, West CJ (2000) Naturalization and invasion of alien plants: concepts and definitions. Divers Distrib 6:93–107
Rosenzweig ML (2001) The four questions: what does the introduction of exotic species do to diversity? Evol Ecol Res 3:361–367
Sakai AK, Allendorf FW, Holt JS, Lodge DM, Molofsky J, With KA, Baughman S, Cabin RJ, Cohen JE, Ellstrand NC, McCauley DE, O'Neil P, Parker MI, Thompson JN, Weller SG (2001) The population biology of invasive species. Annu Rev Ecol Syst 32:305–332
Sax DF, Stachowicz JJ, Gaines SD (2005) Species invasions: insights into ecology, evolution and biogeography. Sinauer Associates Incorporated, Sunderland
Shigesada N, Kawasaki K (1997) Biological invasions: theory and practice. Oxford University Press, Oxford
Thomas CD, Palmer G (2015) Non native plants add to the British flora without negative consequences for native diversity. Proc Natl Acad Sci U S A 112:4387–4392
Villagrán C, Le-Quesne C, Aravena JC, Jiménez H, Hinojosa F (1998) El rol de los cambios de clima del Cuaternario en la distribución actual de la vegetación de Chile central-sur. Bamberger Geographische Schriften 15:227–242
Vitousek PM, D'Antonio CM, Loope LL, Westbrooks R (1996) Biological invasions as global environmental change. Am Sci 84:468–478
Webb SD (1991) Ecogeography and the Great American interchange. Paleobiology 17:266–280
Wilcove DS, Rothstein D, Dubow J, Phillips A, Losos E (1998) Quantifying threats to imperiled species in the United States. Bioscience 48:607–615
Williamson M (1996) Biological invasions. Chapman & Hall, London

Chapter 2
Biological Invasions in the Anthropocene

2.1 Preamble

For millennia human beings have served as dispersing agents for other organisms (Elton 1958). Thus, it is not surprising that the patterns and processes associated with human movement across geographical, biogeographical, and political boundaries are reflected in the current distribution of plants, animals, and microorganisms (McNeely 2001; Crosby 2004). In fact, the history of human movement and its relation with exchange of goods have provided new means and routes of dispersal by which organisms mobilize (Kowarik 2003; Pyšek and Prach 2003; Perrings et al. 2005; Meyerson and Mooney 2007). For this reason, the consideration of contextual aspects, both historical and geographical, associated with human beings contribute to improving our understanding of biological invasions today (di Castri 1989). This situation is particularly true for South America, because as history has shown, this region was colonized by Europeans from the sixteenth century, a fact that led to the introduction of a number of species for domestic use that allowed the first settlers to settle more easily.

Perhaps the importance of analyzing the history and geography of invasions can be illustrated by the following example. Several authors have found that European plant species tend to be successful colonizers in the regions where they have been introduced, sometimes surpassing the representation of native or exotic species from other regions (di Castri 1989). Some authors also agree that this success would be guaranteed by the presence of traits "pre-adapted" to colonization, which evolved in response to the environmental changes experienced in Europe after the glaciations and the development of agriculture (di Castri 1989; Mooney and Hobbs 2000; Mooney and Cleland 2001). Daehler (2003) evaluated this hypothesis based on a review of studies carried out in various locations around the globe, where it was possible to compare the performance of native and European plant species. This author found that the proposed hypothesis was not supported, since native species exhibited equal or better biological performance than exotic species, evaluating this

F. M. Jaksic, S. A. Castro, *Biological Invasions in the South American Anthropocene*, https://doi.org/10.1007/978-3-030-56379-0_2

performance as competitive ability or colonization capacity. Daehler (2003) offered an alternative explanation, stating that the over-representation of European plants in recipient regions could be more a consequence of the historical exchange relations maintained with that continent and not of biological attributes inherent to these species.

An important lesson from the above example is that examination of the historical and geographical context of invasions provides a fundamental background for understanding this type of phenomenon. In particular, Ruiz et al. (2000) emphasized that many of the patterns documented in invasion ecology may be only apparent or spurious if their previous history is unknown.

2.2 Historical Imprint

The transport and exchange of goods and services have promoted biological invasions (Williamson 1996; Mack et al. 2000). However, this relationship has varied qualitatively and quantitatively over the last 500 years of history, especially in South America. In pre-Columbian periods, for example, the pepper seed trade (*Schinus molle*) would have allowed the expansion of its range into the interior of the region, from Peru to Mexico in the north and Chile in the south (Latcham 1936). More recently, *Schinus molle* has crossed the Pacific and Atlantic oceans, having been introduced in both Australia and South Africa, where it is considered an invasive species (Henderson 2001; Iponga et al. 2008). Thus, in this case, the provision of faster and more extensive means of transport facilitated wider geographical expansion.

The relevance of the historical context to understand biological invasions is highlighted by some archaeological studies that have contributed to establishing trade routes and, therefore, dispersion for species in pre-Columbian times, with surprising findings. For example, a DNA sequencing analysis performed on samples of bone tissue from rodents found on Easter Island showed that they corresponded to the Polynesian rat, *Rattus exulans* (Barnes et al. 2006). The Polynesian rat has an original distribution corresponding to Southeast Asia but is currently not found on Easter Island. The reduced genetic variation found in these samples suggests that the *Rattus exulans* was introduced on Easter Island in a small number of events, a fact associated with Polynesian colonization (Barnes et al. 2006).

More controversial is the finding documented by Storey et al. (2007). Analyzing pre-Columbian archaeological remains obtained in the Arauco Peninsula (37°22′15’ S; 73°36′45’ W), these authors found bone remains of a bird. Based on molecular evidence of DNA sequences and radiocarbon dating, the authors came to determine that it was the species *Gallus gallus*, a taxon that until then was thought to have been introduced by Europeans to America. Storey et al. (2007) managed to date their registration between 1300 and 1450 A.D. According to these authors, the presence of bones of *Gallus gallus* in this site evidences the pre-Columbian contact that would have taken place between South America and Polynesia, the original distribution region of this bird.

Similarly, in Europe there are those species transported under the protection of continuous migrations and human invasions, which served as support for their dispersion from and to Africa, Asia, and America (see di Castri 1989). Two examples of the above are the black rat (*Rattus rattus*), native to the Indochina peninsula, and the house mouse (*Mus musculus*), native to the Middle East, species that probably arrived in the Mediterranean basin 5000 or 6000 years ago. In contrast, the invasion of the gray rat (*Rattus norvegicus*) to the Old Continent would have occurred more recently, in the late nineteenth century from China (di Castri 1989).

For obvious reasons, for pre-Columbian times there is no abundant historical record documenting which species were traded or exchanged from one region to another and for what reason. However, it is reasonable to assume that this type of event must have involved a small number of species, and relatively limited geographic scales, in contrast to the massive and globalized character in which subsequent invasions began to occur (Williamson 1996). In effect, it is estimated that as a consequence of the discovery of America in 1492, there was a substantial increase in the rate of species exchange and in the geographical distances involved (Crosby 2004; Williamson 1996).

Hulme (2009) illustrated this idea conveniently by indicating that, in the last 500 years, there would have been two global historical phases in which invasion rates increased. It is important to advance that, although in general terms other authors agree on the establishment of these periods (see Perrings et al. 2005; Meyerson and Mooney 2007), they should rather be understood as referential milestones. The first of these phases would have begun with the discovery of America, associated with the development of global navigation and exploration, the expansion of European colonialism, and the beginning of changes in agriculture, commerce, and industry (see Mack et al. 2000; Crosby 2004; Preston et al. 2004). On the other hand, the second phase would have occurred as a result of the Industrial Revolution, a period in which trade and territorial connectivity increased as a result of advances in transport. In addition, some geopolitical changes during the nineteenth century also had repercussions, such as the emancipation of the colonies in America, which allowed the New World republics to broaden their range of trade relations with other regions of the globe (see Crosby 2004).

On the basis of the antecedents exposed, it is possible to illustrate through a graphical model the historical trend that would have experienced the accumulation of species introduced to a certain territory (Fig. 2.1), a pattern that should be more or less common among the areas that have been exposed to exchange. In simple terms, it is accepted that until about 1500, the number of species translocated by human assistance was relatively small. However, from that date the rate would have increased, albeit slightly (Fig. 2.1). By the year 1800 (as a reference landmark), the level of global interconnection reached as a result of the improvement and intensification of maritime transport would have led to a second change in the dynamics of invasions, implying an increase in the number of geographically relocated species, a situation that has been maintained to the present day (Fig. 2.1). As we will see in the next section, this pattern has been documented by various studies in different countries and/or regions.

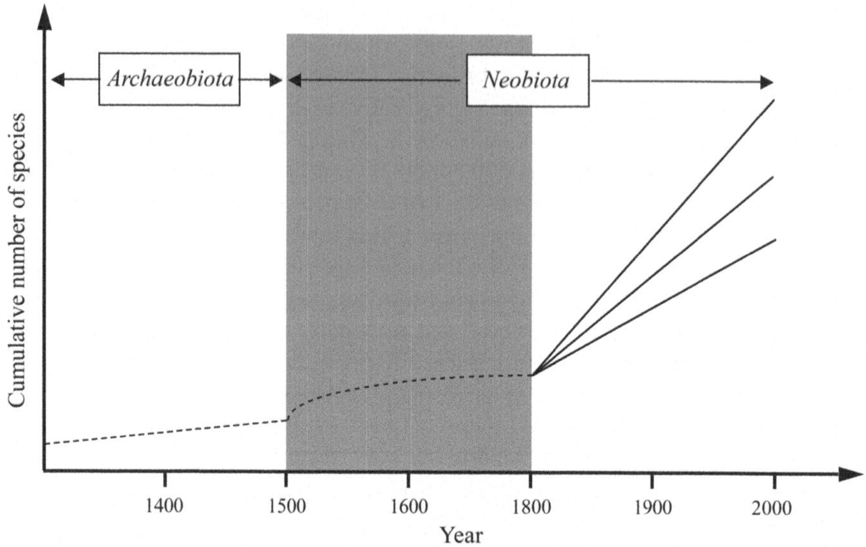

Fig. 2.1 Exotic species accumulation dynamics for hypothetical area. A distinction is made between Archaeo and Neobiota, to differentiate species that have been introduced before and after the year 1500, respectively (see text)

2.3 Species Accumulation Curves

The historical events mentioned in the previous section have had their imprint on patterns of global invasion and local diversity. Through the use of accumulation curves (Pyšek and Prach 2003), various studies have shown how rates of species introduction have changed over the last 500 years.

For the United States, for example, accumulation curves have been established for various groups of economically and ecologically important exotic organisms, such as plant pathogens and terrestrial vertebrates (Fig. 2.2a), fish and mollusks (Fig. 2.2b), and insects and plants (Fig. 2.2c). Independent of the taxonomic group, these curves exhibit a growing increase in the number and rates of introduction over the last 200 years. Along with the above, it is also possible to note that this trend remains high today, with no signs of decreasing or stopping (OTA 1993).

Perhaps the accumulation curves obtained for plants are the most frequently documented in the literature. In Chile, Castro et al. (2005) reconstructed the accumulation curve for a sample of 428 naturalized plant species, based on historical evidence and records of collections deposited in herbaria (Fig. 2.3a). Interestingly, these authors showed that the beginning of the introduction of plants to Chile can be traced as early as the beginning of the sixteenth century, coinciding with the process of Spanish colonization. In those years, the presence of at least four species was already registered: *Cardamine hirsuta*, *Medicago polymorpha*, *Spartium junceum*, and *Bromus hordeaceus*, all these herbs widely distributed in the territory today.

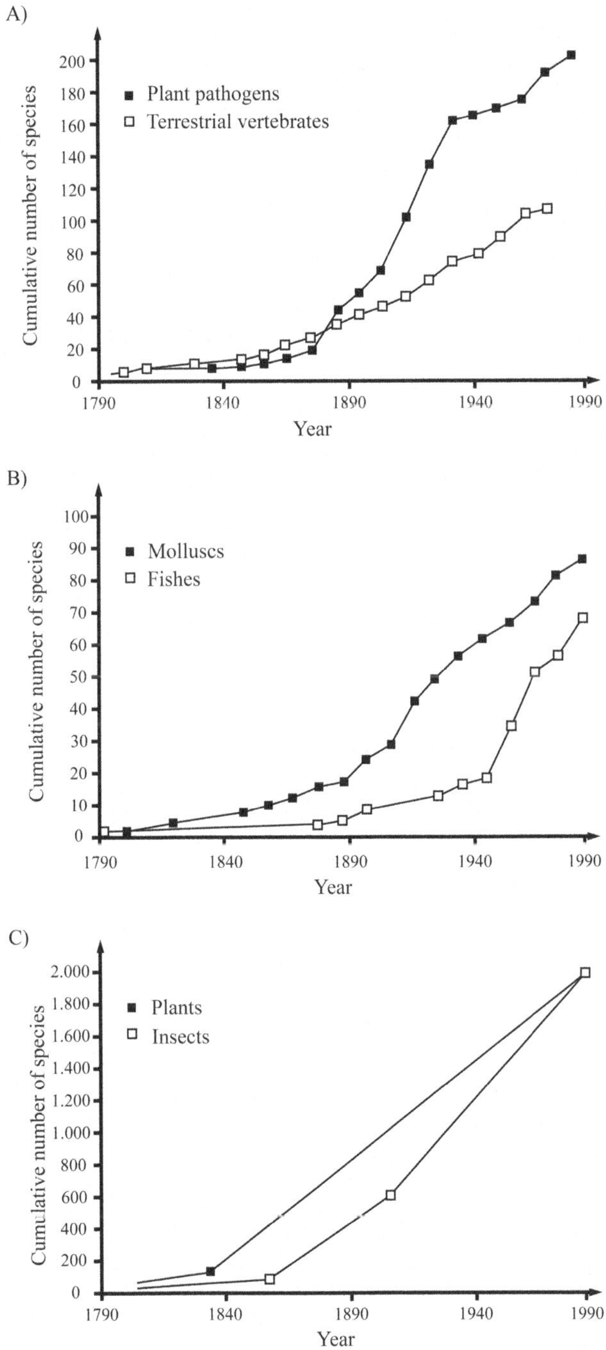

Fig. 2.2 Cumulative number of non-native species introduced into the United States between 1790 and 1990, according to the Office of Technology Assessment (OTA 1993). (**a**) Pathogens of terrestrial plants and vertebrates; (**b**) mollusks and fish; (**c**) plants and insects

Fig. 2.3 Cumulative number of naturalized plant species in three different areas. (**a**) Chile (Castro et al. 2005); (**b**) California (Raven 1988; Rejmánek et al. 1991); (**c**) Europe (Lambdon et al. 2008). In B, the continuous and segmented lines after the 1988 "Jepson" represent possible trends in the rate of species accumulation, depending on the functioning of rules that control the entry of new species into California. In C, "European Origin" represents species that are translocated within the continent, while "Exotic to Europe" includes species translocated from outside the continent

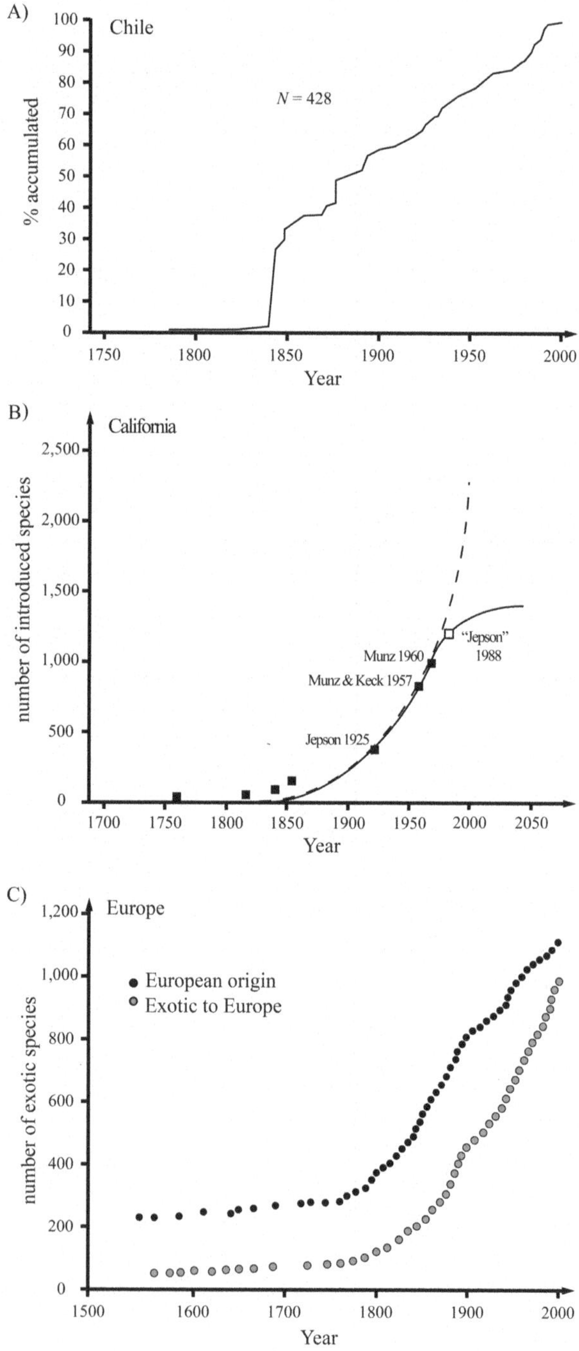

Although the most continuous and complete record of collections began in 1850, clearly the fraction of species introduced up to that year already reached 30% of the current richness (Fig. 2.3a). From there, the number of naturalized plants has grown at an estimated rate of between two and three species per year (Castro et al. 2005).

In addition to the above, Fuentes et al. (2007) examined the accumulation curves for 71,764 specimens of native (56,776) and naturalized (14,988) plants in Chile, all deposited in the herbarium of the Universidad de Concepción (CONC); the latter species are represented mainly by weed species that grow in association with crops and have been collected together with the natives in different sampling sites in Chile. Based on this information, these authors managed to recognize different phases of territorial expansion in which both native and naturalized flora have managed to extend their range of distribution in the last century (i.e., twentieth century). An important contribution of this study was to show the relationship between the processes of expansion of native and naturalized flora with different phases of development that agriculture has experienced in the country. In this context, the authors distinguished two phases. The first phase would have occurred between the years 1910 and 1940, coinciding with the growth in area extension of agriculture, while a second phase would have occurred between 1980 and 2000, coinciding with a stage of conversion of agricultural land to forest and increased efficiency and/or agricultural production (Fuentes et al. 2007; Martín-Forés et al. 2012).

Raven (1988) and Rejmánek et al. (1991) analyzed the dynamics of accumulation of naturalized plants in the state of California (Fig. 2.3b), finding that between 1700 and 1860, some 134 species were naturalized in the territory, while between 1861 and 1988 this value increased to 1116 taxa. Depending on the effectiveness of the zoonotic and phytosanitary policies implemented in California, it is a matter of speculation whether this number will continue to grow, maintain itself, or decline (Rejmánek et al. 1991).

For its part, the situation of the European continent is particularly interesting, as it appears as the main source of exotic organisms for other regions of the globe (Crosby 2004). However, Europe has also received species from other continents (di Castri 1989), being one of the best known regions in terms of historical record of introductions. As in the cases mentioned above, records obtained from terrestrial plants and vertebrates show a sustained increase in the number and rate of species introduction over the last 200 years (Fig. 2.3c), a trend that has been maintained to date at an estimated speed of between four and six naturalized species per year (Genovesi et al. 2009; Kark et al. 2009; Lambdon et al. 2008).

It is important to point out that the accumulation curves we have commented have been developed for a very particular and reduced set of exotic species, the so-called naturalized ones (see Table 1.1). Therefore, the estimation of both numbers and rates of introduction of alien species sensu lato are clearly underestimated as they do not include the entire set of exotic taxa, i.e., those species that have not yet been naturalized or are in an incipient state of naturalization. Despite this, the trends documented in the literature show a sustained growth of these values (in number and rate), confirming that the translocation of species assisted by humans is a ubiquitous process, which does not seem to have a brake. However, it is important to note that the patterns illustrated show a linear trend (see Figs. 2.2 and 2.3), a fact that is far from some appreciations that describe the phenomenon of biological invasions as exponential in nature. This fact is striking because the linear trend shown suggests a constant rate of naturalization, in contrast to an accelerated rate that would result in exponential dynamics. More research is needed to clarify this point.

Accumulation curves have not only shown their usefulness in recognizing historical processes that have led to invasions but also in estimating rates and future trends. As will be seen later (Chap. 6), this information is important to establish control and management actions of exotic species that become harmful.

2.4 Minimum Residence Time

The historical circumstances surrounding the invasion of a species are relevant variables, both for their theoretical and applied value (Shigesada and Kawasaki 1997). This information can help reconstruct and understand how the invasion has taken place, as well as to make predictions about its future dynamics. However, for most alien species, there is little history of invasion. Basically, there are two reasons for this. On the one hand, in the past numerous species were inadvertently introduced and were only recognized as exotic when they were already widely distributed. On the other hand, the interest in the study of invasions is relatively recent in ecology, as it has not been more than 20 years since publications on the subject have begun to appear systematically (Kolar and Lodge 2001; Pyšek et al. 2008a, b).

The year of introduction and the time elapsed between this moment and the present time are important attributes for understanding invasive dynamics. However, given the uncertainty and incompleteness of the biological records, many authors have used auxiliary information to estimate the year of introduction or to have a relative idea of the time elapsed. For example, some studies have been based on observations made by travelers and naturalists, official records of the import or export of goods, or the presence of collections of specimens deposited in biological collections (Pyšek and Prach 2003). Thus, the concept of "minimum residence time" (Rejmánek 2000) has been established as a rather conservative approach, which seeks to estimate the time elapsed from the introduction of a certain taxon to the present. Strictly speaking, minimum residence time can be defined as the period that has elapsed since the first recording of an exotic species in a territory until the present moment (see Table 1.1).

The minimum residence time has shown important correlations with some characteristics of the invasion. For example, the longer the minimum time of residence, the greater the geographic range acquired by a given species in the recipient area (Pyšek et al. 2003; Castro et al. 2005) or its present population abundance (Pyšek et al. 2003). Recently, Pyšek et al. (2009) showed that it can also affect the probability of naturalization of exotic cultivated species, because a longer minimum residence time increases the lapse (and with it, the probability) in which a greater number of propagules could arrive or the escape of a crop could occur.

2.5 Archaeo and Neobiota

Because the exotic flora and fauna of a region may be associated with different waves of introduction over time (Fig. 2.1), it is sometimes not possible to establish whether a particular species present in a region is native to that region or arrived as a result of a remote invasion, for which there are no records. In fact, continuing wars and human invasions in Europe have intermingled the biota of this continent to such an extent that for many species their native or exotic origin is unclear (Lambdon et al. 2008). A more reliable picture only began to emerge from the discovery of America, which occurred more or less 500 years ago. For this reason, the terms archaeo and neobiota have been coined to distinguish organisms whose minimum residence time is greater than or less than 500 years, respectively (Holub & Jirásek 1967 in Pyšek et al. 2002). Although the decision to use 500 years as a criterion seems arbitrary, it is basically inspired by the rate change experienced by the translocation of species globally after AD 1500 (Figs. 2.2 and 2.3). This decision is also based on the recognition that the process of globalization has more or less recognizable roots in the expansion of Western culture. Consequently, the term archaeobiota – which can be used as archaeoflora or archaeofauna, depending on the type of organism (Table 1.1) – is used to distinguish species presumably introduced before 1500, such as *Rattus rattus* or *Mus musculus* in Europe. In contrast, neobiota – it can be used as a neoflora or neofauna (Table 1.1), depending on the type of organism – is used for species introduced after AD 1500, as in the case of *Rattus norvegicus* in Europe (di Castri 1989).

2.6 Geographical Imprint

If transport relations and the exchange of goods and services are conducive to invasions, then the regions participating in this network should show this traceable footprint in the composition of their exotic biota. In this sense, Europe constitutes a peculiar continent because precisely in the last 500 years of history it has acted as an important source of human and cultural expansion in the era of globalization, especially for South America (Crosby 2004). As a result, the composition of the exotic biota in different parts of the globe shows an important representation – sometimes majority – of species that come from this continent.

The numerical importance of European biota in other continents has been evidenced in floristic studies (Pyšek et al. 2009). For example, Kalin-Arroyo et al. (2000) estimated that the diversity of vascular plants established in Chile (i.e., naturalized) would reach about 690 species, of which 71% comes from the Old Continent (Kalin-Arroyo et al. 2000; see Fig. 2.4). In this country, the European imprint can also be recognized when analyzing the taxonomic composition of some particular vegetational formations, such as those of the Mediterranean region of central Chile. In this area, flora from Europe represents 77% of naturalized species (Montenegro

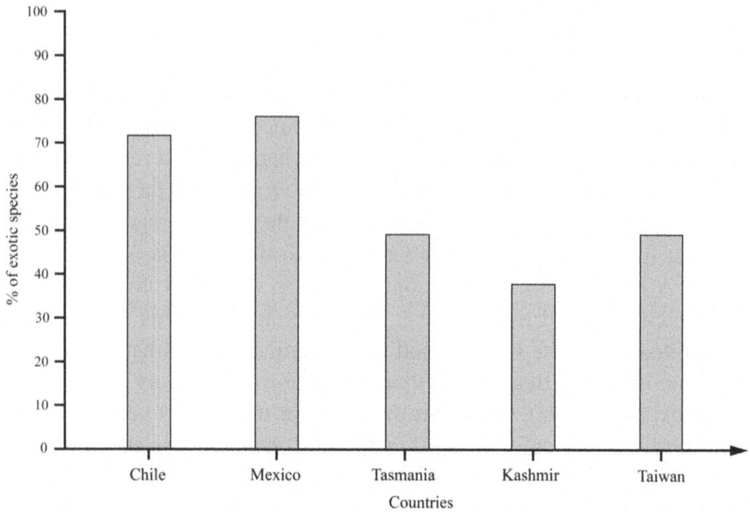

Fig. 2.4 Percentage representation of plant species of European origin in floras from Chile (N = 690; Kalin-Arroyo et al. 2000), Mexico (N = 618; Villaseñor and Espinosa-García 2004), Tasmania (N = 740; Rozebelds et al. 1999), Kashmir (N = 571; Khuroo et al. 2007), and Taiwan (N = 340; Wu et al. 2004)

et al. 1991). This representation may be even higher in formations known as "espinales," where the coverage and richness of species of European origin may exceed 90% (Figueroa et al. 2004, 2011; Martín-Forés et al. 2012).

In a similar analysis in Mexico, Villaseñor and Espinosa-García (2004) documented that 78% of naturalized species come from Europe (Fig. 2.4). In Tasmania (Rozebelds et al. 1999) and Kashmir (Khuroo et al. 2007), an equivalent situation occurs, because although the representation of European flora in these countries (50% and 38%, respectively; see Fig. 2.4) is somewhat lower than that documented for Chile and Mexico, its frequency at the local level is greater compared to the set of plants coming from other regions.

Similarly, of the 340 species of plants naturalized in Taiwan, Wu et al. (2004) found that more than 50% of them come from the American continent, the main source of exotic plants for this large island (Fig. 2.4). Harris et al. (2007) analyzed 179 species of naturalized vines in Australia and found that 23% of them come from South America, the region with the highest contribution of vine species to this country, followed by Europe (21%) and Africa (20%).

Because transport and exchange links can change over time, this fact can also be recognized as an imprint on exotic biota. Costa Rica, for example, currently has 1048 species of naturalized plants, of which only 13% correspond to flora of European origin (Chacón and Saborio 2006). The rest are of Asian and South American origin, with representation of 19% and 18%, respectively (Chacón and Saborio 2006). According to Chacón and Saborio (2006), this situation is due to the recent increase in the introduction of ornamental species, whose naturalization has been favored because these species come from tropical regions that have climatic homology with Costa Rica. According to these authors, until 1821 the representation of European flora in Costa Rica was a majority.

In contrast to terrestrial ecosystems, coastal-marine ecosystems are more open to invasions, manifested in the multiplicity of geographical origins of their exotic biota. For example, the diversity of benthic macro-invertebrates and exotic macroalgae on the coasts of Chile totals 47 naturalized species (Castilla et al. 2005), distributed over a wide taxonomic range. This includes 12 species of algae, 3 annelids, 5 ascidians, 3 bivalves, 2 bryozoans, 4 decapod crustaceans, 4 gastropods, 3 porifers, 10 fish, and even a grass (*Heterozostera tasmanica*). Of this biota, ten taxa are distributed on the eastern Pacific coast, from Mexico to Alaska (Fig. 2.5); seven are distributed on the northwestern Pacific coast, including Japan; six are found on the southwestern Pacific coast comprising Australia, New Zealand, and Tasmania (Fig. 2.5); five are found on the European continent including the North Atlantic and Mediterranean Basin; and three species are located on the southern coast of Africa (see Fig. 2.5). As it can be noticed, the totality of the exotic species present in ecosystems of marine coast of Chile is equally found in one or more different regions of the planet.

Similarly, analysis of the composition of exotic flora and fauna present on the US coast shows the presence of 298 naturalized species (Ruiz et al. 2000). This diversity is represented mainly by crustaceans, mollusks, annelids, and algae, whose native range of distribution corresponds to the Pacific coast Indo-West (30% of species) and West Atlantic (20% of species), while the representation of organisms from Europe and South America was less than 5%, in both cases. Invasions from

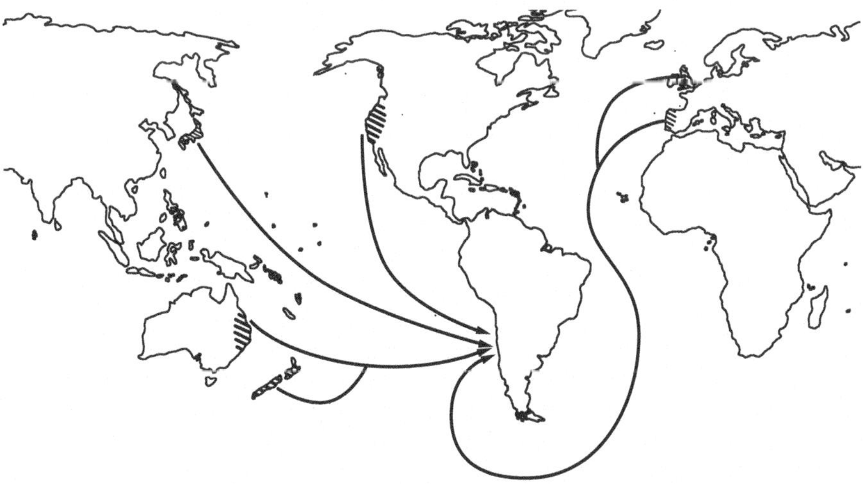

Fig. 2.5 Global geographic origin of the exotic species present in the coastal marine ecosystems of the Chilean coastline. The dwarfed areas represent the original distribution regions for this group of species (see Castilla et al. 2005)

other coastal regions of Asia, Europe, Africa, and Australia had an intermediate representation in the US marine-coastal neobiota, which ranged from 10% to 6%.

It is important to note that in studies of invasions in marine ecosystems, it is rarely possible to unequivocally attribute the region of origin, because – as has already been seen – these species tend to show a wide geographical distribution, with great disjunctions. This is the case of the annelid *Spiophanes bombyx,* first documented off the coast of Chile in 1974 (Castilla et al. 2005). This organism is also present in the coasts of the North and South Atlantic, Mediterranean Basin, South Africa, Gulf of Mexico, Southeast Australia, Falkland Islands or Malvinas, from the west of Canada to the south of California, Japan, and Peru. In view of the fact that Chile maintains or has maintained maritime trade with all these regions, the origin of the invasive population of *Spiophanes bombyx* is not possible to determine. As we will see below, molecular genetic tools that show the genealogical relationships between global populations of this annelid could clarify the origin and provenance of its populations in all the regions it has colonized.

Geographical imprinting can not only be analyzed in terms of the origin or provenance of alien species in a given region, but it is also necessary to pay attention to the types of areas exposed to invasions. This includes remote regions whose environmental conditions are extreme for life or limiting for certain taxa. Due to patterns of global transport and human occupation of territories, there are virtually no areas in the World that do not show the presence of exotic species, and South America is no exception. Oceanic islands, for example, which usually occur hundreds of kilometers off the mainland coast, possess exotic species that have arrived through various mechanisms (Courchamp et al. 2003). Given the volcanic origin of this type of islands, only those species with adaptations to long-distance dispersion had spontaneously arrived there (Lockwood and McKinney 2001; Sax et al. 2005). However, the invasion of new species has now been greatly facilitated, by commerce, species which probably would not have arrived without the aid of human transport (Sax et al. 2002).

Another example is mountain environments, which in light of environmental climate changes become more vulnerable to invasions from lower lands (Alexander et al. 2016). Indeed, it is estimated that mountain habitats cover 5% of the Earth's land area, harboring a high diversity of vascular plant species (Nagy and Grabherr 2009). Their altitude and geographic extent are generally assumed to limit human access, and therefore these environments are poorly exposed to anthropogenic impacts, including the arrival of exotic species (Pauchard et al. 2009; Haider et al. 2018). However, due to the changes the planet is undergoing in the context of the Anthropocene (Malhi 2017), an increasing number of exotic species have begun to show up and expand their distribution as environmental conditions become more suitable for their persistence (Becker et al. 2005; McDougall et al. 2011; Seipel et al. 2012; Morueta-Holme et al. 2015).

For extreme regions, it may be surprising to recognize that polar and subpolar environments, including Antarctica, are no exception to having been colonized by exotic species (Chown et al. 2012). A variety of species of microorganisms, fungi, plants, and animals have established themselves on sub-Antarctic islands and in some parts of the Antarctic continent (Frenot et al. 2005). Most of this diversity of species comes from Europe, and its presence imposes direct and indirect effects on the functioning of Antarctic ecosystems, although little known to date (Frenot et al. 2005). Because the diversity of species resident in these ecosystems is reduced, the presence of exotic species poses a danger to their conservation, which is exacerbated in complementarity with climate change (Frenot et al. 2005; Convey et al. 2012). In fact, Cavieres et al. (2017) have shown that if thermal conditions become more favorable, then the invasive grass species *Poa annua* could reduce the abundance of the only two native species of vascular plants on the Antarctic peninsula, *Deschampsia antarctica* and *Colobanthus quitensis*.

2.7 Cryptogenic Species

Many species have been introduced accidentally and inadvertently. Some of them have been able to establish and expand their secondary range without being noticed. In view of the taxonomic and systematic ignorance of biota in some regions and countries, these species can be considered native without really being so. For those cases where there is suspicion that a particular native taxon is actually an exotic species, the term "cryptogenic species" has been coined; that is, species whose origin (i.e., native or exotic) has not been clearly established (Carlton 1996).

The diversity or richness of cryptogenic species may be relatively high in some ecosystem types, such as in marine-coastal species (Ruiz et al. 2000). There are probably two reasons for this. On the one hand, the importance of inadvertent dispersion, as occurs with ballast water or in the hulls of boats (Roman 2006), and on the other, the action of marine currents that disperse organisms spontaneously (Ruiz et al. 2000; Hillebrand et al. 2001; Minchin 2007). The numerical representation that cryptogenic species can reach is considerable, as some prospective studies have shown. For example, between 2000 and 2003, a team of taxonomists examined floating docks and piers off the coast of New England (United States) to identify the diversity of native and exotic organisms. In this limited but accurate study, 71 exotic species were identified, of which 37 were of cryptogenic origin (Pederson et al. 2003).

A similar panorama is evidenced by studies carried out in South America. For example, Pérez et al. (2007) examined the diversity of exotic and cryptogenic marine species of the coasts of Venezuela, finding 22 non-native species (represented by 2 species of algae, 4 species of mollusks, 8 species of crustaceans, 1 species of ascidian, and 7 species of fish) and 67 other cryptogenic taxa (represented by 23 species of algae, 21 species of mollusks, 5 species of polychaetes, 12 species of crustaceans, 5 species of ascidians, and 1 species of fish) for which there is doubt of their native or exotic origin. In Argentina, the number of species introduced to the coast totaled 35 and the number of cryptogenic species 48 (Schwindt 2008). Among the latter were species of wide taxonomic diversity, such as 1 species of macroalgae, 2 sponges, 6 hydrozoa, 3 anemones, 13 polychaetes, 2 cirripedians, 10 amphipods, 1 mysidacean shrimp, 2 tanaid shrimps, 3 crabs, and 4 bryozoans (Schwindt 2008).

Cryptogenic species challenge the certainty of our knowledge of biodiversity. As indicated above, many species that have been described as native and assigned to a certain systematic position turn out to be exotic, which will produce modifications in the nomenclatural range of the taxonomic diversity of a given territory. It also happens that after determining that a certain cryptogenic taxon is actually exotic, it remains to be seen from where this invasion has come. Although these considerations are intricate, they can be clarified using modern molecular tools. Basically, these tools consist of the analysis of kinship between populations, information that can be obtained from specific regions of the DNA molecule that act as markers (Avise 2004). Two examples illustrating the confusing systematic status of certain species come from mussels of the genus *Mytilus,* those with high phenotypic variability. On the coasts of California, the morphological similarity between the species of the genus has concealed the invasion of the European mussel *Mytilus galloprovincialis,* very similar to the native species *Mytilus trossulus,* whose introduction could only be detected by genetic markers (Geller 1999). Similarly, the covert arrival of *Mytilus galloprovincialis* to southern Chile could only be revealed recently by genetic studies (e.g., Borsa et al. 2012), which also confirmed that, of the two species with which it was commonly confused, *Mytilus chilensis* and *Mytilus edulis,* the first had been wrongly described and both corresponded to the subspecies *Mytilus edulis platensis,* geographically isolated from *Mytilus edulis* in the northern hemisphere.

In northern Chile, Castilla et al. (2002) documented a similar case, after analyzing the status of the tunicate *Pyura stolonifera,* commonly known as "piure de Antofagasta." For a long time, the native or exotic origin of this taxon was subject to controversy (Castilla et al. 2002). *Pyura stolonifera* has been considered a disjunct-distribution taxon, present in Australia (*Pyura stolonifera praeputialis*), South Africa (*Pyura stolonifera stolonifera*), and South America (*Pyura* sp.). On the coast of Antofagasta (23°44' S, 70°26' W), *Pyura stolonifera* is a dominant species in rocky intertidal habitats, although it shows a limited geographic range, occupying an extension of no more than 70 km of coastline (Fig. 2.6a). Using the molecular sequence of the cytochrome oxidase I (COI) marker, Castilla et al. (2002) analyzed sequences of specimens of *Pyura,* finding that its presence in Chile would correspond to a recent introduction from Australia. Consequently, these two populations – the Australian and the Chilean – were the same species, currently called *Pyura praeputialis* (Fig. 2.6b). The results of Castilla et al. (2002) also showed that the South African taxon would represent a second species, now named *Pyura stolonifera* (Fig. 2.6b).

2.8 Biodiversity Catalogues and Databases

It should come as no surprise that the exotic biota resident in a given region or country is not fully known, as interest in studying biological invasions has only gained strength in the last 30 years. Due to the strengthening of national and international

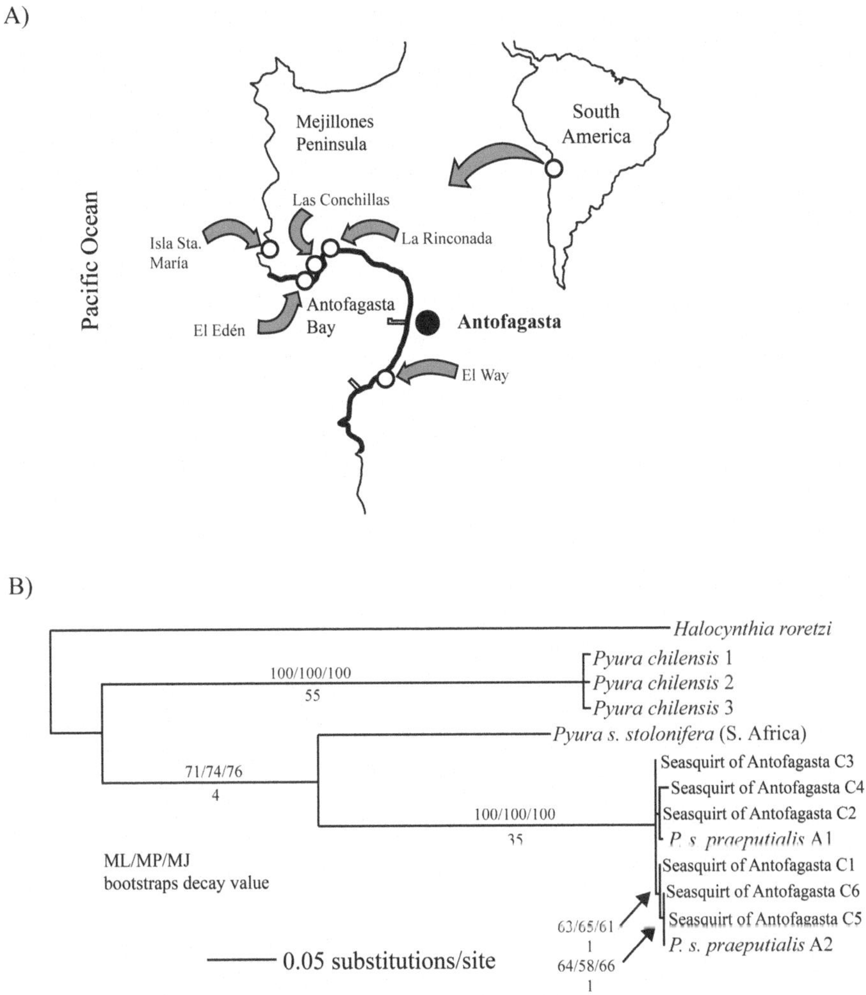

Fig. 2.6 Geographical distribution and phylogenetic analysis for *Pyura* of Antofagasta Bay (see Castilla et al. 2002). (**a**) Location of Antofagasta Bay and Mejillones Peninsula. The blackened coastline represents the distribution of the "Antofagasta piure" (*Pyura* sp.); the main collection sites are indicated in white circles: Isla Santa María, El Edén, Las Conchillas, La Rinconada, and El Way; (**b**) phylogenetic hypothesis obtained for the relations in *Pyura*, supported by the use of three reconstruction methods: maximum likelihood (ML), maximum parsimony (MP), and neighbor-joining (NJ) (Castilla et al. 2002)

computer networks, both countries and entire regions have made efforts to inventory their exotic biota, seeking to establish diagnoses and regulations that limit the entry of unwanted species or to control those already present (Vilà et al. 2010). In this context, the databases published on the Internet constitute an important step forward. In the United States alone, there are at least 319 databases of exotic species

(Crall et al. 2006), covering a wide taxonomic range of organisms, including vascular plants, vertebrates, invertebrates, and fungi.

In Europe, the Delivering Alien Invasive Species Inventories for Europe (DAISIE; www.europe-aliens.org) initiative, which integrates information from 48 countries, has flourished. Through this collaborative network, a complete and updated register of the exotic flora and fauna present in this continent has been created, covering 12,122 species, 79 continental areas, and 57 coastal and marine areas, with the support of 2440 researchers. Their results in flora, for example, show that for a total of 5789 species of exotic plants, 2843 come from non-continental areas, i.e., they have non-European origin; the remaining 2946 taxa correspond to translocations to the interior of the continent (e.g., from one country to another). For those species coming from regions outside Europe, it was determined that most of them come from two regions, Asia (45%) and America (45%), while the remaining 10% come from Australia and Africa. In the case of species translocated within Europe, it has been revealed that they have been extracted from a few countries, such as Belgium, the Czech Republic, and the United Kingdom (Lambdon et al. 2008; Pyšek et al. 2009), while the rest of the countries have contributed a small number of taxa.

In contrast, for Latin America in general and South America in particular, the survey of databases on the Internet is in a lesser state of progress. Very few countries have complete and reliable inventories of their exotic biodiversity, and of these, a minority fraction has been published on the web. In general, the level of completeness of these inventories is heterogeneous. It is also possible to note taxonomic biases where, for example, vascular plants and terrestrial vertebrates tend to be better represented and with a higher level of completeness of information, while those referring to terrestrial or aquatic invertebrates, fungi and other organisms are less well attended.

In Latin America and the Caribbean, these initiatives have been financed by national and international efforts, such as the Inter-American Biodiversity Information Network (IABIN; www.oas.org/en/sedi/dsd/iabin), an organization that has encouraged the compilation of digitized registers and databases in light of its thematic axis of invasive exotic species in Argentina, Brazil, Chile, Colombia, Costa Rica, Guatemala, Jamaica, Paraguay, and Uruguay. It is hoped that these databases will contribute to cementing collaborative efforts between countries, especially in research matters, as well as in the control of pre- and post-frontier species entry.

Recently, the United Nations Programme (UNPD 2017) published the first "Catalogue of Wild/Naturalized Exotic Species in Chile", a pioneering effort in Latin America. An extensive group of specialists participated in its elaboration, among which were botanists and zoologists with extensive trajectory in systematics and taxonomy of each group. The catalogue recognized a total of 1119 species belonging to different taxa. These values included 22 extra-range native species (18 vascular plants, 3 birds, and 1 mammal) that being native to the country, have been introduced into other portions of the territory where they have established expanding populations (UNDP 2017). Among the exotic species, vascular plants were the most diverse with 774 species, 29 nonvascular species, 21 species of algae, 71 fungi, 23 mammals, 13 birds, 2 reptiles, 1 amphibian, 28 fish, 109 insects, 21 species of aquatic invertebrates, and finally 27 species of terrestrial invertebrates. Most of this diversity is distributed in the central portion of Chile, associated with the most important urban development centers (UNDP 2017). 39% of the species recognized as wild or naturalized in the catalogue come from the European continent, 14% from the Eurasian region, and 13% from North America (UNDP 2017). Despite these important advances in terms of recognizing the diversity of naturalized species in Chile, there are still aspects that need to be improved. Among them is the need to standardize the definition of naturalized species as it usually differs between specialists or taxonomic groups. In addition, the biological information available for the different groups is not equivalent; latitudinal range, altitudinal range, year of introduction, life cycle, and biogeographic origin are unknown information fields for many species. Finally, this catalogue is a first step towards an understanding and commensurability of the diversity of exotic species present in Chile, leaving more species to be included, as well as complete taxonomic groups to be reported.

2.9 Biotic Homogenization

Species relocation is changing the distribution and abundance patterns of taxa (Lockwood and McKinney 2001; see Chap. 7). A central problem for the conservation of biodiversity as we know it today is to predict the future magnitude and meaning of these modifications into the future (Vitousek et al. 1997; Chapin III et al. 2000; Sala et al. 2000). Within this area of research, efforts have been made to establish the possible scenarios of diversity or species composition that we will face (Chapin III et al. 2000; Sala et al. 2000).

It is accepted by the community of ecologists that biological invasions constitute one of the main drivers of global change in biodiversity, a dimension called global biotic change (Rahel 2000, 2002; McKinney 2004a, 2004b; Olden et al. 2004). Thus, the increase in the rate of introduction of exotic species throughout the planet seems to lead to a state of homogenization of biodiversity (McKinney and Lockwood 1999). Biotic homogenization can be defined as a process of increasing compositional similarity between ecological units (such as communities or ecosystems) or territorial units (such as islands, watersheds, cities), originally more dissimilar (McKinney and Lockwood 1999, 2005). This homogenization process is given by an increase in the number of species shared between areas as a result of the relocation of species. Fig. 2.7a illustrates this situation by comparing two areas that include a different species composition. After a new species is introduced in both regions simultaneously (Fig. 2.7a), the taxonomic similarity, which was initially null or small, is now visualized as larger as this new taxon constitutes a species common to both areas.

Although biotic homogenization is the most expected trend, other compositional results are also possible. Indeed, as Olden and Poff (2003) pointed out, the compositional trend of the areas being compared depends on the initial distribution of diversity and on what and where introduced species are inoculated. Fig. 2.7b, for

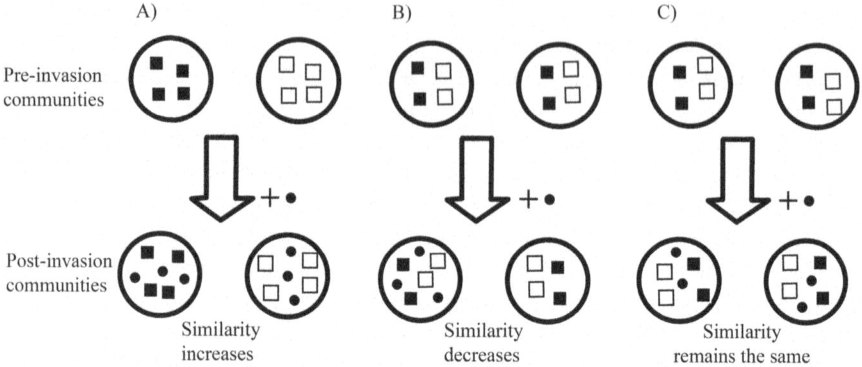

Fig. 2.7 Hypothetical scenarios of homogenization, differentiation, and invariability in compositional similarity, promoted by biological invasions. (**a**) An exotic species is introduced and colonizes simultaneously two communities with different composition, promoting the increase of the compositional similarity (= biotic homogenization); (**b**) an exotic species colonizes only one of the previous communities, promoting a decrease of the compositional similarity (= biotic differentiation); (**c**) a species colonizes simultaneously two communities with similar composition, maintaining the original level of similarity (= biotic tracking)

example, illustrates two areas that maintain similar species composition, so introducing a species into only one of these areas reduces the fraction of shared species; in other words, regions differ in their composition by decreasing their taxonomic similarity (Olden and Poff 2003). In a third possible scenario, the compositional similarity patterns between the two areas being compared could remain relatively unchanged, despite the occurrence of new species introduction events (Fig. 2.7c). This situation may be due to the fact that the introduction of these new taxa does not modify the levels of compositional similarity, so that the similarity remains unchanged over time (Castro and Jaksic 2008).

Homogenization studies have been implemented in various taxa (fish, vascular plants, birds, invertebrates, etc.; Olden and Poff 2003; McKinney 2006; Knop 2016) and community types (basins, islands, urban areas; etc.; Olden 2004; McKinney 2006; Castro et al. 2010; Lososová et al. 2012), as well as emphasizing their temporal dynamism (Clavero and García-Berthou 2006). However, depending on the nature of the distributional information, different indices of diversity can be used, such as those based on composition (α-diversity) or spatial change (β-diversity), which fit more appropriately to discrete or continuous spatial units, respectively (Collins et al. 2002). The α-diversity indicators are based on the quantification of species composition (as intuitively shown in Fig. 2.7, quotation), whose value is expected to increase with homogenization. Meanwhile, the β-diversity indices are based on the measurement of distributional turnover under a more continuous concept of space (Collins et al. 2002); in this case, biotic homogenization can be evidenced through a reduction in spatial turnover when comparing different areas (Baselga 2010).

Our model, which illustrates the compositional changes experienced by a two-area system (Fig. 2.7), can face some complications when the number of areas being compared is greater than two. This would lead to paired contrasts between the multiple areas of interest. Baselga et al. (2007) pointed out that the use of diversity indices based on paired contrasts may yield somewhat spurious values given the lack of independence (Baselga et al. 2007). Baselga (2010) has developed a series of indices that can be implemented in systems of multiple areas (or sites) of interest to compare. These indices not only control the lack of statistical independence between sites but also factor β-diversity into their nesting and spatial replacement components (Vellend 2001; Anderson et al. 2011). The spatial nesting corresponds to the component of the β-diversity due to random changes by gain or loss of species (Vellend 2001; Anderson et al. 2011; Legendre 2014). Meanwhile, spatial change responds to the replacement of species between biota, as a result of established environmental and geographic limitations (Baselga 2010; Chase et al. 2011).

As a text of biological invasions, we have emphasized the role of the introduction of species as a causal agent of biotic homogenization or differentiation. However, an equally important complementary role assists species extinctions or extirpations (McKinney and Lockwood 1999; Olden and Poff 2003, 2004). In the understanding that the native species that make up the initial biodiversity of the areas being compared may disappear, the taxonomic composition and compositional similarity between areas shall be modified (Olden and Poff 2003). This is particularly critical when those species that disappear from the system are endemics, because their local extinction (or extirpation) is equivalent to the extinction of the taxa. Native species tend to grant distinction or differentiation between areas, so their disappearance will have an effect on indices of similarity or spatial change (Olden and Poff 2003). According to different studies, extinctions have indeed contributed to modify the patterns of compositional similarity between areas, but apparently to a lesser extent, since the number of introduced species tends to have a much greater effect on indicators (Smart et al. 2006; Spear and Chown 2008; Winter et al. 2009; Castro et al. 2010).

2.10 Genetic, Phylogenetic, and Functional Homogenization

The concept of "homogenization" has expanded beyond taxonomic or compositional diversity and has been applied in other contexts (Olden 2006). Indeed, the homogenizing effect induced by exotic species in the recipient area may additionally affect genetic (Olden 2006), phylogenetic or evolutionary (Winter et al. 2009), and functional diversity (Devictor et al. 2007).

Genetic homogenization can be defined as an increase in the similarity of the gene pool of initially separate or isolated populations, or an increase in genetic similarity with an exotic taxon, resulting from intraspecific and interspecific hybridization. A clear example is the hybridization between the fishes *Salmo trutta* and *Salmo salar*, which represents a threat to the conservation of the latter species, as it has small populations and therefore hybridization contributes to the loss of its identity and genetic heritage (Verspoor 1988).

Phylogenetic homogenization implies an increase in the similarity of the component of evolutionary information contained in a community, area, or region, over time. This is a consequence of the fact that the presence of a new species also implies the presence of a lineage, which, depending on the recipient community, may constitute an already represented or completely new entity (Knapp et al. 2008a; Pavoine and Ricotta 2014). Carvallo and Castro (2017) stressed that the arrival of 756 species of angiosperms on six oceanic islands of Chile has not significantly altered their diversity and phylogenetic similarity, because most of the introduced species belong to lineages equivalent to taxonomic genera and families, which were already present on the islands.

In its functional dimension, homogenization is conceived as a consequence of the establishment of species with ecological roles that tend to dominate the recipient community, which may or may not be represented in the extant community (Olden and Rooney 2006). At the same time, resident species that have specific or specialized roles should reduce their importance, both in number and abundance (Olden and Rooney 2006; Devictor et al. 2008). Given the complexity of quantifying the role of species in a given community, an auxiliary approach considers evaluating the presence of morphofunctional traits as proximal variables (a proxy) that enables to have an idea of which processes or functional characteristics of the communities have been in transit to homogenization. For example, Pool and Olden (2012) selected ecological and life history features to analyze functional homogenization in the assembly of fishes that inhabit the Lower Colorado River Basin. In this basin the fish composition was 62 species, 23 native and 39 introduced. Among the traits analyzed in this study were the preferred temperatures of each species, its trophic habit, maximum body length, and age at reproductive maturity, among other characteristics. Interestingly, the authors recorded that in functional terms, i.e., the similarity of common functional features in the rivers of the basin increased by 6%. This value was higher than that obtained by taxonomic homogenization indicators that showed an increase of only 1%. Still, these authors showed that functional homogenization and taxonomic homogenization were positively correlated (Pool and Olden 2012).

2.11 Biological Invasions and Urbanization

The relationship between urbanization and biological invasions is close, partly because cities function as centers of introduction of new species (McKinney 2008; Haeuser et al. 2018). This is particularly true for plant species as the composition and distribution of flora within cities is the result of a set of anthropic processes of different nature, operating at different spatial and temporal scales (Marco et al. 2010). On the one hand, through a series of decisions, human beings determine what and how many species we wish to have in our urban environment; and on the other, the environmental and habitat conditions of cities can affect this diversity, facilitating or impeding the survival of these species (Kowarik 1995, 2011; Lososová et al. 2012). As a result, urban floras will be made up of a set of species of different geographic and evolutionary origins, including native species remnants of the original vegetation (Knap et al. 2008a, b; Kowarik 2011; Haeuser et al. 2018). The balance between native and exotic species composition can be a first step in understanding the consequences of urbanization for biodiversity, as well as a first indicator of the effect of urbanization on biodiversity.

So far, a low proportion of the world's cities have been studied in terms of plant species composition. Most of these have been carried out in European cities, in countries that have enjoyed a long botanical tradition (e.g. Gaston et al. 2005a, b; Kühn et al. 2004; Kowarik 2011; Kowarik et al. 2013; La Sorte and Pyšek 2009; Celesti-Grapow et al. 2013). Interestingly, in these cities the representation of native species still tends to exceed the representation of exotic species both at the level of entire cities (Lososová et al. 2012) and in the different habitats within them, such as squares, parks, or streets (Lososová et al. 2012). Lososová et al. (2012), for example, analyzing the composition and floristic distribution in 32 Central European cities, found that the composition of native plants was 48%, of exotic archaeoflora (plant species introduced before AD 1500) was 16%, and of exotic neoflora (plant species introduced after 1500) was 33%. These authors also reported that in different urban habitat types, such as parks, residential areas, blocks, etc., native species diversity always surpassed archaeoflora and neoflora components. On the other hand, on the basis of floristic information obtained for 85 European and 25 non-European cities, La Sorte et al. (2014) and Aronson et al. (2014) documented that the representation of native species registered a median of 52% in the 110 cities and that the representation of these was greater in non-European cities, which was close to 70%.

Both the study by La Sorte et al. (2014) and the study by Aronson et al. (2014) have shown that urban floras located in different localities of the globe seem to be in the process of biotic homogenization. Indeed, in quantitative terms, these studies indicated that the composition of native plant species shows less similarity between cities, while when comparing the set of exotic plants (introduced), the similarity was significantly greater (Aronson et al. 2014; La Sorte et al. 2014).

The state of floristic knowledge in European cities contrasts with what we know in other continents or regions, including South America. In fact, for South America, attempts to describe the diversity of urban plant species are still incipient. For example, Méndez (2005) documented that 68% of the species present in the interior of Luján (Argentina) were exotic, while native species were represented by 32%. For the city of Mendoza (Argentina), Martínez-Carretero (2010) analyzed the composition of the city's synanthropic flora, finding that for a total of 487 registered taxa, 98% of them were exotic. Córdova-Stroobandt (2013) studied the diversity of woody species (trees and shrubs) present in Cochabamba (Bolivia) and found that 90% of them were exotic and only 10% native. Moro and Farías-Castro (2015) reported for the city of Fortaleza (Brazil) a diversity of 219 tree species of which 72% were exotic taxa.

Perhaps Chile is the country that exhibits the greatest number of urban flora studies, at least to date. At the national level, for example, Hoffmann (1998) recorded 105 species of trees and shrubs characteristic of the country's cities in a field guide, although without indicating their distribution; of these, 93% were exotic species, while only 7% were native. In a similar effort, Rodríguez et al. (2005) identified 283 species of trees and shrubs in Chile, 43% of which were native and 57% exotic. According to these authors, the majority of exotic species are represented by those that are not naturalized but that can reach a large abundance inside cities. De la Maza et al. (2002) carried out one of the first systematic surveys in the city of Santiago, although focused only on the tree component. This information was summarized by Hernández and Villaseñor (2018), who also visited the same sites sampled by de la Maza et al. (2002) reaching similar conclusions as the latter authors: of the approximately 118 tree species found, over 90% were exotic. In a study that included all life forms present in the interior of the city of Santiago, Figueroa et al. (2016) found that the floristic composition was represented by 661 species, of which 85% were exotic and the remaining 15% were native.

In an effort to broaden geographic understanding of floristic composition in cities, Santilli et al. (2018) analyzed the composition of woody species for five cities in central Chile, including La Serena, Valparaíso, Santiago, Rancagua, and Talca. These authors recorded that of the 302 species recorded in streets, squares, and parks of these cities, about 85% were exotic and 15% native, values similar to those obtained for Santiago by Figueroa et al. (2016). The former authors suggested that these cities are probably in transit of floristic homogenization, although they did not provide quantitative indicators.

Another taxonomic group widely studied in cities is birds. In general terms, it has been indicated that the composition of avifauna in cities depends closely on the biogeographic region in which the city is located (Aronson et al. 2014; Marzluff 2016), as well as the number of exotic species that have been introduced and naturalized in the country (Lim and Sodhi 2004; Aronson et al. 2014; Sanz and Caula 2015). A generalized pattern in urban ecosystems shows that the number of native species exceeds the number of exotic species (Marzluff 2016). Likewise, it has been documented that the avifauna usually shows a distribution in abundance that diminishes from the perisphere towards the interior of the city (Clergeau et al. 2006). In terms of their functional traits, generalist species tend to be more represented within the city (McKinney 2006). In trophic terms, for example, the most frequent species are omnivorous (Marzluff 2016). In contrast, carnivorous and insectivorous birds are less diverse and abundant (Chace and Walsh 2006; Kark et al. 2007; Blair and Johnson 2008; Macgregor-Fors 2008; Ortega-Álvarez and MacGregor-Fors 2009; Sanz and Caula 2015).

Urban fauna studies have been less developed in South America. In particular, for Santiago de Chile, Fernández and Simonetti (2013) have studied the effect of the fragmentation of peri-urban vegetation in the urban-rural boundary of the city on the diversity and abundance of small mammals. These authors noted a decline in the abundance of native rodents from the rural area to the interior of the city, while the exotic species *Rattus rattus* and *Mus musculus* exhibited an inverse pattern.

Gutiérrez-Tapia et al. (2018) analyzed the diversity of birds in Santiago, attempting to characterize their geographic origin and distribution patterns. These authors found that the avifauna in the interior of Santiago was represented by 46 species, of which 41 were native. This majority representation of native species does not differ from that found in other cities (Aronson et al. 2014; Marzluff 2016). The former authors also found that general nesting and feeding habits were the predominant functional traits in the set of native and exotic species (Gutiérrez-Tapia et al. 2018).

The results obtained in floristic and faunal studies in the cities carried out in South America challenge us to understand how the process of urbanization occurs in this region of the planet, and if it shares, or not, the tendencies of homogenization recognized for European cities. So far, it is possible to conclude that South American flora have a higher proportion of exotic species than Europe's urban flora, thus greater efforts are needed to integrate those cities with the native biota of the urban environment for conservation and sustainable development purposes (Santilli et al. 2018).

References

Alexander JM, Lembrechts J, Cavieres LA, Daehler C, Haider S, Kueffer C, Liu G, McDougall K, Milbau A, Pauchard A, Rew LJ, Seipel T (2016) Plant invasions into mountains and alpine ecosystems: current status and future challenges. Alp Bot 126:89–103

Anderson MJ, Crist O, Chase JM, Vellend M, Inouye BD, Freestone AL, Sanders NJ, Cornell HV, Comita LS, Davies KF, Harrison SP, Kraft NJB, Stegen JC, Swenson NG (2011) Navigating the multiple meanings of β-diversity: a roadmap for the practicing ecologist. Ecol Lett 14:19–28

Aronson MF, La Sorte FA, Nilon CH, Katti M, Goddard MA, Lepczyk CA, Dobbs C (2014) A global analysis of the impacts of urbanization on bird and plant diversity reveals key anthropogenic drivers. Proc R Soc Lond B 281:20133330

Arroyo MTK, Marticorena C, Matthei O, Cavieres L (2000) Plant invasions in Chile: present patterns and future predictions. In: Mooney HA, Hobbs RJ (eds) Invasive species in a changing world. Island Press, Washington, DC, pp 385–421

Avise J (2004) Molecular markers, natural history, and evolution, Sinauer, MA

Barnes SS, Matisoo-Smith E, Hunt TL (2006) Ancient DNA of the Pacific rat (*Rattus exulans*) from Rapa Nui (Easter Island). J Archaeol Sci 33:1536–1540

Baselga A (2010) Partitioning the turnover and nestedness components of beta diversity. Glob Ecol Biogeogr 19:134–143

Baselga A, Jiménez-Valverde A, Niccolini G (2007) A multiple-site similarity measure independent of richness. Biol Lett 3:642–645

Becker T, Dietz H, Billeter R, Buschmann H, Edwards P (2005) Altitudinal distribution of plant species in the Swiss Alps. Perspect Plant Ecol Evol System 7:173–183

Blair RB, Johnson EM (2008) Suburban habitats and their role for birds in the urban-rural habitat network: points of local invasion and extinction? Landsc Ecol 23:1157–1169

Borsa P, Rolland V, Daguin-Thiébaut C (2012) Genetics and taxonomy of Chilean smooth-shelled mussels, *Mytilus* spp. (Bivalvia: Mytilidae). C R Biol 335:51–61

Carlton JT (1996) Biological invasions and cryptogenic species. Ecology 77:1653–1655

Carvallo GO, Castro SA (2017) Invasions but not extinctions change phylogenetic diversity of angiosperm assemblage on southeastern Pacific Oceanic islands. PLoS One 12:e0182105

Castilla JC, Collins AG, Meyer CP, Guíñez R, Lindberg DR (2002) Recent introduction of the dominant tunicate, *Pyura praeputialis* (Urochordata, Pyuridae) to Antofagasta, Chile. Mol Ecol 11:1579–1584
Castilla JC, Uribe M, Bahamonde N, Clarke M, Desqueyroux-Faúndez R, Kong I, Moyano H, Rozbaczylo N, Santelices B, Valdovinos C, Zavala P (2005) Down under the southeastern Pacific: marine non-indigenous species in Chile. Biol Invasions 7:213–232
Castro SA, Jaksic FM (2008) How general are global trends in biotic homogenization? Floristic tracking in Chile, South America. Glob Ecol Biogeogr 17:524–531
Castro SA, Figueroa JA, Muñoz-Schick M, Jaksic FM (2005) Minimum residence time, biogeographic origin, and life cycle as determinants of geographic distribution of naturalized plants in continental Chile. Divers Distrib 11:183–191
Castro SA, Daehler CC, Silva L, Torres-Santana CW, Reyes-Betancort JA, Atkinson R, Jaramillo P, Guezou A, Jaksic FM (2010) Floristic homogenization as a teleconnected trend in oceanic islands. Divers Distrib 16:902–910
Cavieres LA, Sanhueza AK, Torres-Mellado G, Casanova-Katny A (2017) Competition between native Antarctic vascular plants and invasive *Poa annua* changes with temperature and soil nitrogen availability. Biol Invasions 20:1597–1610
Celesti-Grapow L, Capotorti G, Del Vico E, Lattanzi E, Tilia A, Blasi C (2013) The vascular flora of Rome. Plant Biosyst 147:1059–1087
Chace JF, Walsh JJ (2006) Urban effects on native avifauna: a review. Landsc Urban Plan 74:46–69
Chacón E, Saborio G (2006) Análisis taxonómico de las especies de plantas introducidas en Costa Rica. Lankesteriana 6:139–147
Chapin FS III, Zavaleta ES, Eviner VT, Naylor RL, Vitousek PM, Reynolds HL, Hooper DU, Lavorel S, Sala OE, Hobbie SE, Mack MC, Díaz S (2000) Consequences of changing biodiversity. Nature 405:234–242
Chase JM, Kraft NJB, Smith KG, Vellend M, Inouye BD (2011) Using null models to disentangle variation in community dissimilarity from variation in α-diversity. Ecosphere 2:1–11
Chown SL, Lee JE, Hughes KA, Barnes J, Barrett PJ, Bergstrom DM, Frenot Y (2012) Challenges to the future conservation of the Antarctic. Science 337:158–159
Clavero M, García-Berthou E (2006) Homogenization dynamics and introduction routes of invasive freshwater fish in the Iberian Peninsula. Ecol Appl 16:2313–2324
Clergeau P, Croci S, Jokimäki J, Kaisanlahti-Jokimäki ML, Dinetti M (2006) Avifauna homogenisation: analysis at different European latitudes. Biol Conserv 127:336–344
Collins MD, Vázquez DP, Sanders NJ (2002) Species–area curves, homogenization and the loss of global diversity. Evol Ecol Res 4:457–464
Convey P, Aitken S, Di Prisco G, Gill MJ, Coulson SJ, Barry T, Jónsdóttir IS, Dang PT, Hik D, Kulkarni T, Lewis G (2012) The impacts of climate change on circumpolar biodiversity. Biodiversity 13:134–143
Córdova-Stroobandt KL (2013) Caracterización de la biodiversidad urbana en la cuenca central de Cochabamba, Bolivia. Acta Nova 6:94–121
Courchamp F, Chapuis JL, Pascal M (2003) Mammal invaders on islands: impact, control and control impact. Biol Rev 78:347–383
Crall AW, Meyerson LA, Stohlgren TJ, Jarnevich CS, Newman GJ, Graham J (2006) Show me the numbers: what data currently exist for non-native species in the USA? Front Ecol Environ 4:414–418
Crosby AW (2004) Ecological imperialism: the biological expansion of Europe, 900–1900. Cambridge University Press, Cambridge
Daehler CC (2003) Performance comparisons of co-occurring native and alien plants: implications for conservation and restoration. Annu Rev Ecol Syst 34:183–211
de la Maza CL, Rodríguez R, Bown H, Hernández J, Escobedo F (2002) Vegetation diversity in the Santiago de Chile urban ecosystem. Arboricult J 26:347–357
Devictor V, Julliard R, Couvet D, Lee A, Jiguet F (2007) Functional homogenization effect of urbanization on bird communities. Conserv Biol 21:741–751

Devictor V, Julliard R, Clavel J, Jiguet F, Lee A, Couvet D (2008) Functional biotic homogenization of bird communities in disturbed landscapes. Glob Ecol Biogeogr 17:252–261

Di Castri F (1989) History of biological invasions with special emphasis on the Old World. In: Drake JA, Mooney HA, di Castri F, Groves RH, Kruger FJ, Rejmánek M, Williamson M (eds) Biological invasions: a global perspective. Wiley, Chichester, pp 1–30

Elton CS (1958) The ecology of invasions by animals and plants. University of Chicago Press, Chicago

Fernández IC, Simonetti JA (2013) Small mammal assemblages in fragmented shrublands of urban areas of Central Chile. Urban Ecosyst 16:377–387

Figueroa JA, Castro SA, Marquet PM, Jaksic FM (2004) Exotic plant invasions to the mediterranean region of Chile: causes, history and impacts. Rev Chil Hist Nat 77:465–483

Figueroa JA, Teillier S, Castro SA (2011) Diversity patterns and composition of native and exotic floras in central Chile. Acta Oecol 37:103–109

Figueroa JA, Teillier S, Guerrero N, Ray C, Rivano S, Saavedra D, Castro SA (2016) Vascular flora in public space of Santiago, Chile. Gayana Botánica 73:44–69

Frenot Y, Chown SL, Whinam J, Selkirk PM, Convey P, Skotnicki M, Bergstrom DM (2005) Biological invasions in the Antarctic: extent, impacts and implications. Biol Rev 80:45–72

Fuentes N, Ugarte E, Kühn I, Klotz S (2007) Alien plants in Chile: inferring invasion periods from herbarium records. Biol Invasions 10:649–657

Gaston KJ, Warren PH, Thompson K, Smith RM (2005a) Urban domestic gardens (IV): the extent of the resource and its associated features. Biodivers Conserv 14:3327–3349

Gaston KJ, Smith RM, Thompson K, Warren PH (2005b) Urban domestic gardens (II): experimental test of methods for increasing biodiversity. Biodivers Conserv 14:395–413

Geller JB (1999) Decline of a native species masked by sibling species invasion: the case of Californian marine mussels. Conserv Biol 13:661–664

Genovesi P, Scalera R, Solarz W, Roy D (2009) Towards an early warning and information system for invasive alien species threatening biodiversity in Europe. Eur Environ Agency 5:1–52

Gutiérrez-Tapia P, Azócar MI, Ignacio M, Castro SA (2018) A citizen-based platform reveals the distribution of functional groups inside a large city from the Southern Hemisphere: e-Bird and the urban birds of Santiago (Central Chile). Rev Chil Hist Nat 91:3

Haeuser E, Dawson W, Thuiller W, Dullinger S, Block S, Bossdorf O, Carboni M, Conti L, Dullinger I, Essl F, Klonner G, Moser D, Münkemüller T, Parepa M, Talluto MV, Kreft H, Pergl J, Pyšek P, Weigelt P, Winter M, Hermy M, van der Veken S, Roquet C, van Kleunen M (2018) European ornamental garden flora as an invasion debt under climate change. J Appl Ecol 55:2386–2395

Haider S, Kueffer C, Bruelheide H, Seipel T, Alexander JM, Rew LJ, Arévalo JR, Cavieres LA, McDougall KL, Milbau A, Naylor BJ, Speziale K, Pauchard A (2018) Mountain roads and non-native species modify elevational patterns of plant diversity. Glob Ecol Biogeogr 27:667–678

Harris CJ, Murray BR, Hose GC, Hamilton MA (2007) Introduction history and invasion success in exotic vines introduced to Australia. Divers Distrib 13:467–475

Henderson L (2001) Alien weeds and invasive plants, Plant Protection Research Institute Handbook No. 12. Agricultural Research Council, Pretoria

Hernández JH, Villaseñor NR (2018) Twelve-year change in tree diversity and spatial segregation in the Mediterranean city of Santiago, Chile. Urban Forest Urban Green 29:10–18

Hillebrand H, Watermann F, Karez R, Berninger UG (2001) Differences in species richness patterns between unicellular and multicellular organisms. Oecologia 126:114–124

Hoffmann A (1998) El árbol urbano en Chile. Ediciones Claudio Gay, Santiago de Chile

Hulme PE (2009) Trade, transport and trouble: managing invasive species pathways in an era of globalization. J Appl Ecol 46:10–18

Iponga DM, Milton SJ, Richardson DM (2008) Superiority in competition for light: a crucial attribute of the invasive alien tree *Schinus molle* (Peruvian pepper tree) in shaping its impact in semi-arid South African savanna. J Arid Environ 72:612–623

Kark S, Iwaniuk A, Schalimtzek A, E Banker (2007) Living in the city: can anyone become an "urban exploiter"? J Biogeogr 34:638–651
Kark S, Solarz W, Chiron F, Clergeau P, Shirley S (2009) Alien birds, amphibians and reptiles of Europe. In: DAISIE (ed) Handbook of alien species in Europe. Springer, Dordrecht, pp 105–118
Khuroo AA, Rashid I, Reshi Z, Dar GH, Wafai BA (2007) The alien flora of Kashmir Himalaya. Biol Invasions 9:269–292
Knapp S, Kühn I, Schweiger O, Klotz S (2008a) Challenging urban species diversity: contrasting phylogenetic patterns across plant functional groups in Germany. Ecol Lett 11:1054–1064
Knapp S, Kühn I, Wittig R, Ozinga WA, Poschlod P, Klotz S (2008b) Urbanization causes shifts in species' trait state frequencies. Preslia 80:375–388
Knop E (2016) Biotic homogenization of three insect groups due to urbanization. Glob Chang Biol 22:228–236
Kolar CS, Lodge DM (2001) Progress in invasion biology: predicting invaders. Trends Ecol Evol 16:199–204
Kowarik I (1995) On the role of alien species in urban flora and vegetation. In: Pyšek P, Prach K, Rejmánek M, Wade M (eds) Plant invasions: general aspects and special problems. PB Academic Publishing, Amsterdam, pp 85–103
Kowarik I (2003) Human agency in biological invasions: secondary releases foster naturalisation and population expansion of alien plant species. Biol Invasions 5:293–312
Kowarik I (2011) Novel urban ecosystems, biodiversity, and conservation. Environ Pollut 159:1974–1983
Kowarik I, von der Lippe M, Cierjacks A (2013) Prevalence of alien versus native species of woody plants in Berlin differs between habitats and at different scales. Preslia 85:113–132
Kühn I, Brandl R, Klotz S (2004) The flora of German cities is naturally species rich. Evol Ecol Res 6:749–764
La Sorte FA, Pyšek P (2009) Extra-regional residence time as a correlate of plant invasiveness: European archaeophytes in North America. Ecology 90:2589–2597
La Sorte FA, Aronson MF, Williams NS, Celesti-Grapow L, Cilliers S, Clarkson BD, Pyšek P (2014) Beta diversity of urban floras among European and non-European cities. Glob Ecol Biogeogr 23:769–779
Lambdon P, Pyšek P, Basnou C, Delipetrou P, Essl F, Hejda M, Jarošík V, Pergl J, Winter M, Andriopoulos P, Arianoutsou M, Bazos I, Brundu G, Celesti-Grapow L, Chassot P, Jogan N, Josefsson M, Kark S, Klotz S, Kokkoris Y, Kühn I, Marchante H, Perglová I, Vilà M, Zikos A, Hulme PE (2008) Alien flora of Europe: species diversity, geographical pattern and state of the art of research. Preslia 80:101–149
Latcham RE (1936) La agricultura precolombina en Chile y los países vecinos. Ediciones de la Universidad de Chile Santiago, Chile
Legendre P (2014) Interpreting the replacement and richness difference components of beta diversity. Glob Ecol Biogeogr 23:1324–1334
Lim HC, Sodhi NS (2004) Responses of avian guilds to urbanisation in a tropical city. Landsc Urban Plan 66:199–215
Lockwood JL, McKinney ML (2001) Biotic homogenization. Kluwer/Academic Press, New York
Lososová Z, Chytrý M, Tichý L, Danihelka J, Fajmon K, Hájek O, Kintrová K, Kühn I, Láníková D, Otýpková Z, Řehořek V (2012) Native and alien floras in urban habitats: a comparison across 32 cities of Central Europe. Glob Ecol Biogeogr 21:545–555
Macgregor-Fors I (2008) Relation between habitat attributes and bird richness in a western Mexico suburb. Landsc Urban Plan 84:92–98
Mack RN, Simberloff D, Lonsdale WM, Evans H, Clout M, Bazzaz FA (2000) Biotic invasions, causes, epidemiology, global consequences, and control. Ecol Appl 10:689–710
Malhi Y (2017) The concept of the Anthropocene. Annu Rev Environ Resour 42:25.1–25.28
Marco A, Barthelemy C, Dutoit T, Bertaudière-Montes V (2010) Bridging human and natural sciences for a better understanding of urban floral patterns: the role of planting practices in Mediterranean gardens. Ecol Soc 15:2

Martínez-Carretero EEM (2010) The synanthropic flora in the Mendoza (Argentina) urban area. Urban Ecosyst 13:237–242

Martín-Forés I, Casado MA, Castro I, Ovalle C, del Pozo A, Acosta-Gallo B, Sánchez-Jardón L, de Miguel JM (2012) Flora of the Mediterranean Basin in the Chilean *espinales*: evidence of colonisation. Pastos 42:137–160

Marzluff JM (2016) A decadal review of urban ornithology and a prospectus for the future. Ibis 159:1–13

McDougall K, Alexander J, Haider S, Pauchard A, Walsh N, Kueffer C (2011) Alien flora of mountains: global comparisons for the development of local preventive measures against plant invasions. Divers Distrib 17:103–111

McKinney ML (2004a) Measuring floristic homogenization by non-native plants in North America. Glob Ecol Biogeogr 13:4–53

McKinney ML (2004b) Do exotics homogenize or differentiate communities? Roles of sampling and exotic species richness. Biol Invasions 6:495–504

McKinney ML (2006) Urbanization as a major cause of biotic homogenization. Biol Conserv 127:247–260

McKinney ML (2008) Effects of urbanization on species richness: a review of plants and animals. Urban Ecosyst 11:161–176

McKinney ML, Lockwood JL (1999) Biotic homogenization: a few winners replacing many losers in the next mass extinction. Trends Ecol Evol 14:450–453

McKinney ML, Lockwood JL (2005) Biotic homogenization and the breakdown of biogeographic barriers. In: Sax DF, Gaines SD, Stachowicz JJ (eds) Exotic species: a source of insight into ecology, evolution and biogeography. Sinauer Associated, Sunderland, pp 365–380

McNeely JA (2001) The great reshuffling: human dimensions of invasive alien species. IUCN, Gland/Cambridge

Méndez E (2005) Flora y vegetación del centro urbano de Luján de Cuyo, Mendoza (Argentina). Revista de la Facultad de Ciencias Agrarias 37:67–74

Meyerson LA, Mooney HA (2007) Invasive alien species in an era of globalization. Front Ecol Environ 5:199–208

Minchin D (2007) A checklist of alien and cryptogenic aquatic species in Ireland. Aquat Invasions 2:341–366

Montenegro G, Teillier S, Arce P, Poblete V (1991) Introductions of plants and animals into the mediterrean-type climate area of Chile. In: Groves RH, di Castri F (eds) Biogeography of mediterranean invasions. Cambridge University Press, Cambridge, pp 103–109

Mooney HA, Cleland EE (2001) The evolutionary impact of invasive species. Proc Natl Acad Sci U S A 98:5446–5451

Mooney HA, Hobbs HA (2000) Invasive species in a changing world. Island Press, California

Moro MF, Farías-Castro ASF (2015) A check list of plant species in the urban forestry of Fortaleza, Brazil: where are the native species in the country of megadiversity? Urban Ecosyst 18:47–71

Morueta-Holme N, Engemann K, Sandoval-Acuña P, Jonas J, Segnitz M, Svenning JC (2015) Strong upslope shifts in Chimborazo's vegetation over two centuries since Humboldt. Proc Natl Acad Sci U S A 112:12741–12745

Nagy L, Grabherr G (2009) The biology of alpine habitats. Oxford University Press, Oxford

Office of Technology Assessment, OTA (1993) Harmful non-indigenous species in the United States, Report OTA-F-565. US Government Printing Office, Washington, DC

Olden JD (2006) Biotic homogenization: a new research agenda for conservation biogeography. J Biogeogr 33:2027–2039

Olden JD, Poff NL (2003) Toward a mechanistic understanding and prediction of biotic homogenization. Am Nat 163:442–460

Olden JD, Poff NL (2004) Ecological processes driving homogenization: testing a mechanistic model using fish faunas. Ecology 85:1867–1875

Olden JD, Rooney TP (2006) On defining and quantifying biotic homogenization. Glob Ecol Biogeogr 15:113–120

Olden JD, Poff NL, Douglas M, Douglas ME, Fausch KD (2004) Ecological and evolutionary consequences of biotic homogenization. Trends Ecol Evol 19:18–24
Ortega-Álvarez R, MacGregor-Fors (2009) Living in the big city: effects of urban land-use on bird community structure, diversity, and composition. Landsc Urban Plan 90:189–195
Pauchard A, Kueffer C, Dietz H, Daehler CC, Alexander J, Edwards PJ, Arévalo JR, Cavieres LA, Guisan A, Haider S, Jakobs G, McDougall K, Millar CI, Naylor BJ, Parks CG, Rew LJ, Seipel T (2009) Ain't no mountain high enough: plant invasions reaching new elevations. Front Ecol Environ 7:479–486
Pavoine S, Ricotta C (2014) Functional and phylogenetic similarity among communities. Methods Ecol Evol 5:666–675
Pederson P, Bullock R, Carlton J, Dijkstra J, Dobroski N, Dyrynda P, Fisher R, Harris L, Hobbs N, Lambert G, Lazo-Wasem E, Mathieson A, Miglietta M-P, Smith J, Smith J III, Tyrrell M (2003) Marine invaders in the Northeast. Rapid assessment survey of non-native and native marine species of floating dock communities. Massachusetts Institute of Technology Sea Grant College Program, Massachusetts
Pérez JE, Alfonsi C, Salazar SK, Mascotay O (2007) Especies marinas exóticas y criptogénicas en las costas de Venezuela. Boletín del Instituto Oceanográfico 46:79–96
Perrings C, Dehnen-Schmutz K, Touza J, Williamson M (2005) How to manage biological invasions under globalization. Trends Ecol Evol 20:212–215
Pool TK, Olden JD (2012) Taxonomic and functional homogenization of an endemic desert fish fauna diversity and distributions. 18:366–376
Preston CD, Pearman DA, Hall AR (2004) Archaeophytes in Britain. Bot J Linn Soc 145:257–294
Pyšek P, Prach K (2003) Research into plant invasions in a cross-roads region: history and focus. Biol Invasions 5:349–356
Pyšek P, Sádlo J, Mandák B (2002) Catalogue of alien plants of the Czech Republic. Preslia 74:97–186
Pyšek P, Sádlo J, Mandák B, Jarosik V (2003) Czech alien flora and the historical pattern of its formation: what came first to Central Europe? Oecologia 135:122–130
Pyšek P, Richardson DM, Pergl J, Jarošik V, Sixtová Z, Weber E (2008a) Geographical and taxonomic biases in invasion ecology. Trends Ecol Evol 23:237–244
Pyšek P, Lambdon PW, Arianoutsou M, Kühn I, Pino J, Winter M (2008b) Alien vascular plants of Europe. In: Drake JA (ed) Handbook of alien species in Europe. Invading nature, Springer Series in Invasion Ecology. Springer, New York, pp 43–61
Pyšek P, Jarošik V, Pergl J, Randall R, Chytry M, Kühn I, Tichý L, Danihelka J, Chrtek J, Sádlo J (2009) The global invasion success of Central European plants is related to distribution characteristics in their native range and species traits. Divers Distrib 15:891–903
Rahel FJ (2000) Homogenization of fish faunas across the United States. Science 288:854–856
Rahel FJ (2002) Homogenization of freshwater faunas. Annu Rev Ecol Syst 33:291–315
Raven PH (1988) The Californian flora. In: Barbour MG, Mayor J (eds) Terrestrial vegetation of California. Sacramento. California Native Plant Society, California, pp 109–137
Rejmánek M (2000) Invasive plants: approaches and predictions. Austral Ecol 25:497–506
Rejmánek M, Thomsen CD, Peters ID (1991) Invasive vascular plants of California. In: Groves RH, di Castri F (eds) Biogeography of mediterranean invasions. Cambridge University Press, Cambridge, pp 81–101
Rodríguez R, Ruiz E, Elissetche JP (2005) Árboles en Chile. Editorial Universidad de Concepción, Chile
Roman J (2006) Diluting the founder effect: cryptic invasions expand a marine invader's range. Proc R Soc Lond B 273:2453–2459
Rozebelds ACF, Cave L, Morris DI, Buchanan AM (1999) The weed invasions in Tasmania since 1970. Aust J Bot 47:23–48
Ruiz GM, Fofnoff PW, Carlton JT, Wonham MJ, Hines AH (2000) Invasion of coastal marine communities in North America: apparent patterns, processes, and biases. Annu Rev Ecol Syst 31:481–531

Sala OE, Chapin FS, Armesto JJ, Berlow E, Bloomfield J, Dirzo R, Huber-Sanwald E, Huenneke LF, Jackson RB, Kinzig A, Leemans R, Lodge DM, Mooney HA, Oesterheld M, Poff NL, Sykes MT, Walker BH, Walker M, Wall DH (2000) Global biodiversity scenarios for the year 2100. Science 287:1770–1774

Santilli L, Castro SA, Figueroa JA, Guerrero N, Ray C, Romero M, Rojas G, Lavanderos N (2018) Composition and diversity patterns of the ornamental flora of Central Chile: implication for urban planning. Gayana Botánica 72:568–588

Sanz V, Caula S (2015) Assessing bird assemblages along an urban gradient in a Caribbean island (Margarita, Venezuela). Urban Ecosyst 18:729–746

Sax DF, Gaines SD, Brown JH (2002) Species invasions exceed extinctions on islands worldwide: a comparative study of plants and birds. Am Nat 160:766–783

Sax DF, Stachowicz JJ, Gaines SD (2005) Species invasions: insights into ecology, evolution and biogeography. Sinauer Associates Incorporated, Sunderland

Schwindt E (2008) Especies exóticas en el Mar Patagónico y sectores aledaños. In: Campagna C (ed) Síntesis del estado de conservación del Mar Patagónico y áreas de influencia. Foro para la Conservación del Mar Patagónico y Áreas de Influencia, Buenos Aires, pp 57–60

Seipel TC, Kueffer C, Rew LJ, Daehler CC, Pauchard A, Naylor BJ, Alexander JM, Edwards PJ, Parks CG, Arevalo JR, Cavieres LA, Dietz H, Jakobs G, Mcdougall K, Otto R, Walsh N (2012) Processes at multiple scales affect richness and similarity of non-native plant species in mountains around the world. Glob Ecol Biogeogr 21:236–246

Shigesada N, Kawasaki K (1997) Biological invasions: theory and practice. Oxford University Press, Oxford

Smart SM, Thompson K, Marrs RH, Le Duc MG, Maskell LC, Firbank LG (2006) Biotic homogenization and changes in species diversity across human-modified ecosystems. Proc R Soc Lond B 22:2659–2665

Spear D, Chown SL (2008) Taxonomic homogenization in ungulates: patterns and mechanisms at local and global scales. J Biogeogr 35:1962–1975

Storey AA, Ramírez JM, Quiroz D, Burley D, Addison DJ, Walter R, Anderson J, Hunt TL, Athens JS, Huynen L, Matisoo-Smith EA (2007) Radiocarbon and DNA evidence for a pre-Columbian introduction of Polynesian chickens to Chile. Proc Natl Acad Sci U S A 104:10335–10339

UNPD (2017) Catálogo de las especies asilvestradas/naturalizadas en Chile. Proyecto GEF/MMA/PNUD Fortalecimiento de los Marcos Nacionales para la Gobernabilidad de las Especies Exóticas Invasoras; Proyecto Piloto en el Archipiélago de Juan Fernández. Santiago de Chile

Vellend M (2001) Do commonly used indices of β-diversity measure species turnover? J Veg Sci 12:545–552

Verspoor E (1988) Widespread hybridization between native Atlantic salmon, *Salmo salar*, and introduced brown trout, *S. trutta*, in eastern Newfoundland. J Fish Biol 32:327–334

Vilà M, Basnou C, Pyšek P, Josefsson M, Genovesi P, Gollasch S, Nentwig W, Olenin S, Roques A, Roy D, Hulme PE, DAISIE Partners (2010) How well do we understand the impacts of alien species on ecological services? A pan-European cross-taxa assessment. Front Ecol Environ 8:135–144

Villaseñor JL, Espinosa-García FJ (2004) The alien flowering plants of Mexico. Divers Distrib 10:113–123

Vitousek PM, D'Antonio CM, Loope LL, Rejmánek M, Westbrooks R (1997) Introduced species: a significant component of human-caused global change. N Z J Ecol 21:1–16

Williamson M (1996) Biological invasions. Chapman & Hall, London

Winter M, Schweigera O, Klotz S, Nentwig W, Andriopoulos P, Arianoutsou M, Basnou C, Delipetrou P, Didžiulis V, Hejda M, Hulme PE, Lambdon PW, Pergl J, Pyšek P, Royl DB, Kühn I (2009) Plant extinctions and introductions lead to phylogenetic and taxonomic homogenization of the European flora. Proc Natl Acad Sci U S A 106:21721–21725

Wu SH, Hsieh CF, Chaw SM, Rejmánek M (2004) Plant invasions in Taiwan: insights from the flora of casual and naturalized alien species. Divers Distrib 10:349–336

Chapter 3
Transport and Introduction

3.1 Primary Dispersal

Various means and mechanisms promote the translocation of species (Mack et al. 2000). However, the onset of primary dispersal does not ensure the successful completion of this stage because in most cases the organisms that initiated it will not survive (Kolar and Lodge 2001). Williamson (1996) estimated that only 10% of the species will make it to the recipient region and that 10% of them will become established. Thus, the expectation of a successful arrival after primary dispersal is a poor bet.

According to our conception of the invasive process (see Chap. 1), primary dispersal depends directly or indirectly on human activities, which in the context of globalization have progressively spread and massed on the planet (Crosby 2004). Some of these practices, such as trade, agriculture, and livestock, are an inherent attribute of our way of life in the Anthropocene (Hodkinson and Thompson 1997). Thus, the various human practices that make the dispersal of organisms possible are grouped into four components known as introduction pathways (Williamson 1996). On the one hand, there are the components related to the causes and purposes that determine that certain species have been transported and introduced into a given territory and, on the other hand, the components represented by vectors and geographical routes of dispersal (Drake and Lodge 2004). At this point it is necessary to point out that different authors have offered different ways of defining and categorizing the mentioned components, for which reason there is no complete consensus on its systematization (Lockwood et al. 2007). In the next section, our conceptual proposal serves the purpose of organizing and orienting this topic in a simple and operational way.

It is important to draw attention to the importance of knowing the routes of introduction as they contain useful information from the point of view of the management of invasions in the territories. Knowing the components associated with those pathways makes it possible to intervene in order to control the introduction of

F. M. Jaksic, S. A. Castro, *Biological Invasions in the South American Anthropocene*, https://doi.org/10.1007/978-3-030-56379-0_3

undesirable species (Mack 2003). In fact, this information is useful for various agencies dedicated to identifying the route of exit or entry of species, improving the effectiveness of early warning and detection procedures (Mack 2003; Wittenberg and Cock 2005). This is particularly important for inadvertent introductions (see below), as by their nature they are not recorded when they arrive, escaping risk analysis (RA) or environmental impact assessment (EIA) systems.

3.2 Causes and Purposes

In simple terms, the cause of introduction refers to human intentionality in producing the transport of a particular species to a region or territory. Because this decision is usually associated with obtaining some kind of good or service derived from the focal species, the need then arises to establish its purpose (i.e., what for). For example, during the nineteenth century some European ecosystems were considered impoverished in species and therefore unattractive from an aesthetic point of view. For this reason, several species of birds, such as the red-vented bulbul (*Pycnonotus cafer*), originating in South Asia and one of the 100 most invasive species in the world (Lowe et al. 2000), were introduced to the European continent. In this case, the cause of introduction can be established as intentional, while its purpose was to obtain the service of aesthetic improvement of the fauna (Kark et al. 2009).

Perhaps a better known example on our continent is the introduction of grasses from the savannah of Central Africa into Central and South America (Parsons 1972; D'Antonio and Vitousek 1992). Indeed, the introduction of *Hyparrhenia rufa*, *Melinis minutiflora*, *Panicum maximum*, and several other species of the genus *Brachiaria* was voluntarily promoted to obtain more nutritious and resistant fodder for intensive cattle raising (Simões and Baruch 1991).

In contrast, other species have been unintentionally introduced. Unintentional introductions, also called involuntary, accidental, or inadvertent (Williamson 1996), involve species that travel as "stowaways", associated with the transport of other species or the exchange of other goods and services. The most characteristic and well-known example of unintentional introduction is the transport of rodents such as *Rattus rattus*, *Rattus norvegicus*, and *Mus musculus* or the cockroach (*Blatta orientalis*), which dispersed in this way, managed to arrive, and successfully establish themselves in five continents. These taxa are currently considered as cosmopolitan species (Williamson 1996). Another case exemplifies the arrival of the brown tree snake (*Boiga irregularis*) to Guam Island (Savidge 1987; Rodda et al. 1997). This species was inadvertently transported in containers, and its unplanned introduction had a number of negative impacts on the island's native diversity. In Guam, *Boiga irregularis* feeds on vertebrates, especially birds, a fact that may have led to the extinction of 10 of the 12 species that originally inhabited the island (Savidge 1987). Obviously, because of the nature of involuntary introductions, it is not possible to attribute purpose in their inoculation.

3.3 Importance of Intentional Introductions

With a few exceptions, in most of the cases studied, intentional introduction is the main reason why species have been moved out of their original range (Pyšek et al. 2008). According to recent estimates of exotic flora in Europe, 63% of the 3749 exotic plant species present on this continent were intentionally introduced, while only 37% were inadvertently introduced (Lambdon et al. 2008). In the case of exotic vertebrates distributed in Europe, the numerical importance of intentional introduction over inadvertent has also been the norm. For example, Genovesi et al. (2009) reviewed the available antecedents for 84 species of naturalized mammals; of these, 64 (76%) came from regions outside Europe, and another 20 (24%) were translocated from one region to another within the continent. These authors found that intentional introduction was the main reason why these mammal species arrived in Europe, an attribute that involved 75% of these taxa.

In South America, the outlook does not appear to be any different. Novillo and Ojeda (2008) reviewed the status of 18 species of wild mammals in Argentina, indicating that most of them were intentionally introduced. In fact, among the purposes associated with these introductions, sport hunting was recognized (e.g., *Axis axis*), together with food source (e.g., wild boar *Sus scrofa*), and fur industry (e.g., beaver *Castor canadensis* and mink *Neovison vison*). Once introduced, these species have been released or escaped from captivity, establishing themselves as self-sustaining populations. A similar panorama for Chile was evidenced by Jaksic (1998), who studied the fauna of exotic terrestrial vertebrates in the country. This author documented that 62% of the exotic vertebrates naturalized in Chile were intentionally introduced; species such as the African clawed frog (*Xenopus laevis*), the Argentinian turtle (*Chelonoidis chilensis*), and the coati (*Nasua nasua*), among other taxa.

Intentional introductions tend to show greater colonization success than unintentional introductions because they correspond to species for which there is a specific interest, and the number of propagules initially introduced and the number of introductory events tend to be sufficiently large to allow their establishment to be secured or enhanced (Reichard and White 2001a, 2001b; Kowarik 2005). Consequently, this increased propagule pressure works for greater probability of arrival and naturalization (Lockwood et al. 2005).

3.4 The Purposes

In the case of organisms that have been intentionally introduced, one aspect that remains to be clarified is the purpose of their introduction. As mentioned above, this purpose is closely linked to the procurement of some good or service.

In plants, for example, diverse purposes are recognized that have elicited the introduction of species in diverse territories, agriculture being one of the most relevant (Kowarik 2005). Since its inception, agricultural practices have promoted the relocation of plants from one region to another. Some of them have reached world importance for their food value, such as rice (*Oryza sativa*), potato (*Solanum tuberosum*), corn (*Zea mays*), as well as numerous other species of vegetables and fruiting trees. Many of these species have not only reached a cosmopolitan distribution as a crop but have also been able to naturalize in the regions where they are cultivated (Randall 2002). Because they are species for agricultural production, they are rarely considered invasive species; however, the fact is that in various regions of the world, these species have established naturalized populations (i.e., wild), an event that in our view of the Anthropocene deserves to be treated as a biological invasion.

The introduction of plants for consumption is not the only purpose recognizable in plants. There are also ornamental, medicinal, forest, and fodder crop species (Richardson 1997). In particular, recent studies have drawn attention to the growing importance of ornamental cultivation as a factor in plant introduction (Mack and Lonsdale 2001; Randall et al. 2001; Reichard and White 2001a, 2001b; Dehnen-Schmutz and Touza 2008), especially in urban centers (Kowarik 1995, 2011; Smith et al. 2006). For example, Reichard (1997) and Reichard and White (2001b) found that in the United States, 82% of 235 woody species identified as crop escapees were introduced first as ornamental species. At present, this market is considered one of the main promoters of species translocation in the northern hemisphere. When quantified, the representation of introduced flora for ornamental purposes has ranged from 50% to 74% in the United States and Czech Republic, respectively (Table 3.1).

Special mention should be made of the forestry industry as the driving force behind the introduction of tree species; since owing to the geographical extension of plantations, they tend to significantly modify the structure and functioning of native

Table 3.1 Percentage representation of plants introduced for ornamental purposes in different countries. This value has been calculated from the total number of exotic plants present in each of the countries.

Country	Percentage	Source
Australia	65% of species naturalized between 1971 and 1995 introduced as ornamentals	Groves (1996)
Belgium	43% of exotic taxa have been deliberately introduced, usually as horticultural plants	Verloove (2006)
Czech Republic	74% of exotic plants deliberately introduced are ornamentals	Pyšek et al. (2002)
England	58% of exotic plants are classified as garden species	Clement and Foster (1994)
Germany	50% of exotic plants are deliberately introduced species	Kühn and Klotz (2002)
Ireland	61% of plants are exotic	Milbau and Stout (2007)
Italy	51% of plants are exotic	Celesti-Grapow et al. (2006)
United States	>50% of naturalized species have been deliberately introduced	Mack and Erneberg (2002)

ecosystems. These effects may be by way of affecting the ecosystems surrounding the plantation or by total replacement of the native ecosystem (Richardson et al. 2014). Diverse species of *Pinus* and *Eucalyptus* are widely distributed as plantations in South America, South Africa, and Australia, affecting a significant environmental impact after naturalization (Richardson 1998; Williams and Wardle 2007). In Argentina, Brazil, and Chile, the cultivation of extensive plantations of *Pinus* is less time-consuming than in Australia and South Africa, so their naturalization status and environmental impacts are a matter of discussion and interest (Richardson et al. 2008; Pauchard et al. 2014). However, regardless of the state of naturalization of these species, given the geographic extension of their crops, it is reasonable to think that these impacts would be at least proportionally extensive to the substitution of the native vegetational matrix (Armesto et al. 2010; Pauchard et al. 2014). In Chile, the intensive plantation of *Pinus radiata* and *Eucalyptus globulus* began in the 1940s, with a marked increase in forest land area from the 1980s (Fig. 3.1), when legislation encouraged this industry through a state subsidy (Camus 2006). According to Armesto et al. (2010), the total land area that these two species accumulate in the form of plantations in Chile reaches three million hectares, under a sustained rate of increase of approximately one thousand hectares per decade in the last 300 years (Fig. 3.1).

Although there are some studies that evaluate the impact of pine plantations on the components of native flora, of particular interest is the study by Lander et al. (2009) in which they studied aspects of the flowering, pollination, and fructification of queule (*Gomortega keule*). This tree, belonging to the monotypic family

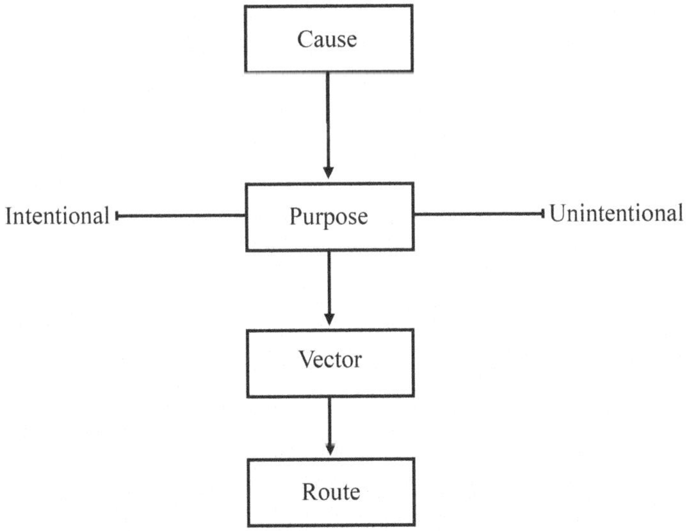

Fig. 3.1 Growth of forest plantations in Chile between 1940 and 2010 (Armesto et al. 2010). This growth is mainly sustained by the planting of *Pinus radiata* and *Eucalyptus globulus*. The dots represent the accumulated area of planted land, and the bars represent the area planted per year. An important turning point came in 1980, when the forestry incentive began as Chile's government policy

Gomortegaceae, is endemic to southern Chile and is currently in danger of extinction (Hechenleitner et al. 2005). Its geographical distribution is limited, restricted only to the western slope of the Cordillera de la Costa between the Region of Maule and the Region of Biobío (35° 44' S - 37° 40' S), in a distribution that occupies approximately 200 km of latitudinal extent (Hechenleitner et al. 2005). This area has experienced massive land use conversion, with forest use currently predominating. According to Lander et al. (2009), the flowering and fructification of *Gomortega keule* was extremely rare in areas of exotic pine plantations, a situation that contrasts with those present in native forest formations, or in less intensively used agricultural lands. The authors suggested that the environmental conditions of pine plantations affect the abundance of syrphid flies, the main pollinating insects of *Gomortega keule*.

An even more dramatic situation has been described for the ruil (*Nothofagus alessandrii*) by Bustamante and Castor (1998). The ruil is a deciduous tree endemic to Chile (Rodríguez et al. 2005), circumscribed to a small area of the Cordillera de la Costa of the Maule Region, between 35° and 36° S (Hechenleitner et al. 2005). Because of the small number of individuals, the destruction of their habitat, and the fragmentation of their populations, the conservation status of the ruil has been declared in critical danger of extinction (Hechenleitner et al. 2005). Due to the extensive and intense forest use of the land in the region, an activity that is fundamentally based on the exploitation of *Pinus radiata*, the ruil forest has experienced intense deforestation and fragmentation. In 1991, there were still 183 fragments of ruil forest, which together totaled 352 ha of forest. For the most part, these fragments were small in size (< 4 ha). According to these authors, between 1981 and 1991, the ruil forest disappeared at a rate of 8% per year, a phenomenon caused by the replacement of native forest by plantations of *Pinus radiata* (Bustamante and Castor 1998). The analysis of the composition of these fragments of ruil groves showed that inside them 13% of the species were exotic (mainly herbs), *Pinus radiata* being the only tree that successfully invaded them.

Other plant species have been introduced to obtain other services, such as soil improvement (i.e., nitrogen fixation), dune stabilization (i.e., root stabilization effect), and/or erosion control. Among the examples that can be cited are different species of the genus *Acacia*, introduced for soil rehabilitation (Gardner 2001). In particular, the silver wattle (*Acacia dealbata*) is used in the mining industry as a phytoremediation species because of its ability to absorb and retain metals from the ground, as well as to stabilize tailings (Sanz-Elorza et al. 2004); this species is native to Australia and Tasmania but has been introduced to the Iberian Peninsula, Southern African cone, and South America (Campos et al. 2004; Lorenzo et al. 2010).

In animals, particularly in the case of vertebrates, the diversity of introductory purposes is also broad. Hunting, recreational fishing, pet trade, and importation as a food source have become important purposes for introducing mammals, birds, reptiles, amphibians, and fish. This scenario has been clearly evidenced at the global level when analyzing the intentionality in the introduction of marine fish, where 17% of the taxa have been inoculated for consumption and/or recreation purposes (Table 3.2). For the United States, the introduction of different species of freshwater

Table 3.2 Percentage (%) and numerical (N) frequency of the different causes and purposes associated with the introduction of species into the world. This percentage was calculated in relation to the total number of mobilized species

Purpose of introduction	Marine fishes worldwide[1]% (N)	Aquatic invertebrates from the United States[2]% (N)	Freshwater fishes from the United States[2]% (N)	Amphibians from the United States[2]% (N)	Reptiles from the United States[2]% (N)	Birds worldwide[3]% (N)	Birds from Australia[3]% (N)	Freshwater fishes from California[1,4]% (N)	Exotic plants from California[5]% (N)
Intentional									
Food and play	17 (9)	11 (8)	42 (205)	14 (8)	13 (8)	25 (103)	22 (20)	69 (49)	13 (?)
Environmental improvement	0 (0)	20 (14)	20 (99)	39 (22)	69 (44)	13 (53)	17 (15)	7 (5)	40 (?)
Biocontrol	0 (0)	0 (0)	3 (14)	11 (6)	0 (0)	4 (18)	6 (5)	3 (2)	
Conservation and science	0 (0)	0 (0)	5 (21)	0 (0)	0 (0)	3 (12)	2 (2)	3 (2)	
Unintentional	73 (41)	41 (29)	26 (132)	11 (6)	0 (0)	28 (117)	22 (20)	18 (13)	5 (?)
Unknown	10 (6)	28 (19)	4 (15)	25 (14)	18 (12)	27 (113)	31 (28)	0 (0)	42 (?)

[1]FishBase (www.fishbase.org/home.htm)
[2]Benson (1999)
[3]Cassey et al. (2004)
[4]Moyle (2002)
[5]Di Tomaso (2000)

fish, amphibians, syrphid flies, and reptiles owes to their use as food, recreation, and environmental "improvement" (Table 3.2), a situation shared with the introduction of birds worldwide in Australia and freshwater fish in California (Table 3.2).

In southern Chile, red deer (*Cervus elaphus*) was introduced into temperate forests for hunting purposes (Jaksic 1998). Recreational fishing has been the main cause of introduction of established exotic fish in freshwater systems in Spain (Clavero and García-Berthou 2006) and Chile. Here, trout and salmon species have initially been introduced for fish farming, but after their numerous escapes, they ended up naturalizing in several hydrographic basins in Chile (Iriarte et al. 2005; Camus and Jaksic 2009). Other species have been introduced for aesthetic purposes, such as the red-bellied squirrel (*Callosciurus erythraeus*) introduced in Argentina in 1973, originally from Southeast Asia (Guichón et al. 2005).

Biocontrol has been the purpose of introducing some vertebrate species in countries such as the United States, where it can reach 11% of cases (see Table 3.2). However, it is important to mention that the need to establish biological control of pest species has also served as a purpose of introducing other species that act as natural enemies. The rationale is that biocontrolling species could reduce the abundance of the unwanted species and thus diminish its effects. This has motivated the introduction of species such as gambusias, toads, various mustelids, cats, foxes, nutrias (*Myocastor coypus*), owls, and passerines (Jaksic and Yáñez 1983; Moors and Atkinson 1984; Sick 1984; King 1990; Common and Norton 1992; Hone 1994; Simberloff and Stiling 1996; Amori and Lapini 1997). Paradoxically and unexpectedly, on some occasions the biocontrolling agent has ended up becoming a problem species as well, since they can start by feeding on native species, without exerting the desired control effect on the invasive species. For example, the coati (*Nasua nasua*) was introduced on Robinson Crusoe Island (Chile) for the purpose of biocontrol. This species inhabits the South American jungle and the southern portion of Central America, and it is absent from the mammal fauna of continental Chile (Muñoz-Pedreros and Yáñez 2000). In 1950, three individuals of coati were introduced to reduce the population of exotic rodents on the island (*Rattus rattus* and *Rattus norvegicus*). However, the coatis reproduced quickly and began to consume eggs from native birds, especially Chilean petrel (*Pterodroma defilippiana*) and the endemic hummingbird of Juan Fernández (*Sephanoides fernandensis*). Currently, the coati is recognized as a pest species, and its hunting is authorized to decrease its population (Danton 2004; Oikonos 2009).

The introduction of the South American gray fox (*Lycalopex griseus*) in Tierra del Fuego Island (shared between Chile and Argentina) is also another example of intentional introduction for the purpose of biological control. According to Jaksic and Yáñez (1983), this canine was introduced from continental Chile to Tierra del Fuego with the intention of controlling European rabbit populations (*Oryctolagus cuniculus*). The landowners considered that the rabbit competed with sheep for forage, so its population control was deemed necessary. However, the fox began to also prey on sheep, which in turn motivated its hunting to control the predator population.

3.5 Involuntary Introduction

The volume and impact of inadvertently introduced species should not be neglected. This makes sense if one considers that the establishment of regulatory mechanisms for the entry of alien species at the border is likely to reduce the number of voluntary introductions but increase the proportion of unintentionally introduced species. This is of concern, because, by their nature, it is common for unintentionally introduced species to be noticed when they have already become naturalized and reached wide distribution in the recipient area (Mack et al. 2000).

Some studies have shown that the number of inadvertently introduced species may be large. For example, in the United States, the Organic Trade Association (OTA 1993) established that 81% of the species interned to this country between 1983 and 1993 entered involuntarily. Another example is the composition of naturalized flora in the Czech Republic. Pyšek et al. (2003) found that the role of unintentional introduction was much more important in the past than it is today, given that 86% of its archaeoflora (i.e., species introduced prior to AD 1500; N = 290 species) were accidentally introduced. In the particular case of marine fish, it is estimated that globally 73% of alien species have been unintentionally introduced, associated with shipping in interoceanic vessels (Table 3.2). In birds, this fraction is lower (26%) but equally significant (Table 3.2).

3.6 Transboundary Diffusion of Species

A special case, where it is not possible to attribute the cause or purpose of introduction, is known as cross-border diffusion. This concept is actually born under a consideration more geopolitical than ecological. Transboundary diffusion occurs when a certain species has been introduced into a certain country and subsequently expands its distribution to neighboring regions thus occupying other countries. As can be seen, cross-border diffusion simply involves the crossing of geopolitical barriers (i.e., borders between countries) as a consequence of secondary dispersal. A well-documented example in Europe is the expansion of estrildid finches that were initially introduced in Portugal but have increased their range to the rest of the Iberian Peninsula in a few years (Martí and Del Moral 2003).

In South America, Jaksic et al. (2002) documented that across the extensive shared border between Chile and Argentina (about 5150 km), there has been transboundary diffusion of eight mammals, including the European rabbit (*Oryctolagus cuniculus*) and the South American gray fox (*Lycalopex griseus*), species that have crossed the border from Chile to Argentina, while the European hare (*Lepus europaeus*), wild boar (*Sus scrofa*), red deer (*Cervus elaphus*), North American beaver (*Castor canadensis*), muskrat (*Ondatra zibethicus*), and mink (*Neovison vison*) were initially introduced in Argentina and expanded to Chile (Fig. 3.2).

It is important to note that one implication of transboundary diffusion is that the effectiveness of population control actions undertaken on a particular alien species

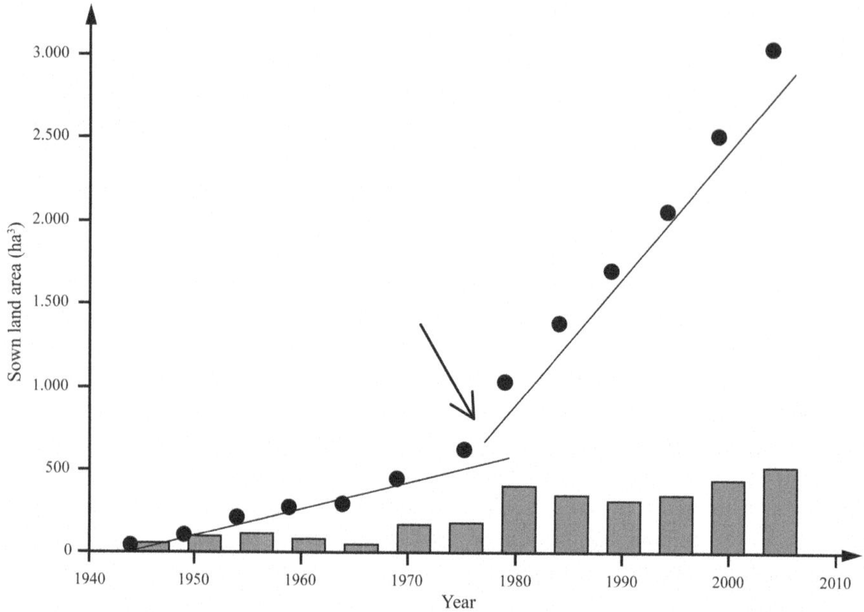

Fig. 3.2 Reciprocal exchange of exotic species between Chile and Argentina through transboundary diffusion (Jaksic et al. 2002). The arrows represent the crosses of (1) European hare (*Lepus europaeus*), (2) European rabbit (*Oryctolagus cuniculus*), (3) beaver (*Castor canadensis*), (4) muskrat (*Ondatra zibethicus*), (5) American mink (*Neovison vison*), (6) wild boar (*Sus scrofa*), (7) red deer (*Cervus elaphus*), and (8) South American gray fox (*Lycalopex griseus*, formerly *Pseudalopex griseus*)

will require the coordinated participation of neighboring countries sharing the invasion, as evidenced by the North American beaver control experience. The invasion of *Castor canadensis* in Tierra del Fuego began with the intentional release of 20 breeding individuals, with the aim of stimulating the fur industry in the Argentine portion of the island (Lizarralde 1993; Massoia and Chébez 1993). The conditions of the region allowed the growth of this beaver population, which since then has expanded throughout the island of Tierra del Fuego, as well as to its surrounding islands, colonizing Navarino Island around 1962, Hoste Island during the 1970s, and Dawson Island in 1989, and reaching the Chilean mainland in 2004 (Sielfeld and Venegas 1980; Skewes et al. 1999; Anderson et al. 2009) (Fig. 3.3). The current population size is estimated between 60 and 90 thousand individuals, which occupy an area of approximately 7,000,000 ha (Lizarralde 1993; Skewes et al. 2006). Currently, coordinated control actions have been carried out in Tierra del Fuego, in an experience that has involved the participation of authorities from Chile and Argentina (Wallem et al. 2014). These actions seem not only necessary but also urgent in light of the economic impact the beaver causes, which is estimated at over US$ 64 billion and projected at another US$ 70 billion, making this rodent one of the seven exotic species with the greatest economic impact in Chile (Table 6.6).

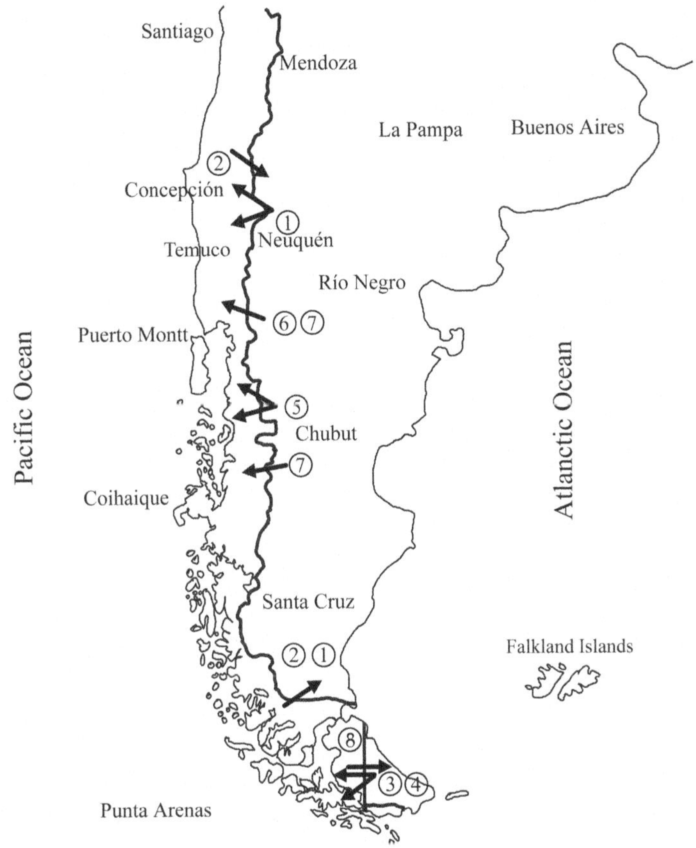

Fig. 3.3 Current geographical distribution of the North American beaver (*Castor canadensis*) in Tierra del Fuego and the continental portion of Chile. The star shows the release point of 10 pairs of breeding beavers whose population has grown and expanded its distribution to the rest of the islands and mainland

3.7 Vectors and Dispersal Routes

Regardless of the causality and purpose of an introduction, many species require initial or secondary dispersal vectors in order to reach from one region to another and then expand, i.e., a physical means of transport that allows them to move. Associated to it, this vector will show a specific geographic route of dispersal, denoting a spatially explicit route. The agents that act as dispersal vectors are different in nature and often show specificity in transporting certain taxa. For example, numerous organisms that inhabit coastal marine environments are dispersed by commercial ships, either in ballast water or attached to the hull, a condition that allows the differential dispersal of sessile or free-living species (Carlton 1985, 1996).

A case exemplifying the distinction between vectors and pathways can be illustrated using the zebra mussel case (*Dreissena polymorpha*), a freshwater bivalve whose original distribution includes the Black Sea and Caspian Sea areas, but which is currently expanding in various locations in Asia, Europe, and North America (Mackie and Schloesser 1996). In the United States, this species appears to have arrived in the ballast waters of commercial vessels, which have been identified as dispersal vectors (Schloesser et al. 1996). Thus, the maritime trade route between Asia, Europe, and North America is the most likely dispersal pathway for this mussel.

3.8 Vector Diversity

The vectors of introduction or expansion of species make up a heterogeneous set of means of transport. Basically, these means could be differentiated between biotic mechanisms mediated by other organisms and abiotic mechanisms mediated by physical transport (i.e., vehicles, ocean currents, etc.).

Perhaps one of the most emblematic models that can be mentioned for the purpose of highlighting the importance of biotic vectors in the spread of invasive species is the case of the expansion of the bubonic plague or black plague in Europe during the fourteenth century. This disease is caused by the bacteria *Yersinia pestis* which was dispersed from Asia to Europe by the black rat (*Rattus rattus*), host species of the bacterium. Using the occurrence of the disease as an information base, Austin-Alchon (2003) configured the distribution of *Yersinia pestis* for the year 1347, which at that time covered the entire Mediterranean Basin (Fig. 3.4). This disease rapidly spread to higher latitudes, covering the whole of Europe by 1351 (Fig. 3.4).

A particularly well-documented case has been the expansion of myxomatosis in introduced rabbit populations (*Oryctolagus cuniculus*). This disease is caused by a virus called *Myxoma*, the natural geographical distribution of which corresponds to Central and South America. This virus causes a mild infection in its natural host, the wild rabbit or tapetí (*Sylvilagus brasiliensis*), but in populations of *Oryctolagus cuniculus*, it causes death. The virus was intentionally introduced into Australia for control purposes in 1950 and into Europe in 1952. Since then its evolution has been monitored periodically (Ratcliffe et al. 1952; Fenner and Ratcliffe 1965; Fenner and Ross 1994; Rogers et al. 1994). This virus was also used for the biological control of rabbits in France, and from there, the disease was transmitted at high speed to neighboring countries, reaching the Iberian Peninsula in 2 years, where it has caused a substantial decline in the population (Calvete 2005; Camps-Rabadà 2008).

Biotic dispersal seems to have been important in island invasion (Traveset and Richardson 2014), not only as an initial arrival vector but also as a secondary dispersal vector. In this context, for example, it is possible to recognize the presence of blackberry (*Rubus ulmifolius*), maqui (*Aristotelia chilensis*), and murtilla (*Ugni molinae*), whose expansion in the Juan Fernández Islands (Chile) has been prompted

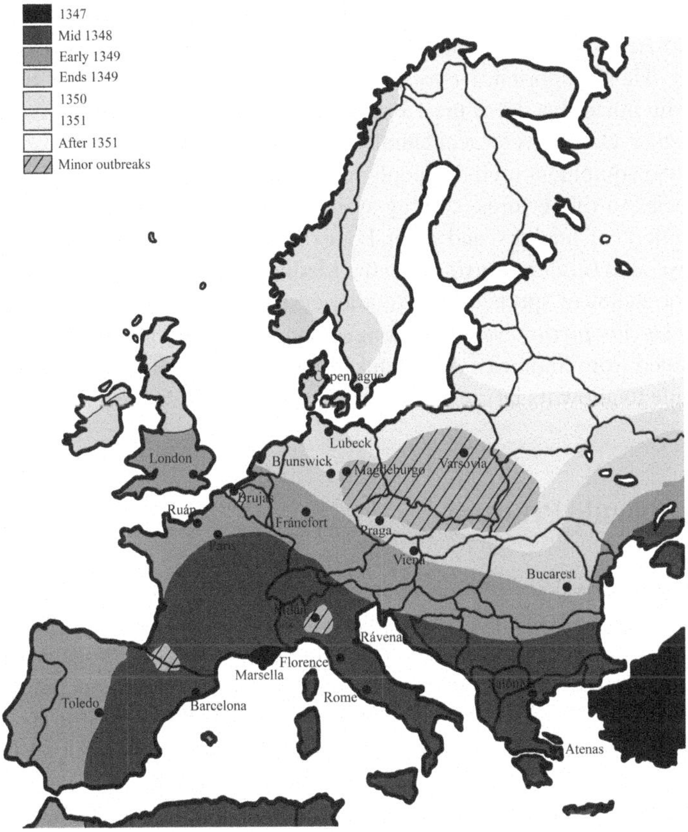

Fig. 3.4 Geographical expansion of the black or bubonic plague in Europe during the fourteenth century (Austin-Alchon 2003). The gray tones represent the geographical extent of this disease at different times, which progressed from southern to northern Europe

by frugivorous birds that consume their fruits and defecate seeds in conditions to germinate. Mora and Smith-Ramírez (2016) studied the germinative response of these three plant species after passing through the digestive tract of the thrush (*Turdus falcklandii*). Their results showed that the germination of *Rubus ulmifolius* and *Ugni molinae* were maintained at around 42% and 50% for the first species, and between 87% and 96% for the second species, while *Aristotelia chilensis* experienced an increase in germination, from 5% in control treatments (with seeds that did not pass through the bird's digestive tract) to 22% in experimental treatment (with seeds that had passed through the digestive tract). These three plant species are considered pests on these islands because once established they tend to quickly take over space, suffocating the growth of native plants (Greimler et al. 2002; Dirnböck et al. 2003; Cuevas et al. 2004; Danton 2004).

On the other hand, among the different abiotic vectors of dispersal are different means of transport linked to human beings, such as motorized vehicles, airplanes, and ships. Their importance may be remarkable, as identified by Christy et al. (2007), who noted that 12 of the 13 species of anurans introduced to Guam Island in the Pacific Ocean were accidentally introduced, where they arrived by boat. Vehicles and equipment used by people in various activities can carry propagules of many species in rubber tires, cooling systems for nautical engines, nets and other fishing equipment, anchors, and so on. For example, the main vectors of dispersal of the invasive alga *Caulerpa taxifolia* in the Mediterranean Sea have been the anchors and anchor chains of sports boats. Another case of interest is that of the tiger mosquito (*Aedes albopictus*), vector of dengue and yellow fever viruses. This dipteran has expanded from Southeast Asia thanks to the transport of tires that keep enough water inside to allow its larvae to survive (Eritja et al. 2005).

3.9 Propagule Pressure

The success of arrival and establishment depends largely on the number of dispersal events combined with the number of organisms participating in each of these events (i.e., propagules). These factors are considered components of propagule pressure, a concept that refers to the number and frequency with which propagules colonize the recipient area (Lockwood et al. 2005; Simberloff 2009). Taylor et al. (2015) have demonstrated the importance of propagule pressure in explaining the establishment of *Pinus contorta* in six globally distributed regions, Río Negro province (Argentina), Aysén and Araucanía regiones (Chile), South Island (New Zealand), and Greater Yellowstone ecosystem (in the United States). The general trend described for *Pinus contorta* was evidenced by a negative correlation between the density of the *Pinus contorta* and distance from the source population (Taylor et al. 2015).

Propagule pressure is an important force that ultimately determines the composition of exotic flora and fauna (Newsome and Noble 1986; Simberloff 1989; Pimm 1991), so its consideration is relevant when evaluating some generalizations in invasion ecology. Along this line, Lonsdale (1999) recognized that among the most accepted generalizations in the study of biological invasions is the one referring to the fact that islands are more invaded than continents (Elton 1958). This fact has received considerable attention, especially because studies suggest that the low diversity of native species present on an island creates conditions conducive to the establishment of exotic taxa (Daehler 2006). However, such differences may appear simply because the islands are subject to a greater influx of propagules per unit area or indirect estimates (Lonsdale 1999; Levine 2000; Daehler 2006).

Propagule pressure has two components: the number of organisms arriving at a given site after a dispersal event and the number of dispersal events occurring in a given time lapse (Lonsdale 1999; Lockwood et al. 2005). Because these components are extremely difficult to measure empirically, much of the evidence available to date rest on observational approaches (Lockwood et al. 2005).

References

Amori G, Lapini L (1997) Le specie di mammiferi introdotte in Italia: il quadro della situazione attuale. Supplement alle Ricerche di Biologia della Selvaggina 27:249–267

Anderson CB, Martínez-Pastur G, Lencinas MV, Wallem PK, Moorman MC, Rosemond AD (2009) Do introduced North American beavers *Castor canadensis* engineer different in southern South America? An overview with implications for restoration. Mammal Rev 39:33–52

Armesto JJ, Manuschevich D, Mora A, Smith-Ramírez C, Rozzi R, Abarzúa AM, Marquet PA (2010) From the Holocene to the Anthropocene: a historical framework for land cover change in southwestern South America in the past 15,000 years. Land Use Policy 27:148–160

Austin-Alchon S (2003) A pest in the land: new world epidemics in a global perspective. University of New Mexico Press, USA

Benson AJ (1999) Documenting over a century of aquatic introductions in the United States. In: Claudi R, Leach JH (eds) Nonindigenous freshwater organisms: vectors, biology, and impacts. Lewis Publishers, Washington, DC, pp 1–31

Bustamante RO, Castor C (1998) The decline of an endangered temperate ecosystem: the ruil (*Nothofagus alessandrii*) forest in central Chile. Biodivers Conserv 7:1607–1626

Calvete C (2005) Modelling the effect of population dynamics on the impact of rabbit hemorrhagic disease. Conserv Biol 20:1232–1241

Campos JA, Herrera M, Biurrun I, Loidi J (2004) The role of alien plants in the natural coastal vegetation in central-northern Spain. Biodivers Conserv 13:2275–2293

Camps-Rabadà J (2008) Los primeros conejos llegaron a América iniciado noviembre de 1493. Información Veterinaria 9:21–24

Camus P (2006) Ambiente, bosques y gestión forestal en Chile (1541–2005). Lom Ediciones, Santiago

Camus P, Jaksic F (2009) Piscicultura en Chile: entre la productividad y el deterioro ambiental (1856–2008). Ediciones de la Pontificia Universidad Católica de Chile, Santiago

Carlton JT (1985) Transoceanic and interoceanic dispersal of coastal marine organims: the biology of ballast water. Oceanogr Mar Biol Annu Rev 23:313–371

Carlton JT (1996) Biological invasions and cryptogenic species. Ecology 77:1653–1655

Cassey P, Blackburn TM, Russell GJ, Jones KE, Lockwood JL (2004) Influences on the transport and establishment of exotic bird species: An analysis of the parrots (Psittaciformes) of the world. Glob Chang Biol 10:417–426

Celesti-Grapow L, Pyšek P, Jarošík V, Blasi C (2006) Determinants of native and alien species richness in the urban flora of Rome. Divers Distrib 12:490–501

Christy MT, Savidge JA, Rodda GH (2007) Multiple pathways for invasion of anurans on a Pacific island. Divers Distrib 13:598–607

Clavero M, García-Berthou E (2006) Homogenization dynamics and introduction routes of invasive freshwater fish in the Iberian Peninsula. Ecol Appl 16:2313–2324

Clement EJ, Foster MC (1994) Alien plants of British isles. Botanical Society of the British Isles, UK

Common MS, Norton TW (1992) Biodiversity: its conservation in Australia. Ambio 21:258–265

Crosby AW (2004) Ecological imperialism: the biological expansion of Europe, 900–1900. Cambridge University Press, Cambridge

Cuevas J, Marticorena A, Cavieres LA (2004) New additions to the introduced flora de of the Juan Fernandez Islands: Origin, distribution, life history traits and, potential of invasion. Rev Chil Hist Nat 77:523–538

D'Antonio CM, Vitousek PM (1992) Biological invasions by exotic grasses, the grass/fire cycle, and global change. Annu Rev Ecol Syst 23:63–87

Daehler CC (2006) Invasibility of tropical islands by introduced plants: partitioning the influence of isolation and propagule pressure. Preslia 78:361–374

Danton P (2004) Plantas silvestres de la Isla Robinson Crusoe guía de reconocimiento. In: Corporación Nacional Forestal, Región de Valparaíso. Chile, Viña del Mar

Dehnen-Schmutz K, Touza J (2008) Plant invasions and ornamental horticulture: pathway, propagule pressure and the legal framework. In: Teixeira da Silva JA (ed) Floriculture, ornamental and plant biotechnology: advances and topical issues. Global Science Books, Isleworth, pp 15–21

Di Tomaso JM (2000) Invasive weeds in rangelands: species, impacts, and management. Weed Sci 48:255–265

Dirnböck T, Geimler J, López SP, Stuessy TF (2003) Predicting future threats to the native vegetation of Robinson Crusoe Island, Juan Fernández archipelago, Chile. Conserv Biol 17:1650–1659

Drake JM, Lodge DM (2004) Global hot spots of biological invasions: evaluating options for ballast-water management. Proc R Soc Lond B 271:575–580

Elton CS (1958) The ecology of invasions by animals and plants. University of Chicago Press, Chicago

Eritja R, Escosa R, Lucientes J, Marques E, Roiz D, Ruiz S (2005) Worldwide invasion of vector mosquitoes: present European distribution and challenges for Spain. Biol Invasions 7:87–97

Fenner F, Ratcliffe FN (1965) Myxomatosis. Cambridge University Press, Cambridge

Fenner F, Ross J (1994) Myxomatosis. In: Thompson HV, King CM (eds) The European rabbit: the history and biology of a successful colonizer. Oxford University Press, Oxford, pp 205–240

Gardner J (2001) Rehabilitating mines to meet land use objectives: bauxite mining in Jarrah forest of Western Australia. Unasylva 52:3–8

Genovesi P, Scalera R, Solarz W, Roy D (2009) Towards an early warning and information system for invasive alien species threatening biodiversity in Europe. Eur Environ Agency 5:1–52

Greimler J, Stuessy TF, Swenson U, Baeza CM, Matthei O (2002) Plants invasions on an oceanic archipelago. Biol Invasions 4:73–85

Groves RH (1996) Recent incursions of weeds to Australia 1971–1995, Technical series 3. CRC Weed Management Systems

Guichón ML, Bello M, Fasola L (2005) Expansión poblacional de una especie introducida en la Argentina: la ardilla de vientre rojo *Callosciurus erythraeus*. Mastozoología Neotropical 12:189–197

Hechenleitner VP, Gardner MF, Thomas PI, Echeverría C, Escobar B, Brownless P, Martínez C (2005) Plantas amenazadas del centro-sur de Chile. Distribución, Conservación y Propagación. Ediciones de la Universidad Austral de Chile y Real Jardín Botánico de Edimburgo

Hodkinson DJ, Thompson K (1997) Plant dispersal: the role of man. J Appl Ecol 34:1484–1496

Hone J (1994) Analysis of vertebrate pest control. Cambridge University Press, Cambridge

Iriarte JA, Lobos GA, Jaksic FM (2005) Invasive vertebrate species in Chile and their control and monitoring by governmental agencies. Rev Chil Hist Nat 78:143–154

Jaksic FM (1998) Vertebrate invaders and their ecological impacts in Chile. Biodivers Conserv 7:1427–1445

Jaksic FM, Yáñez JL (1983) Rabbit and fox introductions in Tierra del Fuego: history and assessment of the attempts at biological control of the rabbit infestation. Biol Conserv 26:367–374

Jaksic FM, Iriarte JA, Jiménez JE, Martínez DR (2002) Invaders without frontiers: cross-border invasions of exotic mammals. Biol Invasions 4:157–173

Kark S, Solarz W, Chiron F, Clergeau P, Shirley S (2009) Alien birds, amphibians and reptiles of Europe. In: DAISIE (ed) Handbook of alien species in Europe. Springer, Dordrecht, pp 105–118

King CM (1990) The handbook of New Zealand mammals. Oxford University Press, Oxford

Kolar CS, Lodge DM (2001) Progress in invasion biology: predicting invaders. Trends Ecol Evol 16:199–204

Kowarik I (1995) On the role of alien species in urban flora and vegetation. In: Pyšek P, Prach K, Rejmánek M, Wade M (eds) Plant invasions: general aspects and special problems. PB Academic Publishing, Amsterdam, pp 85–103

Kowarik I (2005) Urban ornamentals escaped from cultivation. In: Gressel J (ed) Crop ferality and volunteerism. CRC Press, Boca Raton, pp 97–121

Kowarik I (2011) Novel urban ecosystems, biodiversity, and conservation. Environ Pollut 159:1974–1983
Kühn I, Klotz S (2002) Floristischer status und gebietsfremde Arten. In: Klotz S, Kühn I, Durka W (eds) BIOLFLOR: Eine Datenbank zu biologisch-okologischen Merkmalen der Gefaßpflanzen in Deutschland. Bundesamt fur Naturschutz, Bonn, pp 47–56
Lambdon P, Pyšek P, Basnou C, Delipetrou P, Essl F, Hejda M, Jarošík V, Pergl J, Winter M, Andriopoulos P, Arianoutsou M, Bazos I, Brundu G, Celesti-Grapow L, Chassot P, Jogan N, Josefsson M, Kark S, Klotz S, Kokkoris Y, Kühn I, Marchante H, Perglová I, Vilà M, Zikos A, Hulme PE (2008) Alien flora of Europe: species diversity, geographical pattern and state of the art of research. Preslia 80:101–149
Lander TA, Boshier DH, Harris SA (2009) Flower and fruit production and insect pollination of the endangered Chilean tree, *Gomortega keule* in native forest, exotic pine plantation and agricultural environments. Rev Chil Hist Nat 82:403–412
Levine JM (2000) Species diversity and biological invasions: relating local process to community pattern. Science 288:852–854
Lizarralde M (1993) Current status of the introduced beaver (*Castor canadensis*) population in Tierra del Fuego, Argentina. Ambio 22:351–358
Lockwood JL, Cassey P, Blackburn T (2005) The role of propagule pressure in explaining species invasions. Trends Ecol Evol 20:223–228
Lockwood JL, Hoopes M, Marchetti M (2007) Invasion ecology. Blackwell Publishing, Massachusetts
Lonsdale WM (1999) Global patterns of plant invasions and the concept of invasibility. Ecology 80:1522–1536
Lorenzo P, González L, Reigosa MJ (2010) The genus *Acacia* as invader: the characteristic case of *Acacia dealbata* link in Europe. Ann For Sci 67:1–11
Lowe S, Browne M, Boudjelas S, De Poorter M (2000) 100 of the world's worst invasive alien species: a selection from the global invasive species database. Published by The Invasive Species Specialist Group (ISSG) a specialist group of the Species Survival Commission (SSC) of the World Conservation Union (IUCN)
Mack RN (2003) Global plant dispersal, naturalization and invasion: pathways, modes and circumstances. In: Ruiz G, Carlton J (eds) Global pathways of biotic invasions. Island Press, Washington, DC, pp 3–30
Mack RN, Erneberg M (2002) The United States naturalized flora: largely the product of deliberate introductions. Ann Mo Bot Gard 89:176–189
Mack RN, Lonsdale WM (2001) Humans as global plant dispersers: getting more than we bargained for. Bioscience 51:95–102
Mack RN, Simberloff D, Lonsdale WM, Evans H, Clout M, Bazzaz FA (2000) Biotic invasions, causes, epidemiology, global consequences, and control. Ecol Appl 10:689–710
Mackie GL, Schloesser DW (1996) Comparative biology of zebra mussels in Europe and North America: an overview. Am Zool 36:244–258
Martí R, Del Moral JC (2003) La invernada de aves acuáticas en España. Dirección General de Conservación de la Naturaleza-SEO/BirdLife. Organismo Autónomo Parques Nacionales, Ministerio de Medio Ambiente, Madrid, España
Massoia E, Chébez JC (1993) Mamíferos silvestres del archipiélago fueguino. Ediciones LOLA, Buenos Aires
Milbau A, Stout JC (2007) Database of alien plants in Ireland. Irish Biochange Project, Dublin. http://www.biochange.ie/alienplants
Moors PJ, Atkinson IAE (1984) Predation on seabirds by introduced animals, and factors afecting its severity. In: Croxall JP, Evans PGH, Schreiber RW (eds) Status and conservation of the world's seabirds. ICBP Technical Publication, Cambridge, pp 667–690
Mora JP, Smith-Ramírez C (2016) Are birds, wind and gravity legitimate dispersers of fleshy-fruited invasive plants on Robinson Crusoe Island, Chile? Flora 224:167–171
Moyle PB (2002) The decline of anadromous fishes in California. Conserv Biol 8:869–870

Muñoz-Pedreros A, Yáñez JL (2000) Mamíferos de Chile. Centro de Estudios Agrarios y Ambientales (CEA) Ediciones, Valdivia, Chile
Newsome AE, Noble IR (1986) Ecological and physiological characters of invading species. In: Groves RH, Burdon JJ (eds) Biological invasions. Cambridge University Press, Cambridge, pp 1–20
Novillo A, Ojeda RA (2008) The exotic mammals of Argentina. Biol Invasions 10:1333–1344
Office of Technology Assessment, OTA (1993) Harmful non-indigenous species in the United States, Report OTA-F-565. US Government Printing Office, Washington, DC
OIKONOS (2009) Juan Fernández islands conservancy. In: www.oikonos.org/projects/jfic.htm
Parsons J (1972) Spread of African pasture grasses to the American tropics. J Range Manag 25:12–17
Pauchard A, Langdon B, Jiménez A, Cavieres L, Peña E, Núñez MA (2014) Pináceas invasoras en el sur de Sudamérica: patrones, mecanismos e impactos potenciales. In: Jaksic FM, Castro SA (eds) Invasiones biológicas en Chile. Ediciones Universidad Católica de Chile, pp 283–308
Pimm SL (1991) The balance of nature? Ecological issues in the conservation of species and communities. University of Chicago Press, Chicago
Pyšek P, Sádlo J, Mandák B (2002) Catalogue of alien plants of the Czech Republic. Preslia 74:97–186
Pyšek P, Sádlo J, Mandák B, Jarosik V (2003) Czech alien flora and the historical pattern of its formation: what came first to Central Europe? Oecologia 135:122–130
Pyšek P, Lambdon PW, Arianoutsou M, Kühn I, Pino J, Winter M (2008) Alien vascular plants of Europe. In: Drake JA (ed) Handbook of alien species in Europe. Invading nature, Springer Series in Invasion Ecology. Springer, New York, pp 43–61
Randall RP (2002) A global compendium of weeds. RG & FJ Richardson, Melbourne
Randall JM, Benton N, Morse LE (2001) Categorizing invasive plants: the challenge of rating the weeds already in California. In: Groves RH, Panetta FD, Virtue JG (eds) Weed risk assessment. CSIRO Publishing, Collingwood, pp 203–216
Ratcliffe FN, Gay FJ, Greaves T (1952) Australian termites. Commonwealth Scientific and Industrial Research Organization, Melbourne
Reichard SH (1997) Prevention of invasive plant introductions on national and local levels. In: Luken JA, Thieret JA (eds) Assessment and management of plant invasions. Springer, New York, pp 215–227
Reichard SH, White P (2001a) Horticultural introductions of invasive plant species: a north American perspective. In: McNeely JA (ed) The great reshuffling: human dimensions of invasive alien species. IUCN, Gland, pp 161–170
Reichard SH, White P (2001b) Horticultural as a pathway of invasive plant introductions in the United States. Bioscience 51:103–113
Richardson DM (1997) Forestry trees as invasive aliens. Conserv Biol 12:18–26
Richardson DM (1998) Ecology and biogeography of *Pinus*. Cambridge University Press, Cambridge
Richardson DM, van Wilgen BW, Nuñez MA (2008) Alien conifer invasions in South America: short fuse burning? Biol Invasions 10:1387–3547
Richardson DM, Hui C, Núñez MA, Pauchard A (2014) Tree invasions: patterns, processes, challenges and opportunities. Biol Invasions 16:473–481
Rodda G, Fritts T, Chiszar D (1997) The disappearance of Guam's wildlife. Bioscience 47:565–575
Rodríguez R, Ruiz E, Elissetche JP (2005) Árboles en Chile. Editorial Universidad de Concepción, Chile
Rogers PM, Arthur CP, Soriger RC (1994) The rabbit in continental Europe. In: Thompson HV, King CM (eds) The European rabbit: the history and biology of a successful colonizer. Oxford Science Publications, Oxford, pp 22–63
Sanz-Elorza M, Dana ED, Sobrino E (2004) Atlas de las plantas alóctonas invasoras en España. Dirección General para la Biodiversidad. Ministerio de Medio Ambiente, Madrid
Savidge J (1987) Extinction of and island forest avifauna by an introduced snake. Ecology 68:660–667

Schloesser DW, Nalepa TF, Mackie GL (1996) Infestation of unionid bivalves (Unionidae) in North America. Am Zool 36:300–310

Sick H (1984) Ornitologia Brasileira: uma introdução. Editora Universidade de Brasília, Brasil

Sielfeld KW, Venegas C (1980) Poblamiento e impacto ambiental de *Castor canadensis* Kuhl, en Isla Navarino, Chile. Anales del Instituto de la Patagonia (Chile) 11:247–257

Simberloff D (1989) Which insect introductions succeed and which fail? In: Drake JA, Mooney HA, di Castri F, Groves RH, Kruger FJ, Rejmánek M, Williamson M (eds) Biological invasions: a global perspective. Wiley, Chichester, pp 61–76

Simberloff D (2009) The role of propagule pressure in biological invasions. Annu Rev Ecol Evol Syst 40:81–102

Simberloff D, Stiling P (1996) How risky is biological control? Ecol Soc Am 77:1965–1974

Simões M, Baruch Z (1991) Responses to simulated herbivory and water stress in two tropical C4 grasses. Oecologia 88:173–180

Skewes RO, Gonzalez F, Rubilar L, Quezada OM, Olave IT, Vargas V, Avila AC (1999) Investigación, aprovechamiento y control del castor, islas Tierra del Fuego y Navarino. Servicio de Gobierno Regional XII Región. Magallanes y Antártica Chilena, Punta Arenas

Skewes O, González F, Olave R, Ávila A, Vargas V, Paulsen P, König HE (2006) Abundance and distribution of American beaver, *Castor canadensis* (Kuhl 1820), in Tierra del Fuego and Navarino islands, Chile. Eur J Wildl Res 52:292–296

Smith RM, Thompson K, Hodgson JG, Warren PH, Gaton KJ (2006) Urban domestic gardens (IX): composition and richness of the vascular plant flora, and implications for native biodiversity. Biol Conserv 129:312–322

Taylor KT, Maxwell BD, Pauchard A, Nuñez MA, Peltzer DA, Terwei A, Rew LJ (2015) Drivers of plant invasion vary globally: evidence from pine invasions within six ecoregions. Glob Ecol Biogeogr 25:96–106

Traveset A, Richardson DM (2014) Mutualistic interactions and biological invasions. Annu Rev Ecol Evol Syst 45:89–113

Verloove F (2006) Catalogue of neophytes in Belgium (1800–2005). In: Branquart E, Baus E, Pieret N, Vanderhoeven S, Desmet P (eds) . SOS invasions. Conference Abstract Book, Bruselas, Bélgica, pp 10–11

Wallem PK, Soto N, Cabello JL, Castro SA, Jaksic FM (2014) La invasion del castor americano (*Castor canadensis*) en Tierra del Fuego como oportunidad para analizar la valoración del impacto y manejo de vertebrados exóticos invasivos en Chile. In: Jaksic FM, Castro SA (eds) Invasiones biológicas en Chile. Ediciones Universidad Católica de Chile, pp 209–233

Williams MC, Wardle GM (2007) Pinus radiata invasion in Australia: identifying key knowledge gaps and research directions. Austral Ecol 32:721–739

Williamson M (1996) Biological invasions. Chapman & Hall, London

Wittenberg R, Cock MJW (2005) Best practices for the prevention and management of invasive alien species. In: Mooney HA, Mack RN, McNeely JA, Neville LE, Schei PJ, Waage JK (eds) Invasive alien species. Scope, Island Press, Washington/Covelo/London, pp 209–284

Chapter 4
Naturalization

4.1 Preamble

The propagules that have survived the primary dispersion barrier are able to found a colony, initially of reduced population size. From that point on, the newly arrived species may be called exotic (see Fig. 1.7). The colony will be successful if it persists and expands, occupying a larger area in the recipient region. In a broad sense, the persistence or population establishment of an exotic species in the Introduced habitat is called naturalization. This concept (Table 1.1) can be defined as the process in which the individuals that make up the population are able to complete their life cycle, reproduce, and leave fertile offspring, thereby allowing population growth or at least avoiding extinction. Thus, naturalized species correspond to those exotic species whose populations have been established in population terms in the recipient area (Table 1.1), as opposed to non-naturalized exotic species whose fate is either extinction or persistence through human aid (Heywood 1989; Richardson et al. 2000b; Pyšek and Prach 2003).

By the way, these distinctions should be taken with caution for two reasons. The first is that, in the case of newly arrived species, it is not possible to discern whether they are on their way to naturalization or on their way to extinction. This distinction will only be possible as long as sufficient time elapses to ascertain the full course of the phenomenon. A second precaution relates to human interference or participation in the establishment process. Indeed, traditionally the term naturalization has implied that the arrived colonies are able to complete their entire life cycle without requiring human aid (Richardson et al. 2000b). For some species this condition is complex to evaluate, since, as previously noted, numerous exotic species maintain naturalized populations in anthropized environments (e.g., urban weeds), which constitutes a rather indirect human facilitation effect. In the absence of this type of habitat, many exotic species would probably not settle or disappear from the recipient area.

F. M. Jaksic, S. A. Castro, *Biological Invasions in the South American Anthropocene*, https://doi.org/10.1007/978-3-030-56379-0_4

Naturalization and expansion are interrelated population processes but conceptually decoupled (Sakai et al. 2001). To illustrate the difference, let us think that a founding colony can establish itself locally and remain restricted to a certain place, without expanding its range. In this case, the expansion would be null or very reduced. In contrast, a founding population could be established and, in addition, expand beyond the point where it was inoculated, reaching different levels of population abundance (Kolar and Lodge 2001).

Several factors, intrinsic and extrinsic to the population, influence the naturalization process. Here are a few of them.

4.2 Demographic Aspects

One of the most intuitive concepts in population ecology holds that the greater the number of organisms that make up a population, the less likely it will become extinct (Gotelli 2001). From this it is possible to recognize the link between propagule pressure and the likelihood of establishment of exotic species (see Chap. 3). If propagules are introduced repeatedly and they contain a high number of individuals, the probability of establishment in the recipient area will be considerably higher, in contrast to fewer introduction events and fewer inoculated propagules (Lockwood et al. 2005; Daehler 2006; Simberloff 2009).

Several demographic factors are associated with population establishment. For example, when the size of the founding colony is relatively large, its individuals will have greater expectations of finding a mate, and at the same time, there will be less likelihood that environmental events (e.g., disturbances) will extinguish or considerably reduce the founding colony (Dennis 2002). These conditions are manifestations of the so-called Allee effect, characteristic of small populations (Taylor and Hastings 2005).

Logistically growing populations show a negative function between per capita growth rate and population size (Fig. 4.1a). This relationship imposes a denso-dependent dynamic as a consequence of the increase in population size, which negatively affects the per capita growth rate (Gotelli 2001). However, some species under small population size conditions exhibit a different functional relationship, usually modeled as a negative quadratic function (Fig. 4.1b, c). This implies a lower per capita growth rate than expected for its population size. According to Taylor and Hastings (2005), this type of effect can vary between a weak or severe magnitude, depending on the parameters that define the function. In the first case, and despite the small population size, the per capita growth rate always acquires positive values (Fig. 4.1b), leading to a reduction in the population growth rate given a certain population density value. On the other hand, under conditions of severe Allee effect, the per capita growth rate could take negative values under a threshold density (Fig. 4.1c), implying that the small population will decrease to extinction below this point. Because the founding colonies of exotic species are often small, the likelihood of extinction increases.

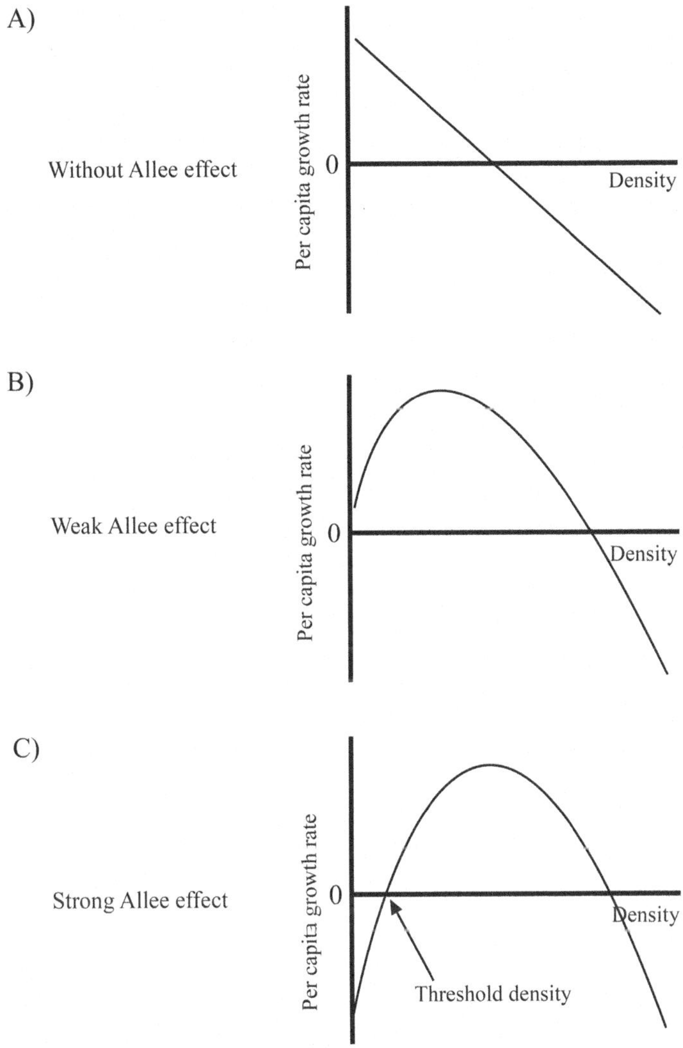

Fig. 4.1 Graphical description of possible functional relationships between per capita growth rate and population density. (**a**) Dynamics without Allee effect in which the per capita growth rate declines with population size; (**b**) dynamics with weak Allee effect in which the per capita growth rate shows a quadratic relationship with population size; (**c**) dynamics with strong Allee effect (see text)

4.3 Genetic Aspects

A field explored in demographic studies of invasive populations is the role of genetic diversity in the establishment or naturalization of the founding colony (Miura 2007). It should be kept in mind that from a population point of view, genetic diversity is an attribute that will allow the population to respond to environmental fluctuations in the recipient area (Lee 2002; Prentice et al. 2008). Therefore, greater genetic diversity is a factor that should facilitate naturalization, in contrast to a condition of lesser genetic diversity. Because the founding colonies usually have a small population size, they contain a minority fraction of the genetic diversity compared to the population distributed in the region of origin (Avise 2004). Therefore, it is likely that this reduced genetic diversity represents an impediment to its population establishment (Prentice et al. 2008). In addition, deleterious genetic effects can lead the founding colony to extinction within a few generations (Hendry et al. 2007), due, for example, to endocrine depression. Endocrine depression occurs when the biological suitability of the descendants of a founding population is impaired by increased homozygosity for deleterious alleles (Ayala 1982; Fonseca et al. 2000; Tsutsui et al. 2000; Golani et al. 2007).

As noted, greater genetic diversity in the founding colony will make its establishment more likely, but at the same time, it will favor spatial expansion, as the new allelic and gene combinations of the progeny will be better prepared to face the new environmental conditions in the recipient region. In angiosperms, polyploidy has been an important evolutionary factor (Adams and Wendel 2005), which has been associated with increased invasive ability by providing hybrid vigor, tolerance, and resistance to environmental stress, adaptive versatility, competitive advantages, and increased phenotypic responsiveness (Baker 1974; te Beest et al. 2011). According to Hurka and Neuffer (1997) and Slotte et al. (2006), the hybridization evolution of the shepherd's purse (*Capsella bursa-pastoris*) would explain its tetrapoliploid endowment (2n = 32), which in turn would be related to its ability to naturalize in various types of environments in the world. Currently, this Brassicaceae is distributed almost all over the globe (Randall 2002). In contrast, the other two species that make up the genus *Capsella* (*Capsella rubella* and *Capsella grandiflora*) are diploid taxa (2n = 16) and exhibit a much more restricted geographic distribution (Slotte et al. 2006).

Although polyploidy or simple ploidy, at least theoretically, can facilitate naturalization, the inverse relationship is not necessarily correct. In other words, a species that occurs in various types of environments or habitats will not necessarily exhibit variation in its ploidy. Indeed, Irimia et al. (2017) analyzed the importance of ploidy in the invasion of *Centaurea solstitialis* in three continents, including the countries of Turkey, Spain, Argentina, Chile, California, and Australia. These authors studied the genetic constitution of 52 populations including their range of origin (Turkey), finding a constancy in their diploidy in these populations (Irimia et al. 2017).

Lizarralde et al. (2007) examined the genetic diversity of North American beaver populations (*Castor canadensis*) in Tierra del Fuego. These authors noted that the beaver population, currently close to 100,000 individuals, descends from a few breeding pairs that were initially introduced for the purpose of stimulating the development of the fur industry in the region. For a total of 30 individuals captured in Tierra del Fuego and 5 individuals from its original range (Alaska), the sequence of three molecular markers (cytochrome b, D-loop, and the 12S region of the rRNA) extracted from different animal tissues (liver, muscle, and spleen) was compared. The results showed that in the South American population, there were ten different lineages, none of which were represented in the original range population. These results reveal the importance of the processes of genetic drift in the generation of diversity, confirming, in addition, a null genetic flow from the original population with the one introduced after its establishment in Tierra del Fuego.

Genetic attributes can determine the success of invasion through the generation of phenotypic plasticity (Daehler 2003; Funk 2008; Davidson et al. 2011), a characteristic that has been widely discussed in plants (Alpert 2000; Hulme 2008). Theoretically, phenotypic plasticity may be advantageous for invasive species by allowing them, for example, to settle beyond the location of the founding colony, to occupy various types of environments or habitats, and to be better able to cope with variations in environmental conditions and resources over space and time (Godoy et al. 2011). However, recent studies have shown that invasive plant species do not necessarily have greater phenotypic plasticity than their native counterparts (Gianoli 2004; Godoy et al. 2011; Dawson et al. 2012). Molina-Montenegro et al. (2018) experimentally evaluated the role of phenotypic plasticity in the success of *Taraxacum officinale*, a perennial herb, whose range of origin corresponds to the European continent. In Chile, this species exhibits a wide tolerance to stressful environmental conditions associated with an extensive range of geographical distribution, which reaches approximately 4000 km latitude (Molina-Montenegro and Naya 2012). In these experiments, the performance of *Taraxacum officinale* was compared to *Hypochaeris thrincioides*, a native species of similar distribution to *Taraxacum officinale* and of the same taxonomic family (Asteraceae). Under controlled greenhouse conditions, the plants were confronted with different types of environmental conditions as a function of changes in temperature, light intensity, and water available in the substrate. The results showed that *Taraxacum officinale* showed a greater capacity for survival and biomass accumulation than that of *Hypochaeris thrincioides*. Additionally, it showed greater phenotypic plasticity for several ecophysiological traits (Molina-Montenegro et al. 2018), thus revealing the importance of this trait in the binding ability.

Certainly, naturalization is determined by genetic and phenotypic attributes that are expressed throughout development. When these attributes are compared between species of different lineages, it is possible to wonder about their importance in determining naturalization. Some of the fundamentals and advances in this topic will be discussed below.

4.4 Morphofunctional Attributes Favoring Naturalization

Naturalization requires the existence of preadapted morphofunctional traits to cope with a novel environment. For example, it is unlikely that tropical plant species could colonize temperate or desert areas, where environmental conditions of soil, humidity and temperature are not appropriate for their survival. These types of considerations have stimulated two lines of research seeking to predict which species could become naturalized and expand their distribution by becoming invasive. On the one hand, there are those studies focused on the search for morphofunctional attributes associated with invasive success and, on the other hand, those that evaluate environmental correlations between the areas of origin and arrival.

The search for morphofunctional attributes that facilitate naturalization was initiated by Baker (1974). Baker developed this concept thinking mainly about weeds of cultivated fields, by means of a comparative analysis of species that were catalogued as such versus those who were not. Although Baker's interest was centered in the agronomic field, his ideas extended to the ecology of invasions.

Various characteristics present in organisms favor or make naturalization more likely. In vascular plants, for example, these characteristics are often related to rapid growth and reproduction, reproductive systems that combine sexual and asexual modes, and abundant production of flowers and seeds (Noble 1989). Additional traits are recognized in seed dispersal (e.g., seed size, dispersal mode), the ability to fix nitrogen, and the biogeographic origin or latitudinal band that invasive species occupy in their original range (Rejmánek 1995; Rejmánek and Richardson 1996). Some of these attributes are phylogenetically associated with certain lineages, so that these lineages have intrinsic characteristics that make them more likely to naturalize once they have been relocated (Pyšek 1998).

For his part, Ehrlich (1989) applied these ideas to vertebrates, wondering why certain vertebrate species are particularly good colonizers and others are not? Are there traits that can help explain this difference? This author found that, in general terms, it is possible to recognize in those species of invasive reputation a set of attributes that a priori seem to ensure their success (see Table 4.1). These attributes include high mobility, a generalist diet, short generation time, high genetic variability, and plasticity in life history attributes (e.g., r and K strategies; Table 4.1), among others. This author also highlighted several cases in which phylogenetically related

Table 4.1 Attributes associated with successful and unsuccessful invasive vertebrate species (Ehrlich 1989)

Successful invaders	Unsuccessful invaders
Wide native range	Small native range
Abundant in its original range	Rare in its original range
Nomad	Sedentary
Generalist diet	Specialist diet
Generational time short	Generational time long
Able to switch between strategies *r* and *K*	Unable to change
High genetic variability	Low genetic variability
Gregarious	Solitary
Female able to colonize on her own	Female unable to colonize on her own
Bigger than most of your relatives	Smaller than most of your relatives
Associated with *Homo sapiens*	Not associated with *Homo sapiens*
Wide environmental tolerances	Narrow environmental tolerances

species, such as the house sparrow *Passer domesticus* and its congener *Passer montanus* were exhibiting differences in their invasiveness despite sharing several of these attributes. It is now clear that not all successful naturalized or colonizing species – whether plants or animals – possess all these invasive traits and that, at the same time, possessing them does not guarantee their naturalization (Crawley 1987).

Studies analyzing the role of invasive traits in determining success usually offer many discrepancies in their conclusions (Pyšek and Richardson 2007). Some of them emerge as a result of the way that different researchers approach their analyses. On the one hand, the so-called naturalization success is a concept that calls for confusion as it has been operated in various ways. Some focus on it as success in the colonization process, others measure it as establishment, while others focus on the process of population expansion (Pyšek and Richardson 2007). Therefore, traits that confer advantage at one stage of the invasive process do not necessarily confer advantage at another stage (Perrings et al. 2005; Pyšek and Richardson 2007). Associated with the above, studies of invasive traits can be grouped into two categories. There are those that compare taxa related at a taxonomic or phylogenetic level, for example, contrasting the performance of species belonging to the same genus or family (Pyšek and Richardson 2007). Others have a more eclectic approach at the time of including species, oriented to the comparison of floras or complete assemblages (Pyšek and Richardson 2007). The studies using the first approach tend to identify with enough precision the traits of interest, since, being related species, it is possible to control the effect of confusing traits (Pyšek and Richardson 2007). In contrast, studies focusing on the comparison of floras are less-detailed approximations, which often introduce more "noise" into the analysis (Pyšek and Richardson 2007). According to Pyšek and Richardson (2007), both types of studies can be informative in validating the role of invasive traits, so their results should be considered complementary.

4.5 Range Environmental Correlations

As mentioned above, the traits that facilitate naturalization are of diverse nature and they are associated with phylogeny or taxonomy (Pyšek 1998). Thus, the different attributes that promote colonization on their own can be found distributed in different lineages, giving rise to the existence of successful colonizers of different evolutionary origins (Pyšek and Richardson 2007). This point can be illustrated by examining the role of vegetative reproduction in vascular plants, a feature that cannot be a condition of success for vertebrates, since this type of reproduction is not their own. This situation has led to the search for other types of mechanisms to explain the success of invasive vertebrates, for example, those based on climate analogy.

The climatic analogy supposes that organisms that in their range of origin are exposed to certain environmental conditions (e.g., climate) should exhibit a high probability of becoming naturalized in areas of similar climate, acquiring a range similar to that shown in the region of origin (Colwell and Rangel 2009). Indeed, this variable has been shown to be a good predictor of naturalization and geographic extent of the secondary range, partly because it assumes that naturalization does not depend exclusively on the morphofunctional configuration of organisms but also on their interaction with the environment (Hierro et al. 2005). For example, Sax (2001) compared the latitudinal limits of the distribution of introduced birds and mammals reciprocally between the continents of North America and Eurasia (Fig. 4.2). The analysis showed statistical significance for the upper latitudinal limit estimated as the highest latitude of the range (birds, $r^2 = 0.6$, $P < 0.01$; mammals, $r^2 = 0.38$, $P < 0.001$; see Fig. 4.2a, b), although this relationship was weaker, but equally

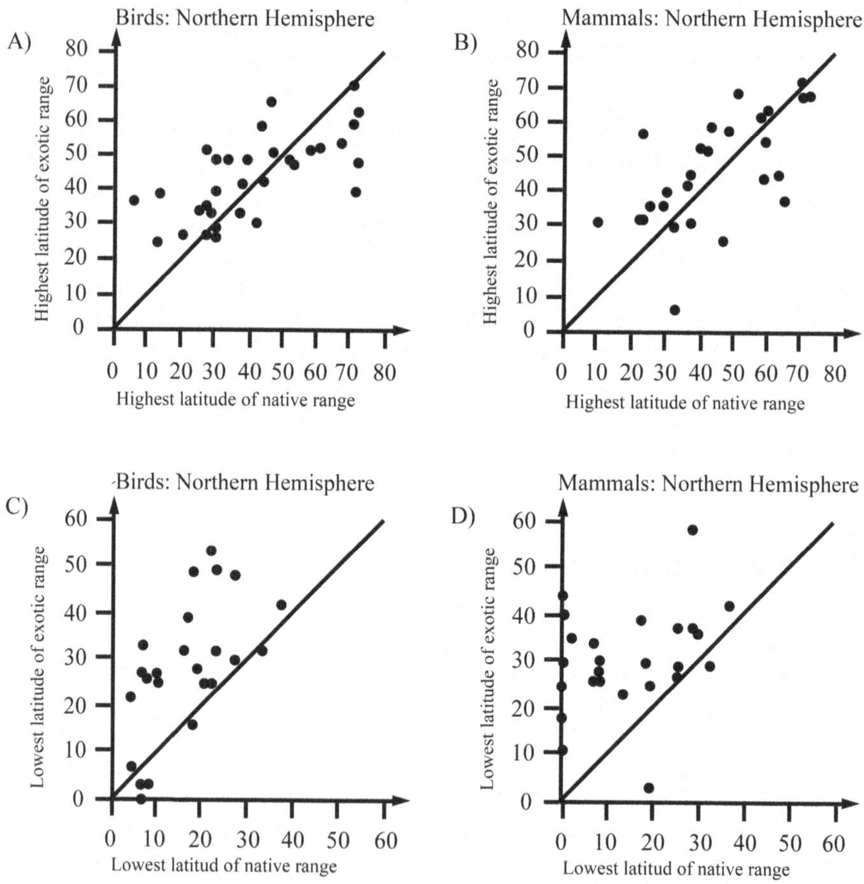

Fig. 4.2 Distribution of geographic range limits for bird and mammal species in northern hemisphere continents (Sax 2001). The relationship between the maximum latitudes of the native range and the maximum latitudes of the exotic range for birds (A) and mammals (B) and the relationship between the minimum latitudes of the native range and the minimum latitudes of the exotic range for birds (C) and mammals (D) are shown

significant for the lower limit, estimated as the lowest latitude of the range (birds, $r^2 = 0.13$, $P < 0.001$; mammals, $r^2 = 0.16$, $P < 0.001$; see Fig. 4.2c, d).

In plants, Casado et al. (2015) analyzed geographic distribution patterns for 645 species of naturalized herbs in Chile; 75% of these species come from the Iberian Peninsula, so they were shared with Spain and Portugal. These authors found correspondence between the prevailing climate in the source region (the Iberian Peninsula) and the recipient (Chile), concluding that these coincident distributional patterns between range of origin and container can be originated by the effect of environmental filters that favor a similar set of species, pre-adapted to the climatic and habitat conditions (Casado et al. 2015). This would be the case of the species members of the families Asteraceae, Brassicaceae, Fabaceae, and Poaceae, which not only exhibit a similar geographic distribution according to the climatic situation, but their diversity in central Chile is represented by a fraction similar to that exhibited in the Iberian Peninsula (Martín-Forés et al. 2012). Similar results were reached by Arianoutsou et al. (2013), after comparing the contingent of naturalized species in the five Mediterranean regions of the world. This study included 1627 established species in the five biomes, climatic equivalents of Mediterranean type, documenting that the greatest representation of naturalized species of the analyzed set was in the southwest of Australia, which, in turn, showed a greater taxonomic similarity between their different habitats (Arianoutsou et al. 2013). These authors indicated that from the point of view of its taxonomic composition in naturalized species, this region would be the most internally homogenized, while the Mediterranean Basin would be the least internally homogenized. These authors ruled out that the species with the widest niche (i.e., generalist species), which allows them to colonize a greater number of habitats within each region, would contribute to the internal homogenization of each of them.

In general terms, these results show that there is a good correspondence between native and naturalized range limits, both for vascular flora and for vertebrate fauna. This phenomenon may be related to the restrictions imposed by the environmental conditions associated with latitude on the morphofunctional traits that support the performance of organisms.

4.6 Population Interactions as Determinants of Naturalization

Population interactions have also been shown to be important in the naturalization process. In effect, once they arrive, exotic species establish interactions with members of the native community or with other previously established exotic taxa, implicating relations of consumption, co-use of resources, pollination or dispersion of seeds, etc., which may well facilitate or impede their establishment. Based on this, several hypotheses have been raised about the mechanisms that promote naturalization, so we will examine some of them.

4.7 Antagonistic Interactions

Charles Darwin was one of the first to address the causes of naturalization, although only recently have some of his ideas been reformulated as hypotheses and formally tested. One of them is known as the "Darwin's naturalization hypothesis" (Daehler 2001). To formulate this idea, Darwin relied on the observations of the Swiss botanist de Candolle (1855), who had noted that the exotic plants present in numerous localities around the world, especially on island communities, belonged to genera not represented in the recipient communities. From this simple observation, Darwin hypothesized that invasions could be favored when the invasive species does not share evolutionary kinship with members of the recipient community, since this evolutionary distance would result in a reduction of the intensity of interspecific competition (Darwin 1859). On the contrary, the invasion of a species belonging to a genus already represented in the recipient community would be faced to a more intense competition because related species may have greater similarity in the requirements of environmental conditions and resources.

For more than a century, these ideas went unnoticed. Analyzing six regional floras from United States, Mack (1996) documented that naturalized species were more frequently distributed in exotic genera not previously represented in such areas, finding support for Darwin's proposal. Rejmánek (1996, 1998) found that most naturalized grass species in California belong to genera not represented in the region's flora. These results also support Darwin's hypothesis. Particularly interesting is the study by Strauss et al. (2006), who, using a phylogenetic tree of the Poaceae family, found the invasiveness of weedy grass species in California was associated with genera with greater phylogenetic distance to native grasses, compared to non-invasive exotic grasses that were more related to native taxa (see Fig. 4.3). That is, the most invasive exotic grasses in California belong to evolutionarily more distant lineages than the least invasive grasses (Strauss et al. 2006). As can be concluded from the evidence analyzed, these antecedents support Darwin's naturalization hypothesis.

Other studies, on the other hand, have not found favorable evidence for the hypothesis. For example, Daehler (2001) – who formalized these ideas under the terms of "Darwin's naturalization hypothesis" – found that in the naturalized flora of Hawaii, the exotic genera (i.e., taxa completely novel on the islands) did not show overrepresentation of naturalized species when compared to genera with exotic species but already represented on the islands. Duncan and Williams (2002) obtained similar results for the naturalized flora in New Zealand, while Ricciardi and Mottiar (2006) analyzing genera of freshwater fish in 11 lakes in the northern hemisphere did not find support for the hypothesis either.

An interesting study conducted in Chile compared the extent of the geographic range occupied by different species of naturalized plants (Escobedo et al. 2011). In this group there are species that belong to completely new genera to the national flora, while others are ascribed to already represented genera. The same applies to higher taxonomic categories, such as family and order. Thus, these authors

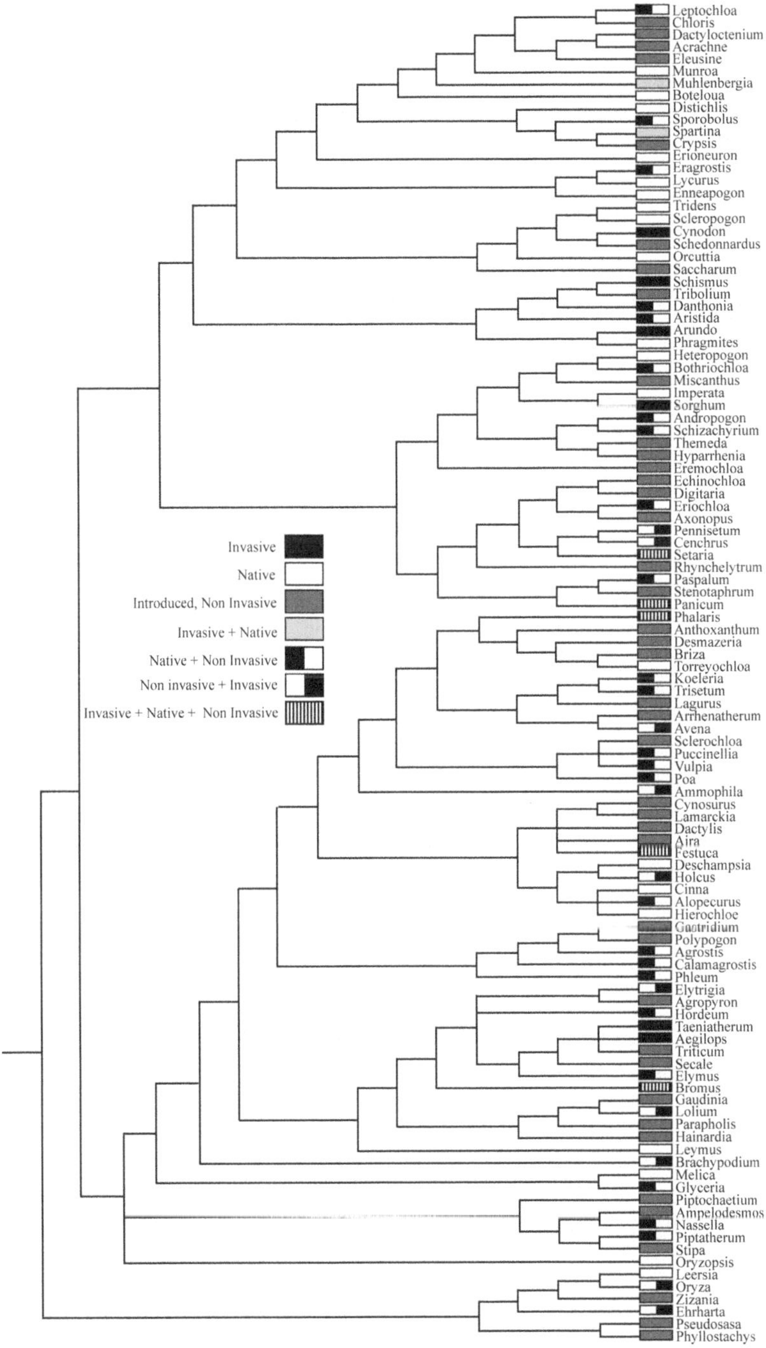

Fig. 4.3 Supertree for native and exotic grass genera (Poaceae) present in California (Strauss et al. 2006). As described by these authors, the species of the Poaceae family that have achieved invasive (weed or plague) status in California are less closely related to the native community than non-invasive species. Boxes of different colors and shrinkage accompany genera that contain native, invasive, and non-invasive species, alone or in combination

distinguished for the naturalized flora, three levels of kinship: near, intermediate, and distant. The "near" level was represented by naturalized taxa belonging to genera already represented in the native flora of Chile; the "intermediate" naturalized flora was composed of species belonging to genera not represented in the native flora but represented at the taxonomic category of family; finally, at the "distant" level of kinship were grouped the species of taxonomic families completely new to the country. Escobedo et al. (2011) found that range size (evaluated as latitudinal extent, in km) was significantly associated with kinship level, although inversely to that predicted by Darwin's naturalization hypothesis. That is to say, the species of close kinship had ranges on average more extensive than those of more distant kinship taxa. These authors suggested that greater kinship between the exotic species and the recipient community may facilitate (and not inhibit) naturalization, since a taxonomic or phylogenetic closeness indicates the possession of homologous, preadapted traits that facilitate the expansion of the relative flora (Escobedo et al. 2011).

The facilitating role of phylogenetic kinship over naturalization was a fact also foretold by Darwin (1859). In fact, this paradox is called *conundrum* (Diez et al. 2008; Thuiller et al. 2010). Although several studies have attempted to evaluate the importance of phylogenetic kinship in facilitating or inhibiting naturalization success (e.g., Jones et al. 2013; Marx et al. 2016), most of them come from evidence obtained from analysis of compositional patterns of communities, in the form of natural experiments (Ma et al. 2016). Approaches based on manipulative experiments, performed under laboratory or field conditions, are scarcer. Among these, we can mention one by Jiang et al. (2010) in which they evaluated Darwin's naturalization hypothesis, using bacterial communities grown in culture media. The communities were assembled with four species of bacteria (*Bacillus cereus*, *Bacillus pumilus*, *Frigoribacterium* sp., and *Serratia marcescens*) and a predatory protist (*Tetrahymena pyriformis*). As an invasive taxon, a specific strain of *Serratia marcescens* was used. The recipient communities were organized into monoculture treatments and different combinations of species, which were inoculated with *Serratia marcescens*. The results showed that the establishment of *Serratia marcescens*, evaluated as colony density, was determined by the level of phylogenetic kinship to the community; thus, the colony density reached by *Serratia marcescens* was greater in the treatments in which the phylogenetic kinship with the recipient community was greater. Peay et al. (2012), using a methodological approach equivalent to the previous one, studied the composition of communities of yeasts that grow in flowers of *Mimulus aurantiacus*. In this case, there were six species belonging to the genera *Candida*, *Starmerella*, and *Metschnikowia* which formed different communities and monoculture containers for the inoculated species. Again, the results of Peay et al. (2012) showed that colonization was more successful in treatments where the phylogenetic distance between the community and the inoculated species was greater. Finally, Morales et al. (2016) analyzed the establishment of the fungus. *Trichoderma* cf. *harzianum* in two types of communities of microorganisms, which were assembled in two treatments of differing phylogenetic distance with respect to *Trichoderma* cf. *harzianum*. The results obtained by these authors showed that the abundance reached by the inoculated species was greater in the treatment in which the community showed less phylogenetic kinship.

In general terms, these studies made by Jiang et al. (2010), Peay et al. (2012), and Morales et al. (2016) have shown evidence in support of Darwin's naturalization hypothesis. They have in common, not only the fact of manipulative experimentation but also by using microorganisms as subjects of study. As far as we know, only Castro et al. (2014) have evaluated Darwin's naturalization hypothesis in experimental terms, using vascular plants as study subjects. These authors conducted experiments evaluating the lettuce establishment (*Lactuca sativa*) in previously established vascular plant communities. The experiments were implemented in greenhouses, where irrigation and soil, temperature, and humidity conditions were controlled. The experimental assemblages consisted of five species considered as resident assemblages, which were extracted from a pool of 14 possible assemblages. Thus, five different container assemblies were established in their composition, but at the same time, located in a gradient of phylogenetic kinship to the inoculated species. In contrast to previous studies, the results showed that the establishment of *Lactuca sativa* was not related to phylogenetic kinship, which then did not support Darwin's naturalization hypothesis.

Another three hypotheses explaining successful naturalization and which are widely disseminated in the literature are (a) natural enemy release, (b) novel weapon, and (c) increased competitive ability. The first was also raised by Darwin (1859) and further refined by Elton (1958); he proposes that host communities usually do not contain natural enemies (i.e., predators, parasites, pathogens), which are present in the community of origin. Therefore, organisms introduced into a new area are favored by being freed from the negative effect of natural enemies.

The natural enemy release hypothesis has been proposed to explain the invasion of the European rabbit (*Oryctolagus cuniculus*) in central Chile. Using a comparative approach between Mediterranean communities in Chile and Spain, Jaksic and Soriguer (1981) documented an under-representation of the exotic lagomorph in the diet of native Chilean predators. In fact, for 29 predator species present in Spain, the region of origin of *Oryctolagus cuniculus*, the consumption of European rabbit represented 20% of total prey, while in central Chile, this representation was 2% for 16 species of predators. These authors determined that the predator sizes between the two Mediterranean ecosystems were similar and hypothesized that the apparently lower efficiency of Chilean predators in capturing the rabbit could explain their escape from predation.

It is interesting to note that while the natural enemy release hypothesis constitutes the conceptual basis of pest and weed control practices, widely applied in biological conservation and productive systems (Huffaker and Messenger 1976; Holt and Lawton 1994; Zavaleta et al. 2001), the empirical evidence supporting it is only partial. In plant-herbivore systems, for example, the distinction between generalist and specialist herbivores is relevant, since every recipient community probably contains generalist herbivores capable of consuming a broad spectrum of plants or their tissues, including exotic species (Crawley 1987, 1997; Maron and Vilà 2001). However, several studies indicate that in their communities of origin, plants are rarely controlled by their natural enemies, so that the eventual colonization of recipient communities does not necessarily agree with the metaphor of escape. At the same time, in the recipient communities, it is also possible to find herbivores capable of negatively affecting exotic plants, so that this situation again does not coincide with the expectations of the present hypothesis (Maron and Vilà 2001; Keane and Crawley 2002; Parker et al. 2006). According to Meijer et al. (2016), many studies have evaluated predictions derived from natural enemy release hypothesis, using a community approach (where the performance of native and non-native species in the same habitat is compared) or a biogeographic approach (where the performance of a species in its native and non-native range is compared); these authors pointed out that together, these studies have offered heterogeneous and discrepant results among themselves (Keane and Crawley 2002; Jeschke et al. 2012). Focusing the analysis on studies carried out on herbivorous plant systems, it is evident that the hypothesis tends to be supported because the diversity (i.e., number of species) and its abundance of herbivorous insects reach lower levels in exotic than in native plants. However, these authors pointed out that very few studies apply both approaches simultaneously on a reciprocal basis, which constitutes a more appropriate and complete approach to test the natural enemy's escape hypothesis.

The novel weapon hypothesis was originally proposed to explain the naturalization of plants that exude allelopathic compounds and that by this mechanism displace native species with which they compete (Thompson 1991; Callaway and Ridenour 2004). Specifically, this hypothesis proposes that the release of such allelochemical compounds by the exotic plant may act as a factor inhibiting or excluding the growth of native plants in the resident community. This would favor naturalization by reducing native competitors and could probably also discourage herbivores. Allelopathy has long been suggested as one of the factors affecting vegetational dynamics in Mediterranean ecosystems and the ability to invade them (Müller 1966; Ballester and Vieitez 1979). However, in South America this effect has been poorly evaluated in field conditions (Montenegro et al. 1978; Fuentes et al. 1984, 1987; Kruse et al. 2010). Still, there are a few examples that demonstrate the importance of allelopathy for the success of the invasion in Mediterranean ecosystems, such as the case of pastures made of *Elytrigia repens* and *Vulpia myuros*, both species of European origin (Friebe et al. 1995; An et al. 1997), or of *Centaurea diffusa*, an invader of Eurasia whose allelopathic effects are expressed on western North American herbs (Hierro and Callaway 2003). Particularly severe seems to be the effect of lantana (*Lantana camara*), which from tropical America and South America has invaded Mediterranean forests of Australia, exerting allelopathic effects (Gentle and Duggin 1997). Experimental evidence indicates that the exudation of allelopathic compounds by lantana is capable of suppressing germination and reducing the growth rate and survival of two native species (Gentle and Duggin 1997).

The silver wattle (*Acacia dealbata*) has been recognized as an exotic species that may affect the germination and growth of native species established under or in the vicinity of its canopy. This effect would be determined by the exudation of allelochemical compounds whose identity has not been fully determined (Aguilera et al.

2015). Two native species, one tree (*Quillaja saponaria*) and an herb (*Helenium aromaticum*), were experimentally exposed to exudates from *Acacia dealbata* extracted from leaves, flowers, pods, and seeds. The germination and early growth of seedlings of *Quillaja* saponaria and *Helenium aromaticum* were significantly affected with respect to the control treatments (Aguilera et al. 2015). However, the role of this mechanism in promoting the invasion of *Acacia dealbata* still requires further examination, as intraspecific allelopathic effects (towards seedlings of the same species) have also been reported (Aguilera et al. 2017).

Finally, the hypothesis of increased competitive ability is based on the fact that organisms are subject to compromises in the allocation of nutrients and available energy that allow them to keep up with physiological functions such as reproduction, growth, and defense (Tilman 1988; Stearns 1992). Assuming that there are few natural enemies in host communities, this hypothesis predicts that a net amount of nutrients and energy will be reallocated from defensive functions, for example, to growth and reproduction. This hypothesis implies that invasive species are relatively exempt from physiological commitments, the result of which is better and more competitive performance against native species (Daehler and Strong 1997; Willis et al. 1999).

4.8 Facilitating Interactions

Most hypotheses and theoretical explanations regarding naturalization emphasize antagonistic interactions, whether of competition, predation, or parasitism. However, some studies have shown that positive interactions within the community (i.e., mutualism, facilitation) can be very important in promoting naturalization (Simberloff and von Holle 1999; Richardson et al. 2000a). These positive interactions can be between exotic species that mutually favor their naturalization (Simberloff and von Holle 1999; Richardson et al. 2000a; Christian 2001; Grosholz 2005) or between native species that favor the naturalization of exotic species (Richardson et al. 2000a). One of the most cited examples in Chile involves the presumed effect of both domestic (e.g., horses, cows) and wild (e.g., rabbits, hares) exotic herbivores, which after being introduced into the central portion of the territory would have facilitated the expansion of exotic Eurasian herbs naturalized in shrub formations (Kruger et al. 1989; Kalin-Arroyo et al. 2000; Fuentes et al. 2007). Indeed, these herbivores can consume and defecate intact seeds, especially herbs that tend to be very small and resistant to digestion. To date, however, no quantitative studies have been carried out on this phenomenon.

Facilitating interactions exerted by native species may be important to allow the establishment of exotic species, especially in extreme environments (Molina-Montenegro et al. 2012). The colonization of exotic plant species in high mountain environments is a clarifying case. Temperature, substrate, and humidity conditions in mountain areas constitute a barrier to the colonization of plant species accustomed to less extreme conditions (Kikvidze et al. 2015). However, given the changes in the average temperature of the atmosphere, mountain environments have become susceptible to colonization (Kikvidze et al. 2015). In this scenario, Cavieres et al. (2007, 2008) have shown that the diversity of cushion plant species facilitates the colonization of exotic species that normally grow at lower altitudes (Kikvidze et al. 2015). Gundale et al. (2016) examined the role of native endophytic fungi in the establishment of *Pinus contorta* in various locations around the globe. These authors found that the diversity of native endophytic fungi favors the establishment of *Pinus contorta*.

4.9 Biodiversity as Biotic Resistance to Naturalization

Elton (1958) observed that communities richer in species usually contain fewer exotic taxa than less diverse communities. Based on this background, Elton proposed that the susceptibility of a community to invasion is negatively related to community richness, so that as richness increases, the number or proportion of exotic species established in the community is reduced (Fig. 4.4a). MacArthur (1970) later gave hypothesis status to Elton's observation, arguing that the mechanism implicit in this relationship was interspecific competition. Invoking the niche

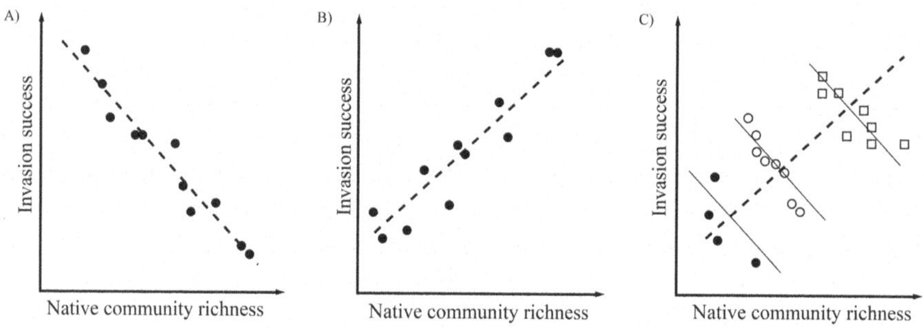

Fig. 4.4 Hypothetical relationships between community diversity (made up of native species) and the success of biological invasions assessed as the number of exotic species capable of colonizing the community. (**a**) Negative relationship describing a decrease in the probability of invasion success as the species richness of the recipient community increases; the biotic resistance mechanism would be interspecific competition (Elton 1958); (**b**) positive relationship showing an increase in the probability of invasion success as the species richness in the recipient community increases; this pattern would be mediated by an increase in spatial heterogeneity and resource availability as diversity increases (Levine 2000); (**c**) "conciliatory" vision describing the positive and negative trends observed in field studies; this model proposes that discrepancy emerges as an effect of the spatial scale at which this functional relationship has been studied (Shea and Chesson 2002); C) 'conciliatory' vision describing the positive and negative trends observed in field studies; this model proposes that discrepancy emerges as an effect of the spatial scale at which this functional relationship has been studied (Shea and Chesson 2002)

theory, MacArthur argued that greater species diversity in the recipient community allows most resources to be occupied. As the community enriches itself, it also saturates the niche axes, decreasing the availability of resources for new settlers, which negatively affects the likelihood of establishment. According to their richness, communities would be able to host a certain number of exotic species, which is why they are said to exhibit invasiveness or differential resistance to invasion (Rejmánek 1989; Levine and D'Antonio 1999; Lonsdale 1999).

Inspired by the approaches of Elton and MacArthur, numerous studies have been conducted over the past 25 years to determine the functional relationship between host community diversity and invasiveness or resistance to invasion. A controversial fact emerged when it was stated that, on a regional scale (a region includes a group of communities), the richness of resident species showed a positive – not negative – relationship with the number of established exotic species (Fig. 4.4b; Fox and Fox 1986; Crawley et al. 1999; Lonsdale 1999; Stohlgren et al. 1999; Levine 2000). The line of argument that explains these findings assumed that the diversity of resident species generates new available resources, which can be used by exotic species, allowing their establishment. These findings opened up a number of avenues for discussion. One of them focused on the role of diversity as an effective and efficient mechanism for repelling new species and the importance of interspecific competition as a factor of resistance to invasion (Levine and D'Antonio 1999).

By using a mechanistic approach, Davis et al. (2000) and Shea and Chesson (2002) have provided conceptual tools to explain and eventually predict naturalization. Davis et al. (2000) proposed a graphic model that emphasizes the importance of available resources for an exotic species to establish itself (Fig. 4.5). According to the model, the availability of resources is determined by the rates of provision or supply of resources (in axis x; Fig. 4.5) and consumption (in axis y; Fig. 4.5), which define a consumption/supply balance isocline. Exotic species could only be established if the rate of provision, consumption, or both increases the availability of resources for them. Although this proposition seems trivial, the contribution lies in the need to evaluate competitive resistance to invasion based on the availability of resources, and not just as a phenomenological response to the number or abundance of the invader.

Shea and Chesson (2002) developed a conceptual framework to explain the success of biological invasions based on the concept of niche opportunity, which describes the set of favorable conditions found in a community that promote the establishment of exotic species. Thus, the level of available resources is part of the niche opportunity, and at the same time, it is the absence of predators, the presence of safe sites and dispersing agents, and so on. Although the authors did not foresee a necessary relationship between niche opportunity and the success of the invasion, their analytical and experimental results reveal that more diverse communities tend to host fewer exotic species compared to less diverse ones.

From a regional perspective, communities are exposed to a natural exchange of species (β-diversity) resulting from the heterogeneity of environmental (climate, disturbances), geographic, and resource conditions (Ricklefs 1987). Thus, the set of communities in a region form habitats conducive to an increasing number of exotic

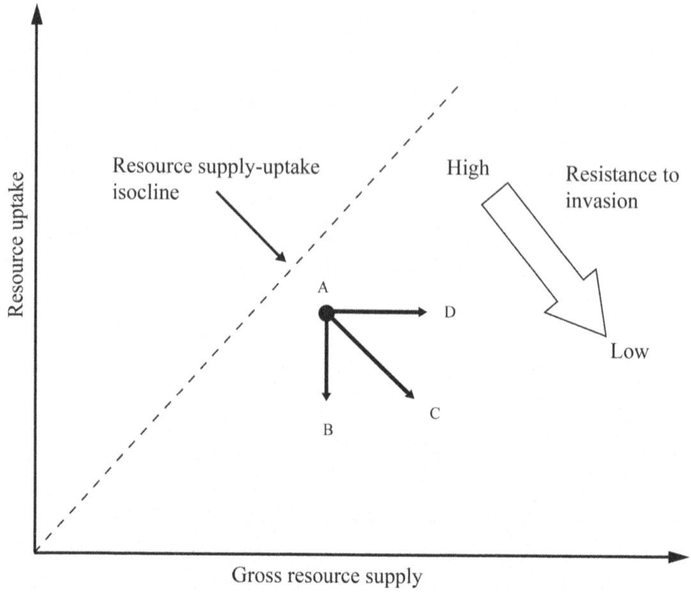

Fig. 4.5 Naturalization model based on the balance of resource supply and consumption by the community (Davis et al. 2000). The segmented line represents the isocline of resource availability in which the supply and consumption by the recipient native community are balanced. Due to this balance, the points close to the isocline constitute an area of maximum resistance to invasion, as there will be no resources available for the exotic species. Point A represents any condition in which a community finds itself with respect to its isocline. Therefore, by increasing the supply of resources to the community (moving from point A to point D) or decreasing the consumption of resources by the community (moving from point A to point B) or both (moving from point A to point C), the community will move into areas where resistance to invasion is diminished

species (Naeem et al. 2000; Cornwell and Grubb 2003). However, this situation alone does not make it possible to recognize whether and how richness effectively promotes resistance to invasion. To unravel this question, field and microcosm experiments have been carried out, which have brought clarity to the discussion. In simple terms, these experiments control the number and/or abundance of native species by adding variable numbers of exotic species (or their propagules). They then assess whether the initial community richness affected the number, abundance, or biomass of the established exotic elements (Tilman 1997). Interestingly, most of these experiments effectively document a negative relationship between diversity and invasiveness or, what is equivalent, a positive relationship between diversity and resistance (Stohlgren et al. 1999; Naeem et al. 2000; Kennedy et al. 2002; Byers and Noonburg 2003; Zavaleta and Hulvey 2004). Thus, the idea that biodiversity itself constitutes a barrier to invasion has gained strength. Shea and Chesson (2002) have attempted to reconcile the discrepancies observed at regional and community level through a graphical model that reconciles both visions (Fig. 4.4c). This model describes the relationship between successful invasion and species richness at different spatial scales. In a local or spatially reduced view, the increase in native species will generate a decrease in the niche opportunity for invasive species attempting to colonize the environment, thus generating a negative relationship between the number of native species and the number of exotic species (Fig. 4.4c). However, extrinsic factors vary considerably as the spatial scale increases, heterogeneity that may favor establishment and coexistence between exotic and native species. In this way, a positive relationship will be generated between the number of native species and the number of invasive species due to an increase in the spatial scale (Fig. 4.4c). However, the debate still persists. A recent review of experimental and observational studies revealed that resistance mediated by interspecific native-exotic interactions, while it may reduce invasion, does not reverse or prevent it (Levine et al. 2004). This has opened up a new front of debate, this time related to the efficiency or permeability of the community resistance barrier.

In Chile, Sax et al. (2002) studied the correlations between specific richness and abundance of native and exotic herbs in coastal scrubland and xeric scrubland communities, comparing them with similar communities in California. Under the different spatial scales used (1–400 m^2), the results were consistent in showing that the specific richness of native and exotic herbs correlated positively, while their respective abundances did not show a significant relationship. According to Sax (2002), these results could be explained by two mechanisms: firstly, the importance of positive interactions (over negative ones) in the communities, which allows the coexistence of native and exotic elements. Secondly, mechanisms that facilitate diversity in native species also seem to favor exotic species. Because Sax et al. (2002) conducted his samplings in sites affected by fires in different years, the results also showed that naturalized species tend to be more abundant in recently disturbed sites (i.e., burned), while these differences tend to fade over time as the site recovers from the disturbance. This author suggested that this would be the result of a native herbaceous flora poorly adapted to fires (Gómez-González et al. 2009), so human disturbances (i.e., fires, grazing) would help to facilitate coexistence between native and exotic herbs (see also Holmgren et al. 2000; Figueroa et al. 2004).

References

Adams KL, Wendel JF (2005) Polyploidy and genome evolution in plants. Curr Opin Plant Biol 8:135–141

Alpert P (2000) The discovery, scope, and puzzle of desiccation tolerance in plants. Plant Ecol 151:5–17

Aguilera N, Guedes LM, Becerra J, Villaseñor-Parada C, González L, Hernández V (2015) Allelopathic effect of the invasive *Acacia dealbata* link (Fabaceae) on two native plant species in south-central Chile. Gayana Botánica 72:231–239

Aguilera N, Guedes LM, Becerra J, González L (2017) Is autotoxicity responsible for inhibition growth of new conspecific seedlings under the canopy of the invasive *Acacia dealbata* link? Gayana Botánica 74:1–14

An M, Pratley JE, Haig T (1997) Phytotoxicity of vulpia residues: I. Investigation of aqueous extracts. J Chem Ecol 23:1979–1995

Arianoutsou M, Delipetrou P, Vilà M, Dimitrakopoulos PG, Celesti-Grapow L, Wardell-Johnson G, Henderson L, Fuentes N, Ugarte-Mendes E, Rundel PW (2013) Comparative patterns of plant invasions in the Mediterranean biome. PLoS One 8:e79174

Arroyo MTK, Marticorena C, Matthei O, Cavieres L (2000) Plant invasions in Chile: present patterns and future predictions. In: Mooney HA, Hobbs RJ (eds) Invasive species in a changing world. Island Press, Washington, DC, pp 385–421
Avise J (2004) Molecular markers, natural history, and evolution, Sinauer, MA
Ayala FJ (1982) Population and evolutionary genetics: a primer. Benjamin/Cummings, Menlo Park
Baker HG (1974) The evolution of weeds. Annu Rev Ecol Syst 5:1–24
Ballester A, Vieitez E (1979) The allelopathic potential of *Erica australis* L. and *E. arborea* L. Bot Gaz 140:433–436
Byers JE, Noonburg EG (2003) Scale dependent effect of biotic resistance to biological invasion. Ecology 84:1428–1433
Callaway RM, Ridenour WM (2004) Novel weapons: invasive success and the evolution of increased competitive ability. Front Ecol 2:436–443
Casado MA, Acosta-Gallo B, Sánchez-Jardón L, Martín-Forés I, Castro I, Ovalle C, del Pozo A, de Miguel JM (2015) Interactive effects of source and recipient habitats on plant invasions: distribution of exotic species in Chile. Divers Distrib 21:609–619
Castro SA, Escobedo VM, Aranda J, Carvallo GO (2014) Evaluating Darwin's naturalization hypothesis in experimental plant assemblages: phylogenetic relationships do not determine colonization success. PLoS One 9:e105535
Cavieres LA, Badano EI, Sierra-Almeida A, Molina-Montenegro MA (2007) Microclimatic modifications of cushion plants and their consequences for seedling survival of native and non-native herbaceous species in the high Andes of central Chile. Arct Antarct Alp Res 39:229–236
Cavieres LA, Quiroz CL, Molina-Montenegro MA (2008) Facilitation of the non-native *Taraxacum officinale* by native nurse cushion species in the high-Andes of central Chile: are there differences between nurses? Funct Ecol 22:148–156
Christian CE (2001) Consequences of a biological invasion reveal the importance of mutualism for plant communities. Nature 413:635–639
Colwell RK, Rangel TF (2009) Hutchinson's duality: the once and future niche. Proc Natl Acad Sci U S A 106:19651–19658
Cornwell WK, Grubb PJ (2003) Regional and local patterns in plant species richness with respect to resource availability. Oikos 100:417–428
Crawley MJ (1987) What makes a community invasible? In: Gray AJ, Crawley MJ, Edwards PJ (eds) Colonization, succession and stability. Blackwell Scientific Publications, Oxford, pp 429–453
Crawley MJ (1997) Plant ecology. Blackwell Science, Oxford
Crawley MJ, Brown SL, Heard NS, Edwards GR (1999) Invasion-resistance in experimental grassland communities: species richness or species identity? Ecol Lett 2:140–148
Daehler CC (2001) Darwin's naturalization hypothesis revisited. Am Nat 158:324–330
Daehler CC (2003) Performance comparisons of co-occurring native and alien plants: implications for conservation and restoration. Annu Rev Ecol Syst 34:183–211
Daehler CC (2006) Invasibility of tropical islands by introduced plants: partitioning the influence of isolation and propagule pressure. Preslia 78:361–374
Daehler CC, Strong DR (1997) Reduced herbivore resistence in introduced smooth cordgrass (*Spartina alterniflora*) after a century of herbivore-free growth. Oecologia 110:99–108
Darwin C (1859) The origin of species. John Murray, London
Davidson AM, Jennions M, Nicotra AM (2011) Do invasive species show higher phenotypic plasticity than native species and, if so, is it adaptive? A meta-analysis. Ecol Lett 14:419–431
Davis MA, Grime JP, Thompson K (2000) Fluctuating resources in plant communities: a general theory of invisibility. J Ecol 88:528–534
Dawson W, Rohr RP, van Kleunen M, Fischer M (2012) Alien plant species with a wider global distribution are better able to capitalize on increased resource availability. New Phytol 194:859–867
De Candolle ALP (1855) Géographie botanique raissoné. V. Masson, Paris
Dennis B (2002) Allee effects in stochastic populations. Oikos 96:389–401
Diez JM, Sullivan JJ, Hulme PE, Edwards G, Duncan RP (2008) Darwin's naturalization conundrum: dissecting taxonomic patterns of species invasions. Ecol Lett 11:674–681

Duncan RP, Williams PA (2002) Darwin's naturalization hypothesis challenged. Nature 417:608–609
Ehrlich PR (1989) Attributes of invaders and the invading processes: vertebrates. In: Drake JA, Mooney HA, di Castri F, Groves RH, Kruger FJ, Rejmánek M, Williamson M (eds) Biological invasions: a global perspective. John Wiley & Sons, Chichester, pp 315–328
Elton CS (1958) The ecology of invasions by animals and plants. University of Chicago Press, Chicago
Escobedo VM, Aranda JE, Castro SA (2011) Hipótesis de Naturalización de Darwin evaluada en la flora exótica de Chile continental. Rev Chil Hist Nat 84:543–552
Figueroa JA, Castro SA, Marquet PM, Jaksic FM (2004) Exotic plant invasions to the mediterranean region of Chile: causes, history and impacts. Rev Chil Hist Nat 77:465–483
Fonseca DM, LaPointe DA, Fleischer RC (2000) Bottlenecks and multiple introductions: population genetics of the vector of avian malaria in Hawaii. Mol Ecol 9:1803–1814
Fox MD, Fox BJ (1986) The susceptibility of natural communities to invasion. In: Groves RH, Burdon JJ (eds) Ecology of biological invasions: an Australian perspective. Australian Academy of Science, Canberra, pp 57–66
Friebe A, Schulz M, Kuck P, Schnabl H (1995) Phytotoxins from shoot extracts and root exudates of *Agropyron repens* seedlings. Phytochemistry 38:1157–1159
Fuentes ER, Otaiza RD, Alliende MC, Poiani AJ (1984) Shrub clumps of the Chilean matorral vegetation: structure and possible maintenance mechanisms. Oecologia 42:405–411
Fuentes ER, Espinoza GA, Gajardo G (1987) Allelopathic effects of the Chilean matorral shrub *Flourencia thurifera*. Rev Chil Hist Nat 60:57–62
Fuentes N, Ugarte E, Kühn I, Klotz S (2007) Alien plants in Chile: inferring invasion periods from herbarium records. Biol Invasions 10:649–657
Funk JL (2008) Differences in plasticity between invasive and native plants from a low resource environment. J Ecol 96:1162–1173
Gentle CB, Duggin JA (1997) Allelopathy as a competitive strategy in persistent thickets of *Lantana camara* L. in three Australian forest communities. Plant Ecol 132:85–95
Gianoli E (2004) Plasticity traits and correlation in two population of *Convolvulus arvensis* (Convolvulaceae) differing in environmental heterogeneity. Int J Plant Sci 165:825–832
Godoy O, Valladares F, Castro-Díez P (2011) Multispecies comparison reveals that invasive and native plants differ in their traits but not in their plasticity. Funct Ecol 25:1248–1253
Golani D, Azzurro E, Corsini-Foka M, Falautano M, Andaloro F, Bernardi G (2007) Genetic bottlenecks and successful biological invasions: the case of a recent Lessepsian migrant. Biol Lett 3:541–545
Gómez-González S, Lohengrin LA, Cavieres A, Torres P, Torres-Díaz C (2009) Competitive effects of the alien invasive *Centaurea solstitialis* on two Chilean *Baccharis* species at different life-cycle stages. Gayana Botánica 66:71–83
Gotelli NJ (2001) A primer of ecology. Sinauer Press, Sunderland
Grosholz ED (2005) Recent biological invasion may hasten invasional meltdown by accelerating historical introductions. Proc Natl Acad Sci U S A 102:1088–1091
Gundale MJ, Almeida JP, Wallander H, Wardle DA, Kardol P, Nilsson MC, Fajardo A, Pauchard A, Peltzer DA, Ruotsalainen S, Mason B, Rosenstock N (2016) Differences in endophyte communities of introduced trees depend on the phylogenetic relatedness of the receiving forest. J Ecol 104:1219–1232
Hendry AP, Nosil P, Rieseberg LH (2007) The speed of ecological speciation. Funct Ecol 21:455–464
Heywood VH (1989) Patterns, extents and modes of invasions by terrestrial plants. In: Drake JA, Mooney HA, di Castri F, Groves RH, Kruger FJ, Rejmánek M, Williamson M (eds) Biological invasions: a global perspective. Wiley, Chichester, pp 31–60
Hierro JL, Callaway RM (2003) Allelopathy and exotic plant invasion. Plant Soil 256:29–39
Hierro JL, Maron JL, Callaway RM (2005) A biogeographic approach to plant invasions: the importance of studying exotics in their introduced and native range. J Ecol 93:5–15
Holmgren M, Avilés R, Sierralta L, Segura A, Fuentes ER (2000) Why have European herbs so successfully invaded the Chilean matorral? Effects of herbivory, soil nutrients, and fire. J Arid Environ 44:197–211

Holt RD, Lawton JH (1994) The ecological consequences of shared natural enemies. Annu Rev Ecol Syst 25:495–520
Huffaker CB, Messenger PS (1976) Theory and practice of biological control. Academic, New York
Hulme PE (2008) Phenotypic plasticity and plant invasions: is it all jack? Funct Ecol 22:3–7
Hurka H, Neuffer B (1997) Evolution processes in the genus *Capsella* (Brassicaceae). Plant Syst Evol 206:295–316
Irimia RE, Montesinos D, Eren O, Lortie C, Kristine F, Cavieres LA, Sotes GJ, Hierro JL, Jorge A, Loureiro J (2017) Extensive analysis of native and non-native *Centaurea solstitialis* L. populations across the world shows no traces of polyploidization. PeerJ 14:e3531
Jaksic FM, Soriguer RC (1981) Predation upon the European rabbit (*Oryctolagus cuniculus*) in mediterranean habitats of Chile and Spain: a comparative analysis. J Anim Ecol 50:269–281
Jeschke J, Aparicio LG, Haider S, Heger T, Lortie C, Pyšek P, Strayer D (2012) Support for major hypotheses in invasion biology is uneven and declining. NeoBiota 14:1–20
Jiang L, Tan J, Pu Z (2010) An experimental test of Darwin's naturalization hypothesis. Am Nat 175:415–423
Jones EI, Nuismer SL, Gomulkiewicz R (2013) Revisiting Darwin's conundrum reveals a twist on the relationship between phylogenetic distance and invasibility. Proc Natl Acad Sci U S A 110:20627–20632
Keane RM, Crawley MJ (2002) Exotic plant invasions and the enemy release hypothesis. Trends Ecol Evol 17:164–170
Kennedy TA, Naeem S, Howe KM, Knops JMH, Tilman D, Reich P (2002) Biodiversity as a barrier to ecological invasion. Nature 417:636–638
Kikvidze Z, Brooker RW, Butterfield BJ, Callaway RM, Cavieres LA, Cook BJ, Lortie CJ, Michalet R, Pugnaire FI, Xiao S, Anthelme F, Björk RG, Cranston BH, Gavilán RG, Kanka R, Lingua E, Maalouf JP, Noroozi J, Parajuli R, Phoenix GK, Reid A, Ridenour WM, Rixen C, Schöb C (2015) The effects of foundation species on community assembly: a global study on alpine cushion plant communities. Ecology 96:2064–2069
Kolar CS, Lodge DM (2001) Progress in invasion biology: predicting invaders. Trends Ecol Evol 16:199–204
Kruger FJ, Breytenbach GJ, MacDonald IAW, Richardson DM (1989) The characteristics of invaded Mediterranean-climate regions. In: Drake JA, Mooney HA, di Castri F, Groves RH, Kruger FJ, Rejmánek M, Williamson M (eds) Biological invasions: a global perspective. Wiley, Chichester, UK, pp 181–213
Kruse M, Strandberg M, Strandberg B (2010) Ecological effects of allelopathic plants: a review, Technical report no. 315. National Environmental Research Institute, Silkeborg
Lee CE (2002) Evolutionary genetics of invasive species. Trends Ecol Evol 17:386–391
Levine JM (2000) Species diversity and biological invasions: relating local process to community pattern. Science 288:852–854
Levine JM, D'Antonio CM (1999) Elton revisited: a review of evidence linking diversity and invasibility. Oikos 87:15–26
Levine JM, Alder PB, Yelenik SG (2004) A meta-analysis of biotic resistance to exotic plant invasions. Ecol Lett 7:975–989
Lizarralde MS, Bailliet G, Poljak S, Fasanella M, Giulivi C (2007) Assessing genetic variation and population structure of invasive North American beaver (*Castor canadensis* Kuhl, 1820) in Tierra Del Fuego (Argentina). Biol Invasions 10:673–683
Lockwood JL, Cassey P, Blackburn T (2005) The role of propagule pressure in explaining species invasions. Trends Ecol Evol 20:223–228
Lonsdale WM (1999) Global patterns of plant invasions and the concept of invasibility. Ecology 80:1522–1536
Ma C, Li SP, Pu Z, Tan J, Liu M, Zhou J, Li H, Jiang L (2016) Different effects of invader – native phylogenetic relatedness on invasion success and impact: a meta-analysis of Darwin's naturalization hypothesis. Proc R Soc Lond B 283:20160663
MacArthur R (1970) Species packing and competitive equilibrium for many species. Theor Popul Biol 1:1–11

Mack RN (1996) Predicting the identity and fate of plant invaders: emergent and emerging approaches. Biol Conserv 78:107–121

Maron JL, Vilà M (2001) When do herbivores affect plant invasion? Evidence for the natural enemies and biotic resistance hypotheses. Oikos 95:361–373

Martín-Forés I, Casado MA, Castro I, Ovalle C, del Pozo A, Acosta-Gallo B, Sánchez-Jardón L, de Miguel JM (2012) Flora of the Mediterranean Basin in the Chilean *espinales*: evidence of colonisation. Pastos 42:137–160

Marx HE, Giblin DE, Dunwiddie PW, Tank DC (2016) Deconstructing Darwin's naturalization conundrum in the San Juan Islands using community phylogenetics and functional traits. Divers Distrib 22:318–331

Meijer K, Schilthuizen M, Beukeboom L, Smit C (2016) A review and meta-analysis of the enemy release hypothesis in plant-herbivorous insect systems. PeerJ 4:e2778

Miura O (2007) Molecular genetic approaches to elucidate ecological and evolutionary issues of the biological invasions. Ecol Res 22:876–883

Molina-Montenegro MA, Naya DE (2012) Latitudinal patterns in phenotypic plasticity and fitness-related traits: assessing the climatic variability hypothesis (CVH) with an invasive plant species. PLoS One 7:e47620

Molina-Montenegro MA, del Pozo A, Gianoli E (2018) Ecophysiological basis of the Jack-and-Master strategy: *Taraxacum officinale* (dandelion) as an example of a successful invader. J Plant Ecol 11:147–157

Montenegro G, Rivera O, Bas F (1978) Herbaceous vegetation in the Chilean matorral: dynamic of growth and evaluation of allelopathic effects of some dominant shrubs. Oecologia 36:237–244

Morales MC, Verdejo V, Orlando J, Carú-Marambio M (2016) Fungal communities as an experimental approach to Darwin's naturalization hypothesis. Res Microbiol 167:126–132

Müller CH (1966) The role of chemical inhibition (allelopathy) in vegetational composition. Bull Torrey Pines Botanical Club 93:332–351

Naeem S, Knops JMH, Tilman D, Howe KM, Kennedy T, Gale S (2000) Plant diversity increases resistance to invasion in the absence of covarying extrinsic factors. Oikos 91:97–108

Noble IR (1989) Attributes of invaders and the invading process: terrestrial and vascular plants. In: Drake JA, Mooney HA, di Castri F, Groves RH, Kruger FJ, Rejmánek M, Williamson M (eds) Biological invasions: a global perspective. Wiley, Chichester, pp 1–30

Parker JD, Burkepile DE, Hay ME (2006) Opposing effects of native and exotic herbivores on plant invasions. Science 311:1459–1461

Peay KG, Belisle M, Fukami T (2012) Phylogenetic relatedness predicts priority effects in nectar yeast communities. Proc R Soc Lond B 279:749–758

Perrings C, Dehnen-Schmutz K, Touza J, Williamson M (2005) How to manage biological invasions under globalization. Trends Ecol Evol 20:212–215

Prentice HC, Malm JU, Hathaway L (2008) Chloroplast DNA variation in the European herb *Silene dioica* (red campion): postglacial migration and interspecific introgression. Plant Syst Evol 272:23–37

Pyšek P (1998) Is there a taxonomic pattern to plant invasions? Oikos 82:282–294

Pyšek P, Prach K (2003) Research into plant invasions in a cross-roads region: history and focus. Biol Invasions 5:349–356

Pyšek P, Richardson DM (2007) Traits associated with invasiveness in alien plants: where do we stand? In: Nentwig W (ed) Biological invasions. Ecological studies 193. Springer, Berlin/Heidelberg, pp 97–126

Randall RP (2002) A global compendium of weeds. RG & FJ Richardson, Melbourne

Rejmánek M (1989) Invasibility of plant communities. In: Drake JA, Mooney HA, di Castri F, Groves RH, Kruger FJ, Rejmánek M, Williamson M (eds) Biological invasions: a global perspective. Wiley, Chichester, pp 369–388

Rejmánek M (1995) What makes a species invasive? In: Pyšek P, Prach K, Rejmánek M, Wade M (eds) Plant invasions: general aspects and special problems. SPB Academic Publishing, Amsterdam, pp 3–13

Rejmánek M (1996) A theory of seed plant invasiveness: the first sketch. Biol Conserv 78:171–180
Rejmánek M (1998) Invasive plant species and invadible ecosystems. In: Sandlund OT, Schei PJ, Vilken A (eds) Invasive species and biodiversity management. Kluwer, Dordrecht, pp 79–102
Rejmánek M, Richardson DM (1996) What attributes make some plant species more invasive? Ecology 77:1655–1661
Ricciardi A, Mottiar M (2006) Does Darwin's naturalization hypothesis explain fish invasions? Biol Invasions 8:1403–1407
Richardson DM, Allsopp N, D'Antonio CM, Milton SJ, Rejmánek M (2000a) Plant invasions: the role of mutualisms. Biol Rev 75:65–93
Richardson DM, Pyšek P, Rejmánek M, Barbour MG, Panetta FD, West CJ (2000b) Naturalization and invasion of alien plants: concepts and definitions. Divers Distrib 6:93–107
Ricklefs RE (1987) Community diversity: relative roles of local and regional processes. Science 235:167–171
Sakai AK, Allendorf FW, Holt JS, Lodge DM, Molofsky J, With KA, Baughman S, Cabin RJ, Cohen JE, Ellstrand NC, McCauley DE, O'Neil P, Parker MI, Thompson JN, Weller SG (2001) The population biology of invasive species. Annu Rev Ecol Syst 32:305–332
Sax DF (2001) Latitudinal gradients and geographic ranges of exotic species: implications for biogeography. J Biogeogr 28:139–150
Sax DF, Gaines SD, Brown JH (2002) Species invasions exceed extinctions on islands worldwide: a comparative study of plants and birds. Am Nat 160:766–783
Shea K, Chesson P (2002) Community ecology theory as a framework for biological invasions. Trends Ecol Evol 17:170–176
Simberloff D (2009) The role of propagule pressure in biological invasions. Annu Rev Ecol Evol Syst 40:81–102
Simberloff D, von Holle B (1999) Positive interactions of nonindigenous species: Invasional meltdown? Biol Invasions 1:21–32
Slotte T, Ceplitis A, Neuffer B, Hurka H, Lascoux M (2006) Intrageneric phylogeny of *Capsella* (Brassicaceae) and the origin of the tetraploid *C. bursa-pastoris* based on chloroplast and nuclear DNA sequences. Am J Bot 93:1714–1724
Stearns S (1992) The evolution of life histories. Oxford University Press, Oxford
Stohlgren TJ, Binkley D, Chong GW, Kalkhan MA (1999) Exotic plant species invade hot spots of native plant diversity. Ecol Monogr 69:25–46
Strauss SY, Webb CO, Salamin N (2006) Exotic taxa less related to native species are more invasive. Proc Natl Acad Sci U S A 103:5841–5845
Taylor CM, Hastings A (2005) Allee effects in biological invasions. Ecol Lett 8:895–908
Te Beest M, Le Roux JJ, Richardson DM, Brysting AK, Suda J, Kubešová M, Pyšek P (2011) The more the better? The role of polyploidy in facilitating plant invasions. Ann Bot 109:19–45
Thompson JD (1991) The biology of an invasive plant: what makes *Spartina anglica* so successful? Bioscience 41:393–401
Thuiller W, Gallien L, Boulangeat I, Bello F, Munkemüller T, Roquet C, Lavergne S (2010) Resolving Darwin's naturalization conundrum: a quest for evidence. Divers Distrib 16:461–475
Tilman D (1988) Plant strategies and the dynamics and structure of plant communities. Princeton University Press, Princeton
Tilman D (1997) Distinguishing between the effects of species diversity and species composition. Oikos 80:185
Tsutsui ND, Suarez AV, Holway DA (2000) Reduced genetic variation and the success of an invasive species. Proc Natl Acad Sci U S A 97:5948–5953
Willis J, Thomas MB, Lawton JH (1999) Is the increased vigour of invasive weeds explained by a trade-off between growth and herbivore resistance? Oecologia 120:632–640
Zavaleta ES, Hulvey KB (2004) Realistic species losses disproportionately reduce grassland resistance to biological invaders. Science 306:1175–1177
Zavaleta ES, Hobbs R, Mooney H (2001) Viewing invasive species removal in a whole-ecosystem context. Trends Ecol Evol 16:454–459

Chapter 5
Range Expansion

5.1 Preamble

Once the founding colony is established, it can begin to expand its secondary distribution range (see Fig. 1.7). Such expansion will not only imply the occupancy of a larger distribution area but also an increase in population size (Sakai et al. 2001). Therefore, the concept of range expansion brings together two coupled population components: the dispersion or spatial movement of individuals and population growth. Therefore, when in the ecology of invasions, we speak of an expanding species, we must conceive of a population that increases the number of individuals and at the same time spreads in space.

A great deal of effort has been put into the development of mathematical models to explain and predict the phenomenon of expansion, and some of these approaches will be reviewed in this chapter. In particular, we will focus our attention on reaction-diffusion models, which, despite the ideal nature of their arguments, have shown great power to emulate ecological complexity under field situations.

5.2 Reaction-Diffusion Models

Reaction-diffusion models gained interest and popularity in invasion ecology from the 1950s onwards. However, it is important to note that they originally emerged from the fields of physics and chemistry, where they were developed to study the motion of particles in space. In this type of phenomena, particles can not only move by diffusion but also react to each other, giving rise to new particles (Okubo 1980). In this way, the reaction-diffusion equations make it possible to estimate the concentration of particles at a point in space after a certain time has elapsed since they were released.

F. M. Jaksic, S. A. Castro, *Biological Invasions in the South American Anthropocene*, https://doi.org/10.1007/978-3-030-56379-0_5

Among the simplest reaction-diffusion expressions is the Kolmogorov-Petrovsky-Piscounov equation (KPP; $\partial_t q = R(q) + D\,\Delta q$), which has been applied to the study of the random motion of microscopic particles in a fluid, also known as Brownian motion or random walk. In this expression, the terms represent the concentration of a substance q ($\partial_t q$), R is a function that considers the occurrence of local reactions, D is the diagonal matrix of diffusion coefficients for the diffusing particles that appear by reaction, and Δ denotes the Laplacian operator for the vector $q\,(x, t)$.

In invasion ecology, a more familiar expression of the KPP equation is:

$$\frac{\partial N}{\partial t} = f(N)N + D\left(\frac{\partial^2 N}{\partial x^2} + \frac{\partial^2 N}{\partial y^2}\right)$$

Right here, N represents the size of the population at one point (x, y) anyone; t is the time elapsed since the release of the organisms at the point of origin ($x = 0$, $y = 0$); $f(N)$ is a function that describes the per capita growth rate of the population; and D is the diffusion coefficient.

For its analysis, this equation can be decomposed into two terms. On the one hand, the term $f(N)$ refers to population growth in the point (x, y), which could essentially take two forms: dense-independent or dense-dependent growth. As can be noted, in this part of the equation, the chemical reactivity of particles becomes analogous to the reproduction of organisms.

The second component of Eq. 1 ($D\,(\partial^2 N/\partial x^2 + \partial^2 N/\partial y^2)$) incorporates the spatial propagation of organisms, since it analogues the diffusion of particles to dispersion. Here, the diffusion coefficient D is a constant, which must be determined empirically; D will depend on each species and, simultaneously, will depend on each type of environment (Kareiva 1983).

Equation 1 assumes that individuals are released at a single point, the point of origin ($x = 0$, $y = 0$). These are then dispersed in random directions at random distances from the origin; this behavior is governed by the Gaussian distribution function:

$$N = \left(\frac{No}{4\pi Dt}\right)e^{-\left(\left(x^2+y^2\right)/4Dt\right)}$$

In this expression, N_o corresponds to the initial number of organisms released at the point of origin. Therefore, Eq. 2 allows to describe the number of individuals (N) at different times and distances from the release point.

5.3 Expansion Rate

From a graphical point of view, this process of range expansion can be conceived as the growth of a ring, which, at least theoretically, expands concentrically over time (Fig. 5.1a, b). Under ideal conditions, organisms would be released at the point of

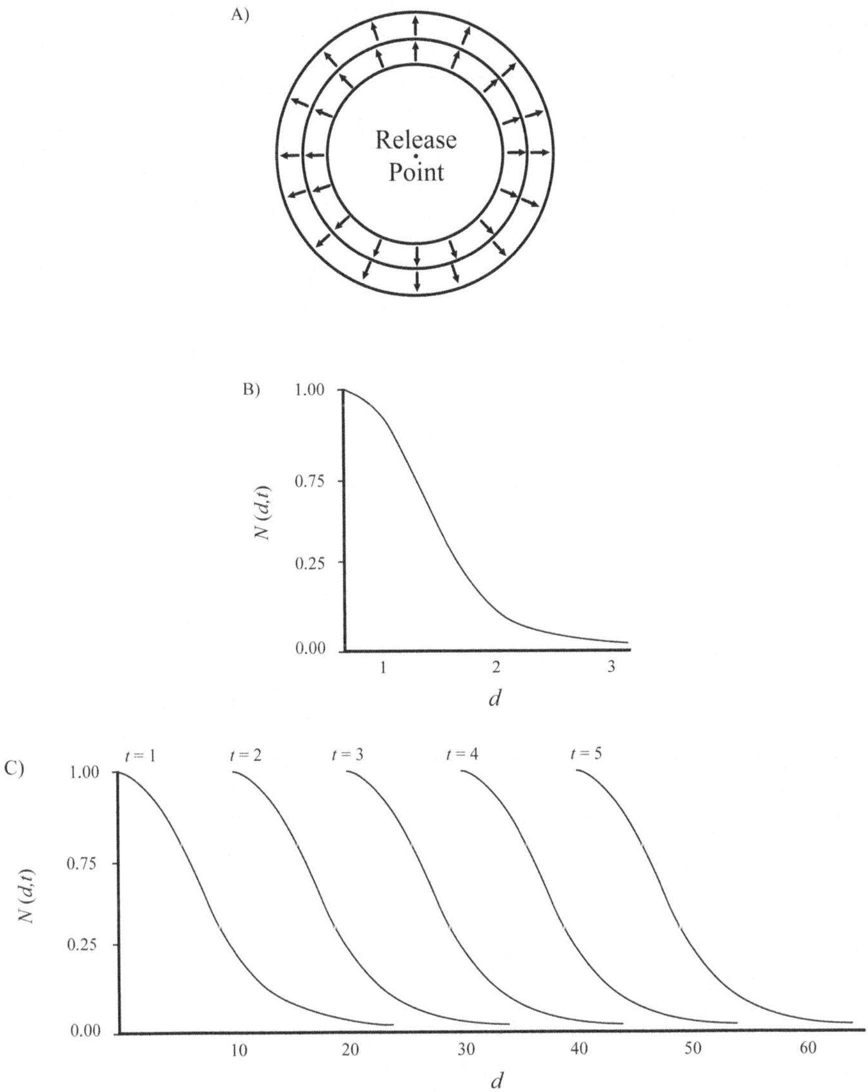

Fig. 5.1 Density function for a hypothetical population whose individuals scatter in space (d). (**a**) Range spatial expansion model; (**b**) density function ($N(d, t)$) for an early stage ($t = 1$); (**c**) density function over time. This pattern is called "wavefront" and is characteristic of the reaction-diffusion model

origin, and from here, they expand their range progressively. Initially, the number of organisms located at the origin would be relatively high, but as time passes, the density function shifts the range concentrically.

Knowing the point of release, as well as the density of organisms at different distances from this point, it is possible to plot the density of organisms depending on distance and time (Fig. 5.1b, c). The function described by this graph is called the density function and allows to set the percentile of the population that is located at different distances from the release point, after a certain amount of time. Again, this curve shows that a majority fraction of individuals should be in the vicinity of the release site. As we move away from this point in any direction, the density of organisms will progressively decline (Fig. 5.1b). With the passing of time ($t = 1, 2, 3, 4, 5$), the shape of the curve will be modified, as the greater density of individuals moves towards the sites furthest from the point of release (Fig. 5.1c). This responds to the expansive colonizing movement of the original population and is a product of local reproduction.

With the terms D and $f(N)$, it is possible to calculate the rate at which the invasion expands (c). This new term, as an expression of velocity, has units of the distance/time type; the simplest expression for c can be derived from Eq. 1, assuming dense-independent growth in the term $f(N)$ (i.e., $f(N) = r$). This assumption is reasonable considering that during the initial phases of the invasion, populations grow exponentially. Thus, the expansion rate will be described by:

$$c = 2\sqrt{rD}$$

Determining the expansion range offers some complexities. This can be accelerated in a variable manner depending on various factors, such as different habitat or environmental types, or the presence of interacting organisms (e.g., predators, competitors, parasites). Shigesada and Kawasaki (1997) have explored and expanded Eq. 1 to include mathematical terms that consider these effects. However, one question that must be asked at this point seems to be the relevance of these models in explaining invasions under field conditions. In other words, can the expansive behavior of exotic species in nature be adequately emulated by such equations?

5.4 Demographic Component

The reaction-diffusion equation (Eq. 1) includes the population growth factor in the expression $f(N)N$. This term may represent a function dependent on or independent of population size (Shigesada and Kawasaki 1997). Despite this, there is a perception that biological invasions correspond to population explosion events where exotic populations reach great abundance (Williamson 1996). In fact, the term "biological pest" is usually applied to superabundant exotic species, which therefore negatively impact the environment in which they are found. However, as Williamson (1996) noted, most exotic species present a densely dependent population growth, reaching relatively low levels of abundance, to the point that many of them go unnoticed. In contrast, invasions with irruptive features are rare (Williamson 1996; Mack et al. 2000).

The population growth of an exotic species is governed by the same demographic processes underlying any other species, that is, densodependence and

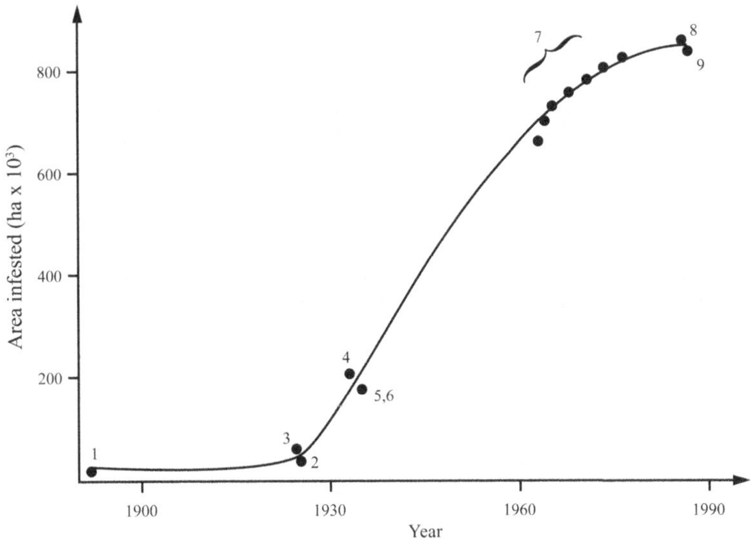

Fig. 5.2 The demographic expansion of the *Opuntia aurantiaca* in South Africa is illustrated, indicating the area of occupation over time (Moran and Zimmerman 1991). After its introduction, the species remains for a long time at low density occupying a reduced area until its population and occupied area increase (around 1930). Around 1960, the population began to decline its rate of territorial occupation to settle around 800,000 hectares

densoindependence (Gotelli 2001). Thus, the characteristic growth of an exotic population is usually characterized in different phases (Fig. 5.2). At first, given the small size of the colony, population abundance may be very small, to such an extent that the probability of extinction would be relatively high. Consequently, if the colony does not become extinct, it may remain small in size for a variable period of time, experiencing little noticeable population growth (Fig. 5.2).

In many cases, as in inadvertently introduced organisms, the abundance during this stage remains so small that the species, although present in an area, is not perceived. The fraction of time in which the exotic species remains at low population levels is called time lag and is associated with the amount of time needed to have a sufficient number of organisms to support further growth (Mack et al. 2000; Sakai et al. 2001).

As mentioned above, during the time lag phase, it is difficult to determine the fate of the founding colony. How many and which of them will become invasive populations or pests represents one of the main focuses of attention in the management and control of invasions. In cases where exotic species become pests, the economic costs for their control are high, and their eradication is virtually unthinkable. In order to avoid future undesired effects, it has been pointed out that it is convenient to detect and control invasions at an early stage, as this considerably facilitates work on small populations with restricted geographical distribution (Pyšek and Richardson 2010).

After this stage, the population will experience population growth giving way to greater abundance (Fig. 5.2). It is at this point that field abundance can exceed a certain threshold, making the presence of the exotic species perceptible or evident. However, as with any population, exponential growth cannot be maintained indefinitely, so eventually the growth rate will be reduced, bringing the population to its carrying capacity, *K* (Gotelli 2001).

Arim et al. (2006) documented the existence of general patterns and mechanisms in the population growth of exotic irruptive species. These authors compared the dynamics of population growth for 30 species, all belonging to taxa as diverse as viruses, plants, invertebrates, fish, amphibians, and birds; their results consistently evidenced the presence of population regulation mechanisms for all these cases. These mechanisms respond to first-order population dynamics, a condition expected when resources are scarce and the magnitude of intraspecific competition increases (Berryman 1999). These authors attributed the time lag in the growth of the expansion front to the fact that this front is generated from a less reproductive population segment, which competes for spaces to establish itself before generating the progeny that will continue to expand the expansion front. These results contrast with those that conceived invasions as phenomena of irruptive, uncontrolled, and idiosyncratic population growth, since the presence of common demographic mechanisms for a wide variety of taxa is evident (Arim et al. 2006).

5.5 Reaction-Diffusion in Nature

So far, modeling based on reaction-diffusion equations has proven to be a useful tool for deriving some properties of the invasive process, at least in theory. But how useful can they be when tested under field conditions?

Skellam (1951) was one of the first researchers to evaluate the fit of the reaction-diffusion equation in relation to the expansion of the invasive range of the muskrat (*Ondatra zibethicus*) in Central Europe. Based on previous surveys conducted by Ulbrich (1930), Skellam produced a map showing the range acquired by the species over time (see Fig. 5.3). The muskrat was introduced into captivity in the Czech Republic for the purpose of stimulating the fur industry; but in 1909, five individuals escaped from the city of Prague, giving rise to one of the best-known point of invasion of this species in Europe (Hengeveld 1989). Skellam (1951) estimated the area occupied by *Ondatra zibethicus* between 1909 and 1928 and, calculating the square root of this value, obtained a constant ratio for the expansion rate (Fig. 5.4), a value close to the growth of the concentric expansion radius visualized in Fig. 5.1a. Beyond this particular case of *Ondatra zibethicus*, Skellam understood that in general terms, the expansion of plants and animals presents a behavior that responds to the theoretical predictions of reaction-diffusion models (Eq. 3). The deviations observed in field conditions from the derived predictions of mathematical models were attributed to ecological particularities of both the expanding species and the habitats through which organisms expand.

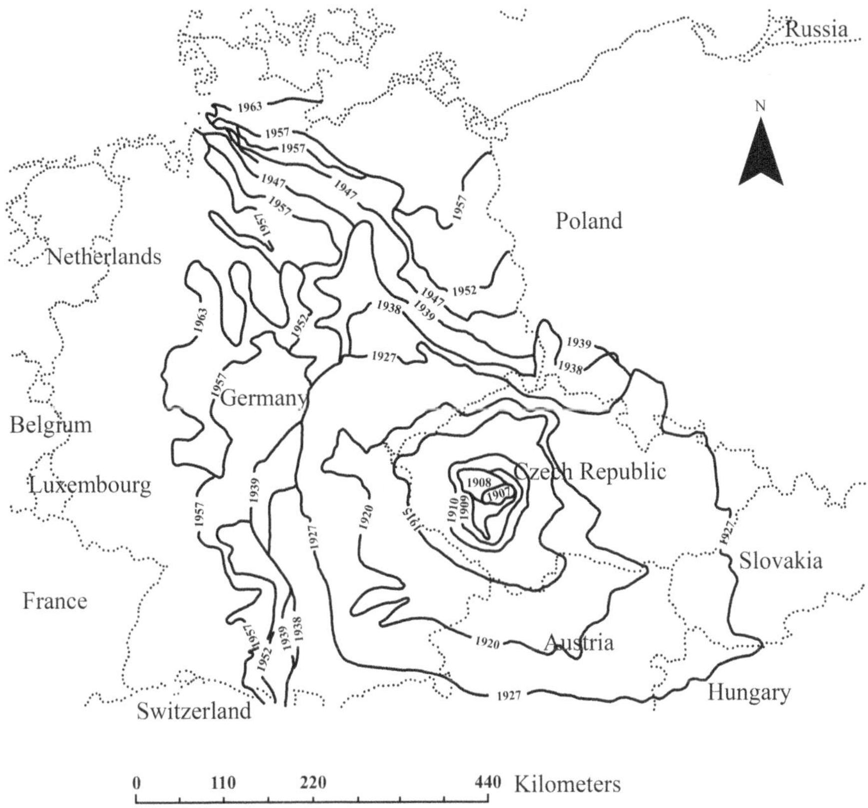

Fig. 5.3 Geographical expansion of the muskrat (*Ondatra zibethicus*) in central Europe, between 1907 and 1963, from the original site of introduction, located in the city of Prague, Czech Republic (Skellam 1951)

In South America, the invasion of some vertebrates has been studied from a rather empirical perspective but inspired by the search for parameters present in the reaction-diffusion equations. Jaksic et al. (2002) collected information on the expansion rate of the European hare (*Lepus europaeus*), the European rabbit (*Oryctolagus cuniculus*), the Canadian beaver (*Castor canadensis*), the muskrat (*Ondatra zibethicus*), and the American mink (*Neovison vison*). Depending on the taxa, expansion rates show wide variation, reaching extremes between 2.7 and 20.0 km per year for the beaver and European hare, respectively (see Table 5.1). This variation was also notable when comparing the estimates obtained for the same species in different localities. For example, for *Oryctolagus cuniculus* the expansion rate has been estimated between 7.0 and 16.0 km/year, while in the case of *Castor canadensis*, values range from 2.7 to 6.3 km/year (Table 5.1). These differences in the expansion rates may respond to inequality in the permeability attributes of the landscape, which in the case of the beaver, for example, could be due to the geographical extension of the micro-watershed network that limits its rate of expansion (Skewes et al. 1999).

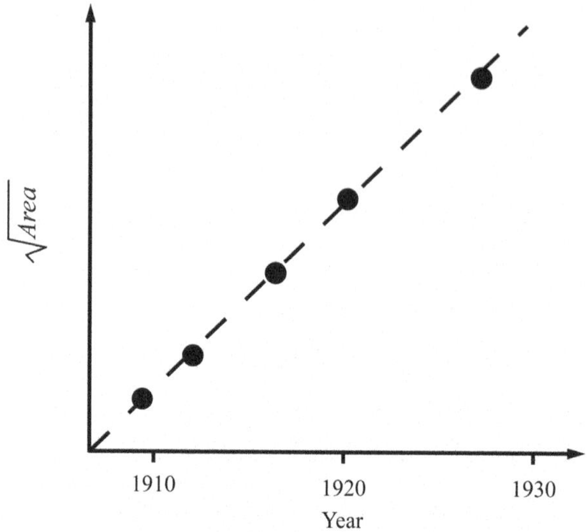

Fig. 5.4 Functional relationship between the time and the area of occupation of the muskrat (*Ondatra zibethicus*) in Central Europe (Skellam 1951)

Table 5.1 Some species of exotic mammals present in Chilean and Argentine Patagonia, for which there is enough information regarding their introduction (Jaksic et al. 2002)

Species rate	Primary source	Secondary source	Current area	Expansion rate, *c* (Km./year)
European hare	Hamburg, Germany (1888)	Cañada de Gómez, Santa Fé Province (Argentina)	Argentinean Pampa	18.6
	Germany (1896 or 1907)	Última Esperanza, region XII (Chile)	Patagonia of Chile and Argentina	20.0
European rabbit	Chillán, VIII region (Chile) Chile (1970)	Neuquén Province (Argentina)	Neuquén and Mendoza provinces (Argentina)	7–16
American beaver	Canada (1946)	Southeast Tierra del Fuego (Argentina)	Central and Southwest Tierra del Fuego (Chile)	2.7–6.3
	Canada (1946)	Southeast Tierra del Fuego (Argentina)	Navarino Island (Chile)	3.1
Muskrat	Ontario, Canada (1948)	Southeast Tierra del Fuego (Argentina)	Central and Southwest Tierra del Fuego (Chile)	10.8
	Ontario, Canada (1948 and 1935)	Southeast Tierra del Fuego (Argentina)	Navarino Island (Chile)	3.9
American mink	Unknown (1946 and 1968)	Cholila, Chubut Province, (Argentina)	Cholila and surroundings (Argentina)	5.5–7.7

Estimates of their expansion rates are given (*c*, km/year). The primary source is that country or region from which mammals were translocated originally; the secondary source is that restricted region or locality from where the mammal began its expansion in the host country or region; the current area is the broadest geographic or political region currently occupied

Fig. 5.5 Current geographical distribution of the European hare (*Lepus europaeus*) in South America (Bonino et al. 2008)

Among these species, particularly interesting is the case of the European hare (*Lepus europaeus*) whose invasion was reviewed by Bonino et al. (2008). The hare was first introduced in Argentina during the nineteenth century, at least in three different locations (Grigera and Rapoport 1983). From here, it expanded to occupy the entire South American Southern Cone, including Chile, Uruguay, Paraguay, Bolivia, and Peru (Fig. 5.5). Studying its expansion rate, Cossíos (2004) recorded a speed of 44.0 and 34.0 km/year in Peru and Bolivia, respectively. In Brazil, Grigera and Rapoport (1983) recorded an average rate of 37 km/year, while in Paraguay it varied between 10.0 and 17.0 km/year (Bonino et al. 2008). According to these results, Bonino et al. (2008) stated that the European hare will continue to expand towards the north of South America, especially in the western portion of the continent. On the other hand, in the central and eastern portion, the Amazon rainforest could represent an insurmountable barrier or one with less permeability to expansion.

5.6 Stratified Dispersion and Long Distance

The contrast between theoretical and empirical studies has led to the conceptual advance of the ecology of invasions. Based on the study of expansion rate, several researchers have documented empirical values that differ from those expected by reaction-diffusion equations. Table 5.2, for example, shows a variety of organisms that denote this difference, as well as others in which the values tend to be comparable (e.g., the muskrat *Ondatra zibethicus* and the otter *Enhydra lutris*). In this area, one of the most striking aspects is that in some cases, the dispersion rate observed under field conditions tends to be greater than the theoretical values expected by simple diffusion, as with the small cabbage butterfly (*Pieris rapae*), the cereal leaf beetle (*Oulema melanopus*), and the gypsy moth (*Lymantria dispar*; see Table 5.2).

A response to this apparent paradox emerges from the observation of the density probability function curves (Fig. 5.1b). As previously indicated, one of the assumptions of the reaction-diffusion models is that the scattering distances around the point of origin follow a Gaussian distribution (Eq. 2). Therefore, a higher expansion rate implies that the tails of this distribution have a higher density of organisms and that, simultaneously, the density around the point of release is lower than expected by this distribution. Consequently, dispersion for some organisms may be governed

Table 5.2 Expansion descriptors for seven invasive taxa

Species	Intrinsic rate of growth, *r* (annual)	Coefficient of diffusion *D* (km[a]/year)	Expansion rate, *c* (km/year)	
			Theoretical	Observed
Musk rat (*Ondatra zibethicus*)[b]	0.2–1.1	51-230	6–32	1–25
Small cabbage butterfly (*Pieris rapae*)[b]	32	2.4–64	9.3-90	15–170
Cereal leaf beetle (*Oulema melanopus*)[b]	1.6–1.9	0.4	1.6–1.7	27–90
Gypsy moth (*Lymantria dispar*)[a]	4.6	<0.34	<2.5	3–20
Otter (*Enhydra lutris*)[c]	0.056	13.5	1.74	1.4
Bacteria of bubonic plague (*Yersinia pestis*)[d]	19	25	720	320–650
Rabies (*Lyssavirus*)[e]	66	40–50	70	30-60

The estimates of the intrinsic rate of population growth (*r*), diffusion coefficient (*D*, km/year), and the theoretical and observed expansion rates estimated from the diffusion model (column of theoretical values) and from occupation maps (column of observed values), respectively, are shown. Expansion rates (*c*) are expressed in km/year units

[a]Liebhold et al. (1992)

[b]Andow et al. (1990)

[c]Lubina and Levin (1988)

[d]Nobel (1974)

[e]Yachi et al. (1989)

by a function other than the normal described in Eq. 2, giving a platykurtic pattern to the density function (Okubo 1980). The platykurtic curve presents a less pronounced decay than the traditional normal curve, denoting a higher expansion rate, c (Kot et al. 1996).

In the biological field, a platykurtic distribution implies the existence of accelerated dispersion mechanisms, beyond what is expected by the reaction-diffusion model. Thus, Hengeveld (1989) coined the concept of stratified dispersion, to discriminate organisms that have bimodal dispersion, referred to as short and long distance dispersion. Short distance dispersion is that which is observed as a consequence of the vagility attributes of each species, in interaction with its dispersing agents. In the absence of interference, under short distance dispersion, the growth of the distributional range should follow a reaction-diffusion dynamic, i.e., a growth with concentric extension with respect to the release point (Fig. 5.1a). If this is the case, the extension of the range would have a growth dynamic at a rate of c, as can be seen in Figs. 5.6a and 5.7a.

However, the occurrence of long-distance dispersion events will contribute to modify the previous form of expansion, accelerating the expansion of range occupancy (Figs. 5.6b, c and 5.7b, c). This mechanism can be facilitated by random or recurring events. From a two-dimensional point of view, the expansion under stratified dynamics is initially visualized as small satellite populations that emerge around the founding colony (Fig. 5.7b, c) and will eventually merge.

The existence of directional events that facilitate long-distance scattering (both biotic and abiotic scattering vectors) can be included in the reaction-diffusion equations. Shigesada and Kawasaki (1997), for example, included an advection factor (ν) in Eq. 1, which could take negative or positive values, either to increase or decrease the rate of expansion. In this way, the range expansion rate could be expressed by the following algorithm:

$$c = 2\sqrt{rD} \pm v$$

Empirical studies have shown numerous examples of accelerated expansion rates for exotic species, yielding long-range dispersal patterns. For example, Suarez et al. (2001) studied the expansion of the Argentine ant (*Linepithema humile*), which has invaded several localities of the globe from its original distribution in South America (Fig. 5.8). This species is dispersed by humans, because its colonies show a high degree of commensalism with *Homo sapiens* (Holway et al. 2002; Roura-Pascual et al. 2006; Silverman and Brightwell 2008). In particular, the analysis of the spatial distribution of this ant in the United States showed an accelerated rate of expansion, from two main geographic focuses: the Pacific coast and the Caribbean (Fig. 5.9). The expansion rates inferred by different models show high levels of displacement, which ranged from 160 to 361 km/year on average (Fig. 5.10), quite high values compared to those from 0 to 200 m/year, calculated for a reaction-diffusion type expansion. The historical records of distribution of this ant indicate that the combined dispersion of long and short distance would be the mechanism responsible for

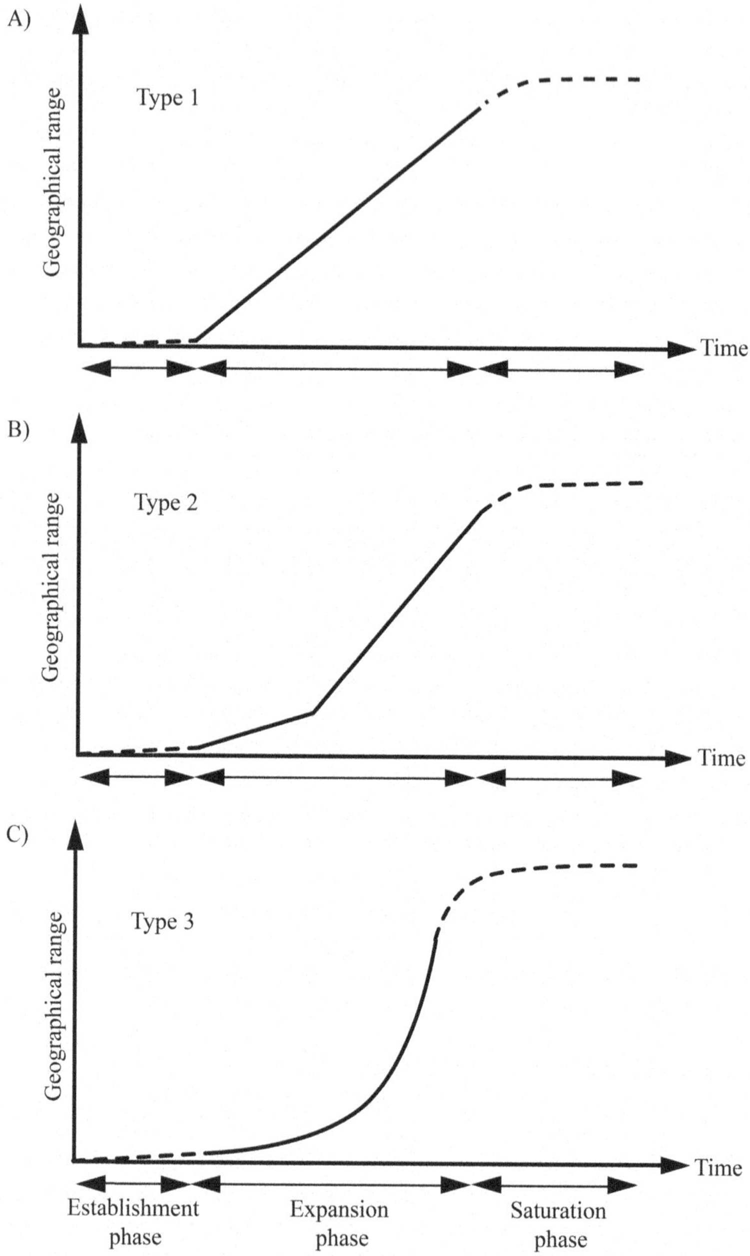

Fig. 5.6 Temporary patterns of geographical range growth. Range growth commonly proceeds in three successive stages: establishment phase (lower segmented line), expansion phase (solid line), and saturation phase (upper segmented line). The expansion phase is classified into three types: (**a**) Type 1, which shows a linear expansion; (**b**) Type 2, which shows a two-phase expansion, with a low initial slope, followed by a steeper linear slope; (**c**) Type 3, where the expansion rate increases continuously and exponentially over time

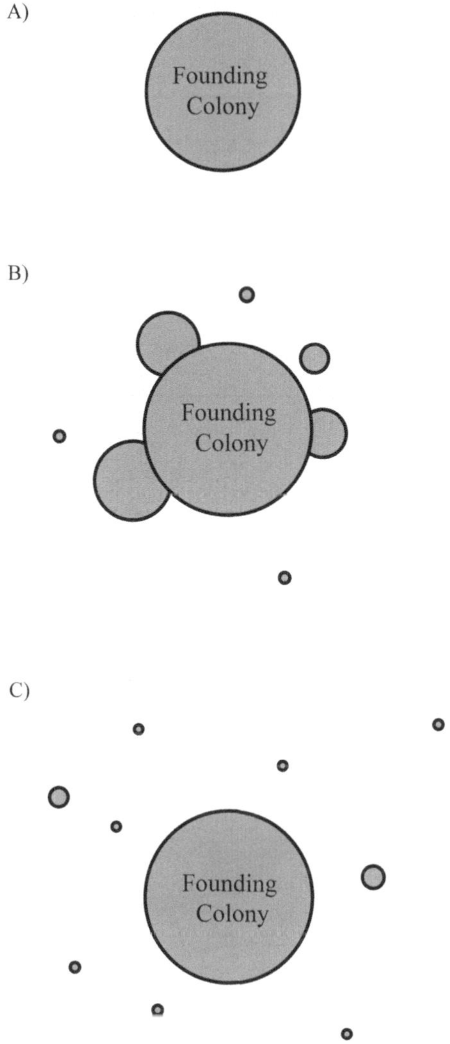

Fig. 5.7 Spatial patterns of range growth from the founding colony (see also Fig. 1.1). (**a**) Hypothetical founding colony that concentrically expands its range; (**b**) satellite colonies are distributed heterogeneously, far enough away from the parental population (founding colony), so that their ranges remain isolated for long periods; (**c**) the colony expands from its periphery, and at the same time satellite colonies are generated produced by dispersants of greater distance that are located near the primary population (founding colony)

Fig. 5.8 Global distribution of the Argentinean ant (*Linepithema humile*). Gray shows the original range in South America, and black indicates the areas where this species has been introduced (Suarez et al. 2001)

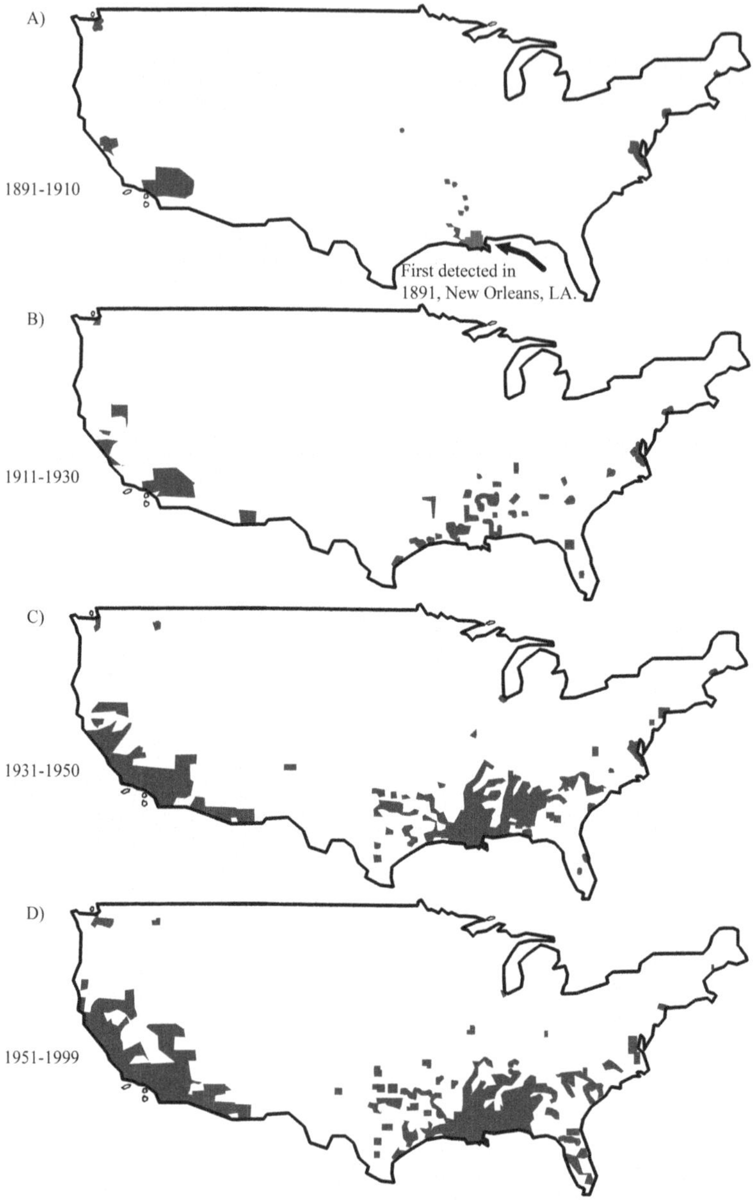

Fig. 5.9 Dynamics of the geographic expansion of the Argentine ant (*Linepithema humile*) in counties of the United States in different periods (Suarez et al. 2001). (**a**) Between 1891 and 1910; (**b**) between 1911 and 1930; (**c**) between 1931 and 1950; (**d**) between 1951 and 1999

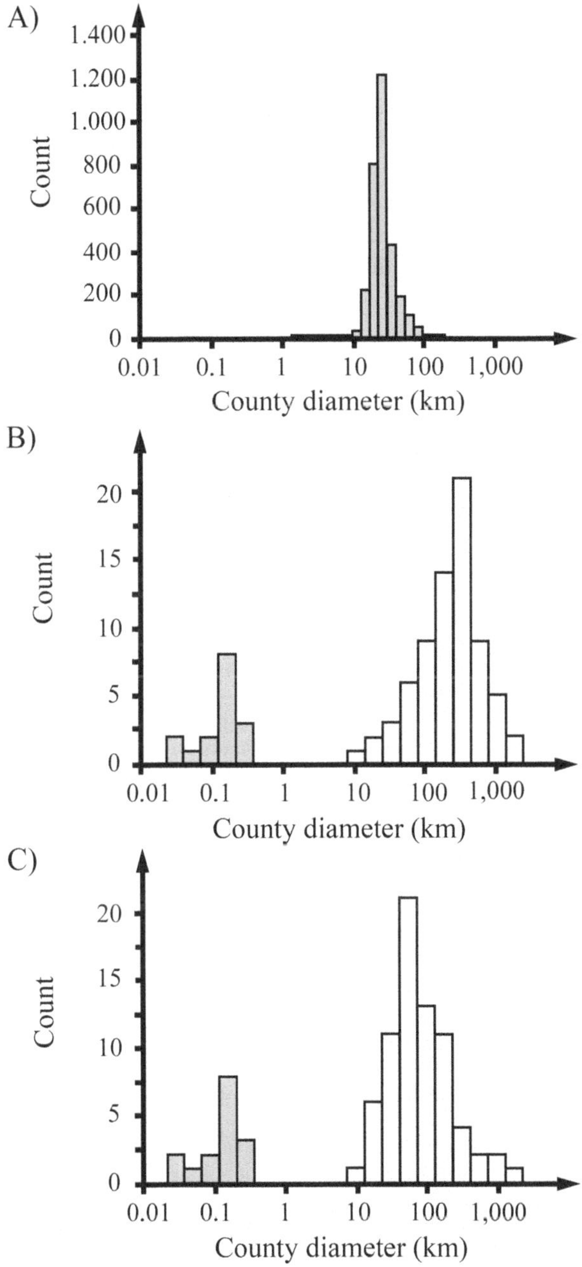

Fig. 5.10 Spatial descriptors for the expansion of the Argentine ant (*Linepithema humile*) in the United States (Suarez et al. 2001). (**a**) Distribution of the diameter of the counties occupied by the species in the United States; (**b**) distribution of the annual expansion distances by short-distance scattering (gray bars) and long-distance scattering (white bars), estimated under the assumption that the starting point was New Orleans; (**c**) distribution of the annual expansion distances by short-distance scattering (gray bars) and long-distance scattering (white bars), estimated under the assumption that the starting point was the nearest county already occupied

the current geographical distribution of *Linepithema humile* in the United States (Suarez et al. 2001; Holway et al. 2002).

Regardless of whether it is stratified or long-distance saltatory dispersal, a small population split from the founding population grows and expands independently. Eventually, after range growth both populations may come into contact and coalesce. From this point on, it is unlikely that the invasive process will be halted, especially if it occupies an increasingly large geographic area. Consequently, the type of management to be applied to this expanding population will depend on the impact that it generates. This aspect will be discussed in the following chapters.

References

Andow DA, Kareiva PM, Levin SA, Okubo A (1990) Spread of invading organisms. Landsc Ecol 4:177–188

Arim M, Abades SR, Neill PE, Lima M, Marquet PA (2006) Spread dynamics of invasive species. Proc Nat Acad Sci U S A 103:374–378

Berryman AA (1999) Principles of population dynamics and their application. Stanley Thornes, Cheltenham

Bonino N, Cossios D, Menegheti J (2008) Dispersión de la liebre europea (*Lepus europaeus*) en Sudamérica. Comunicación Técnica Recursos Naturales 152:1–9

Cossíos ED (2004) La liebre europea, *Lepus europaeus* (Mammalia, Leporidae), especie invasora en el sur del Perú. Rev Peru Biol 11:209–212

Gotelli NJ (2001) A primer of ecology. Sinauer Press, Sunderland

Grigera DE, Rapoport EH (1983) Status and distribution of the European hare in South America. J Mammal 64:163–166

Hengeveld R (1989) Dynamics of biological invasions. Chapman & Hall, London

Holway DA, Suaarez AV, Case TJ (2002) Role of abiotic factors in governing susceptibility to invasion: a test with argentine ants. Ecology 83:1610–1619

Jaksic FM, Iriarte JA, Jiménez JE, Martínez DR (2002) Invaders without frontiers: cross-border invasions of exotic mammals. Biol Invasions 4:157–173

Kareiva P (1983) Local movement in herbivorous insects: applying a passive diffusion model to mark-recapture field experiments. Oecologia 57:322–327

Kot M, Lewis MA, van den Driessche P (1996) Dispersal data and the spread of invading organisms. Ecology 77:2027–2042

Liebhold AM, Halverson JA, Elmes GA (1992) Gypsy moth invasion in North America: a quantitative analysis. J Biogeogr 19:513–520

Lubina JA, Levin SA (1988) The spread of a reinvading species: range expansion in the California Sea otter. Am Nat 131:526–543

Mack RN, Simberloff D, Lonsdale WM, Evans H, Clout M, Bazzaz FA (2000) Biotic invasions, causes, epidemiology, global consequences, and control. Ecol Appl 10:689–710

Moran VC, Zimmerman HG (1991) Biological control of jointed cactus, *Opuntia aurantiaca* (Cactaceae), in South Africa. Agric Ecosyst Environ 37:5–27

Nobel JV (1974) Geographic and temporal development of plagues. Nature 250:276–279

Okubo A (1980) Diffusion and ecological problems: mathematical models. Springer, Berlin

Pyšek P, Richardson DM (2010) Invasive species, environmental change and management, and health. Annu Rev Environ Resour 35:25–55

Roura-Pascual N, Suarez AV, McNyset K, Gomez C, Pons P, Touyama Y, Wild AL, Gascon F, Peterson AT (2006) Niche differentiation and fine-scale projections for Argentine ants based on remotely sensed data. Ecol Appl 16:1832–1841

Sakai AK, Allendorf FW, Holt JS, Lodge DM, Molofsky J, With KA, Baughman S, Cabin RJ, Cohen JE, Ellstrand NC, McCauley DE, O'Neil P, Parker MI, Thompson JN, Weller SG (2001) The population biology of invasive species. Annu Rev Ecol Syst 32:305–332

Shigesada N, Kawasaki K (1997) Biological invasions: theory and practice. Oxford University Press, Oxford

Silverman J, Brightwell RJ (2008) The argentine ant: challenges in managing an invasive unicolonial pest. Annu Rev Entomol 53:231–252

Skellam JG (1951) Random dispersal in theoretical populations. Biometrika 38:196–218

Skewes RO, Gonzalez F, Rubilar L, Quezada OM, Olave IT, Vargas V, Avila AC (1999) Investigación, aprovechamiento y control del castor, islas Tierra del Fuego y Navarino. Servicio de Gobierno Regional XII Región. Magallanes y Antártica Chilena, Punta Arenas

Suarez AV, Holway DA, Case TJ (2001) Patterns of spread in biological invasions dominated by long-distance jump dispersal: insights from argentine ants. Proc Natl Acad Sci U S A 98:1095–1100

Ulbrich J (1930) Die bisamratte. Heinrich, Dresde

Williamson M (1996) Biological invasions. Chapman & Hall, London

Yachi S, Kawasaki K, Shigesada N, Teremoto E (1989) Spatial patterns in propagating waves of fox rabies. Forma 4:3–12

Chapter 6
Impacts on Human Health, Economy and Biodiversity

6.1 Preamble

Once established, exotic species can impact the composition and functioning of the ecosystems they invade (Hulme 2007). Although the magnitude of this impact may vary from imperceptible to conspicuous (Williamson 1996), these species may now be collectively called "invasive" (Table 1.1), and their very presence is conceived negatively (Williamson 1996). It is important to consider that, according to our proposed conceptual framework, an invasive species is an exotic species that is expanding its geographical range (see Table 1.1). However, a complementary definition proposes that an invasive species is an exotic species that causes an undesirable (or negative) effect perceptible to a greater or lesser extent, regardless of geographical expansion in its secondary range. We will use these very similar concepts in an interchangeable manner, because we cannot ascertain for sure whether a given invading species is expanding geographically in its new range. For now, we will mention some examples to illustrate how some exotic species can become problematic.

6.2 Some Examples

The deliberate introduction of salmons (*Salmo* spp.) and trouts (*Oncorhynchus* spp.) in freshwater environments in several countries has caused a serious deterioration on diversity due to its voracity and the breadth of its eating habits. This has led to the local extinction of numerous native fish species in various localities around the globe (Dunham et al. 2002; Kitano 2004; Ricciardi 2004; Revenga et al. 2005). Two species, *Salmo trutta* and *Oncorhynchus mykiss*, both included in a list of the 100 worst invasive species in the world (Lowe et al. 2000), were introduced into lakes and rivers in central and southern Chile (Iriarte et al. 2005). In one of the few

F. M. Jaksic, S. A. Castro, *Biological Invasions in the South American Anthropocene*, https://doi.org/10.1007/978-3-030-56379-0_6

available studies covering 11 lakes and 105 streams, Soto et al. (2007) documented that, on average, 60% of the abundance (Fig. 6.1a) and more than 80% of the biomass of the fish sampled were represented by these exotic species. These results highlight the ecological importance of salmonids in Chilean basins, becoming more abundant species than native fishes. These authors also documented a negative

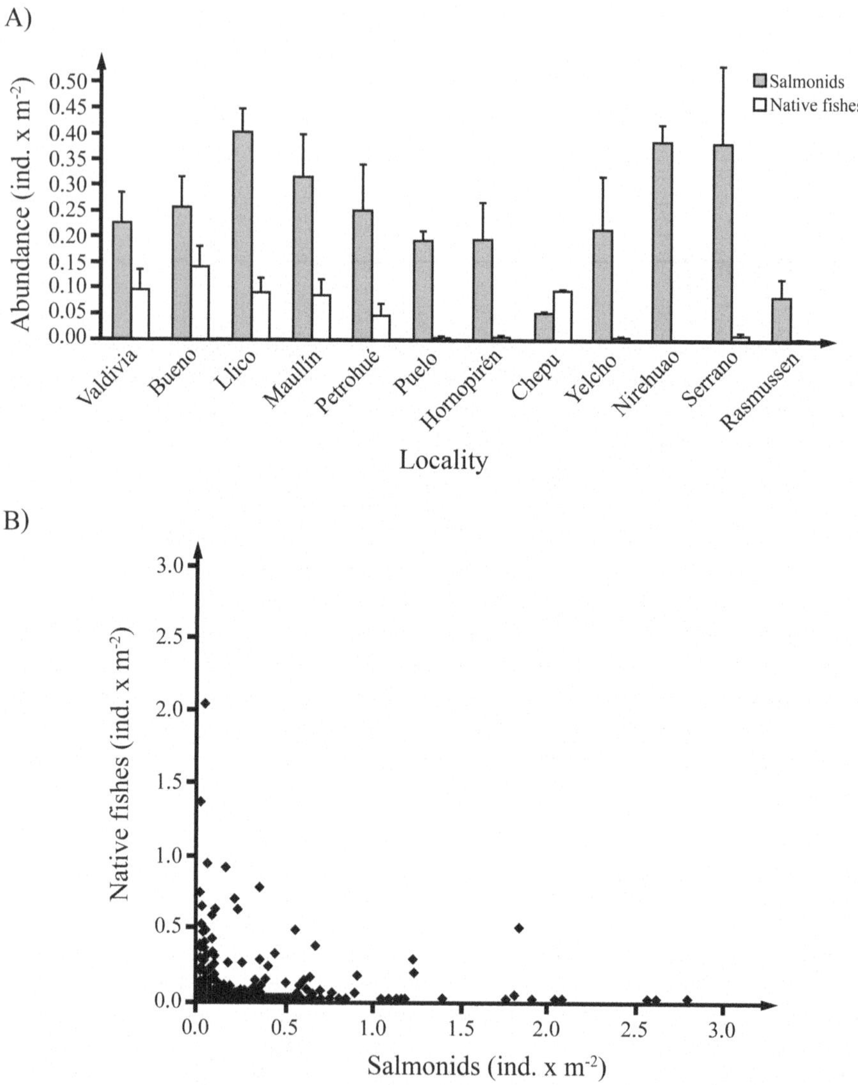

Fig. 6.1 Ecological importance of salmonid invasion in Chile (Soto et al. 2007). (**a**) Mean abundance (individuals × m^{-2} ± 1 ES) of native fishes and salmonids in third to fourth order streams at the 11 watershed studied (watersheds are displayed from north to south in the X axis); (**b**) negative relationship between the average abundance of native fish (all species grouped) and salmonids (all species grouped) in 105 streams, both expressed in individuals × m^{-2}

relationship between the abundance of salmonid species and native ichthyofauna (Fig. 6.1b). This evidence suggests that naturalization of salmonids in freshwater lakes and streams in southern Chile would negatively affect the distribution and abundance of the native fish species assemblage (Soto et al. 2007). The exotic species introduced into Chile, such as trout, salmon, and carp (Habit et al. 2006; Habit el al. 2014; Vila and Habit 2015), are generally larger in size than native fish species and they exhibit general trophic habits at mature stages of development, depredating or competing for habitat and available resources (Soto et al. 2007; Penaluna and Arismendi 2009; Correa et al. 2012; Elgueta et al. 2012).

In terrestrial environments of southern Chile and Argentina, the introduction of the North American beaver (*Castor canadensis*) has progressively modified the appearance of the wooded landscape of Tierra del Fuego, especially in riparian environments (Arismendi et al. 2008). Burrows, built from tree trunks, act as dikes widening rivers and streams, eventually modifying the downstream watercourse (Skewes et al. 1999; Coronato et al. 2003). Consequently, they have modified the physiognomy of the Andean-Patagonian landscape, a fact that can be observed with the naked eye (see Fig. 6.2). These alterations also affect nutrient cycling (Coronato et al. 2003), as well as nearby floristic composition, in response to canopy opening resulting from tree death, flooding, and soil alterations (Jones et al. 1994; Crooks 2002; Parkes et al. 2006; Anderson and Rosemond 2007; Wallem et al. 2010). Being one of the most recent mammal invasions in Chile and Argentina (approximately 70 years old), the presence and ecosystem impact of the beaver are remarkable, so controlling its population expansion requires urgent measures (Anderson et al. 2009).

The examples above are only a sample of the type of effects that invasive species can have. In most cases, however, the effect inflicted is virtually imperceptible or may even have some associated benefits (Williamson 1996). For example, the introduction of the European rabbit (*Oryctolagus cuniculus*) in central Chile was quickly considered a pest (Fuentes et al. 1984; Jaksic and Fuentes 1991; Camus et al. 2008). However, recent evidence has shown that this species yields positive ecosystem services as it has become an important food item for raptors and native foxes, who have increased their rabbit consumption from 20% to 45% (Table 6.1; Jaksic 1998, Pavez et al. 2010); at the same time, *Oryctolagus cuniculus* can act as a seed dispersing agent for native plants, especially in places where the abundance of dispersing frugivorous birds is reduced by anthropogenic effect (Castro et al. 2008). It is important to note that *Oryctolagus cuniculus*, as well as *Salmo trutta* and *Oncorhynchus mykiss*, is listed among the 100 most damaging invasive species in the world (Lowe et al. 2000).

The above evidence unveils a controversial question in invasion ecology: Are biological invasions inherently negative? This point is of great relevance, since as invasive species have occupied government agendas, the need has arisen to finance territorial action policies that regulate international trade and establish population management measures on exotic species already present. Some countries, for example, operate on the logic of species lists, in which globally circulating taxa are recognized and classified according to the level of environmental risk. Clearly the list of the world's 100 most harmful invasive alien species (Lowe et al. 2000) is an

A)

B)

Fig. 6.2 Evidence of the physiognomic impact of beaver (*Castor canadensis*) on the landscape of Tierra del Fuego (Coronato et al. 2003). (**a**) Pond formed by the alteration of the watercourse produced by the beaver; on the margins are dry trees resulting from flooding; (**b**) trees dead as a result of changes in watercourses made by beavers

example of such initiatives. Blacklists bring together unwanted species whose introduction is prohibited or not recommended. On the other hand, there are the so-called white lists, which bring together the taxa whose introduction poses a low environmental risk, and therefore their importation and commercialization are allowed without major reservations. A third set of species is categorized in "gray lists"; these are species that will have an unpredictable environmental impact, so it is advisable to monitor them prior to their admission. The level of environmental risk attributable to a given species will depend on knowledge of the associated negative and positive effects, so assessments with a higher level of integration (recognizing positive and negative impacts) should yield more assertive indicators.

Table 6.1 Temporary changes in the consumption of the European rabbit (*Oryctolagus cuniculus*) by two native predator species (*Lycalopex culpaeus* and *Geranoaetus melanoleucus*) in a site of central Chile (Jaksic 1996)

Predator/prey	Former consumption[a] (%)	Later consumption[b] (%)
Culpeo fox (*Lycalopex culpaeus*)	**1976**	**1984**
Small mammals	70.0	52.0
European rabbit	19.7	48.0
Other vertebrates	10.3	0.0
Black-chested eagle (*Geranoaetus melanoleucus*)	**1973–1974**	**1987–1988**
Small native mammals	75.9	34.3
European rabbit	18.8	43.9
Other vertebrates	5.3	14.0
Insects	0.0	7.8

[a]Simonetti (1986)
[b]Pavez et al. (1992)

Parker et al. (1999) proposed a fairly simple and operational impact measure. This metric, described by the algorithm: $I = R \times A \times E$, gives an account of the total impact (I) of any exotic species and is a function of the size of the occupied geographical range (R in m^2), the average of its abundance along the range (A), and a measure of its effect per individual or per unit of biomass (E). Of these, the evaluation of the individual effect (E) seems to be the most complex parameter to establish, since it implies a comparable quantification for species that impose effects depending on attributes such as their trophic and taxonomic-systematic position and the type of ecosystem invaded, among other factors.

6.3 Types of Impact

It is important to note that the impact of exotic species, whether positive or negative, can go beyond the ecological realm. In fact, it is frequently pointed out in the scientific literature that some exotic species can affect human health, international trade, the production of goods and services, as well as the sociocultural and poverty condition of the human population (Davis 2009). This situation makes the management of exotic species particularly complex, since the control or eradication of some taxa considered environmentally harmful could be discouraged in favor of their role in the local economy (Simberloff 2005). Perhaps one of the cases that best illustrates this commitment between interests is the cultivation of pines (*Pinus* spp.) in Chile. Currently, extensive forest plantations have generated a notorious environmental impact due to the substitution of native vegetation (Armesto et al. 2010), a fact that contrasts with the macroeconomic impact of its exploitation, since the forest industry constitutes the second axis of economic growth of the country after mining

Table 6.2 Annual export of forest products in millions of dollars (data from the Forestry Institute of Chile; INFOR 2010). Annual average composition is expressed in grains/m^3 units

Year	2001	2002	2003	2004	2005	2006	2007	2008	2009
Total	2,205	2,301	2,524	3,396	3,495	3,890	4,952	5,452	4,162

(INFOR 2010). In fact, the evolution of the forest products export market showed an increase from US$ 2200 in 2001 to about US$ 4100 in 2009 (Table 6.2).

Currently, most studies documenting impacts are only partial or incomplete efforts (Vilà et al. 2010). On the one hand, some studies limit themselves to describing effects without quantifying their magnitude, and on the other hand, other studies only consider the negative or undesirable dimension of the impact with little or no attention to the benefits derived from exotic species. Clearly, these facts do not provide adequate decision-making regarding the control or management of introduced species. At present, risk assessments (RA) for species that could potentially be introduced or impact assessment (IA) for exotic species already present in a territory have tended towards more integrated considerations of this multiplicity of factors and variables associated with the invasion (Simberloff 2005).

The need for an integrated approach can be seen from the recent publication by Vilà et al. (2010). These authors have shown that the ecological and economic impacts of vertebrates, invertebrates, and exotic aquatic species in Europe are positively correlated with each other (Fig. 6.3). In other words, those species that cause high ecological damage also determine high economic costs. Thus, in the European case, the control or eradication of those species with the greatest ecological impact would simultaneously reduce their negative economic effect. Although this scenario cannot necessarily be transferred to other regions of the planet, what is important here is the deployment of integrated assessments that support management decision-making.

6.4 Impacts on Human Health

Various species of animals, plants, fungi, and microorganisms can cause or serve as vectors of human disease (Groves 2002; Roques et al. 2008; Canals et al. 2014). Traditional examples are Diptera (mosquitoes), Siphonaptera (fleas), and Phthiraptera (lice). However, one of the most emblematic cases illustrating the impact of a biological invasion on the human population is the spread by contagion of the black plague associated with the black rat (*Rattus rattus*) in Europe. The black rat came to Europe from Asia in the eighth century (Elton 1958), but it was in the fourteenth century when the disease spread, causing an estimated decline between 30% and 60% of the European population; its cause, the bacterium *Yersinia pestis*, is housed in fleas that parasitize the black rat and the human being.

Although not as dramatic, biological invasions can now cause persistent deleterious effects on the human population. For example, there is abundant evidence

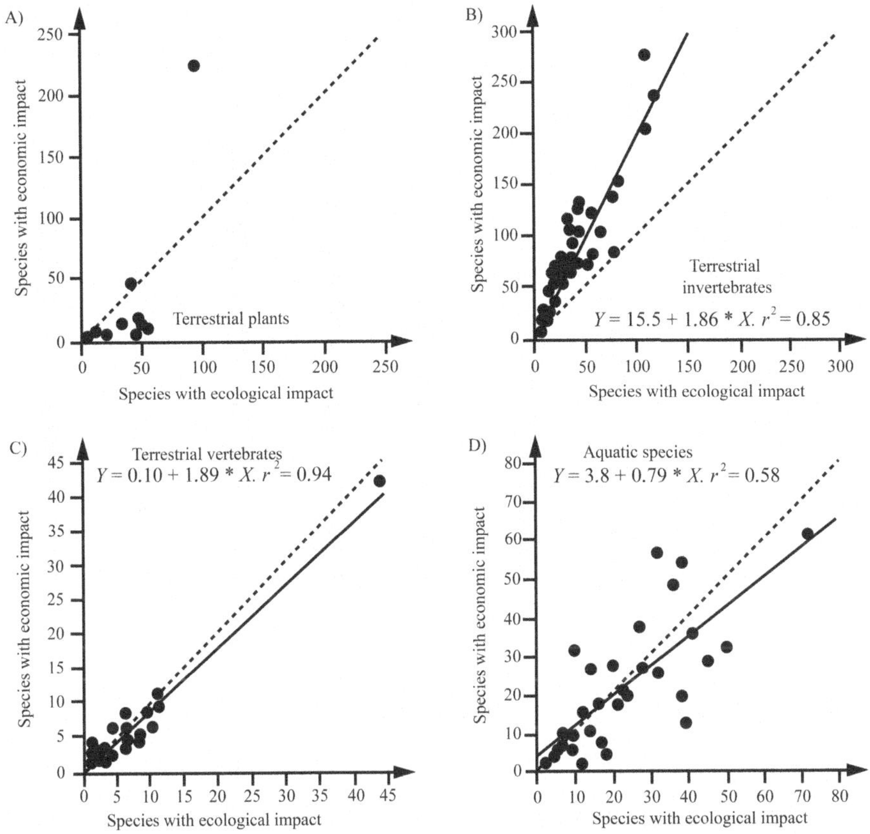

Fig. 6.3 Relationship between the number of exotic species with ecological and economic impact for different taxonomic groups in Europe (Vilà et al. 2010). Each data point represents an individual country, major island, or administrative unit. Dashed line represents the line of equality between the axes. Data from the DAISIE database. (**a**) Terrestrial plants; (**b**) terrestrial invertebrates; (**c**) terrestrial vertebrates; (**d**) aquatic species

showing that pollen release from urban plants causes allergies (Belmonte and Vilà 2004), a condition that can lead to more severe complications. The interesting thing is that an important proportion of the plant species in urban environments of America and Europe correspond to exotic species of ornamental value but with allergenic activity.

Floristic studies documenting a high representation of exotic flora in urban environments have increased in the last decade (e.g., Celesti-Grapow et al. 2006; Fanelli et al. 2006; Ricotta et al. 2009); however, few have explored their actual or potential impacts on the human population, especially those related to pollen allergenicity (Belmonte and Vilà 2004). In Chile some prospective studies are available. For example, Rojas and Roure (2001) studied the composition of atmospheric pollen in Santiago, describing the concentration over 3 years of sampling. Their findings

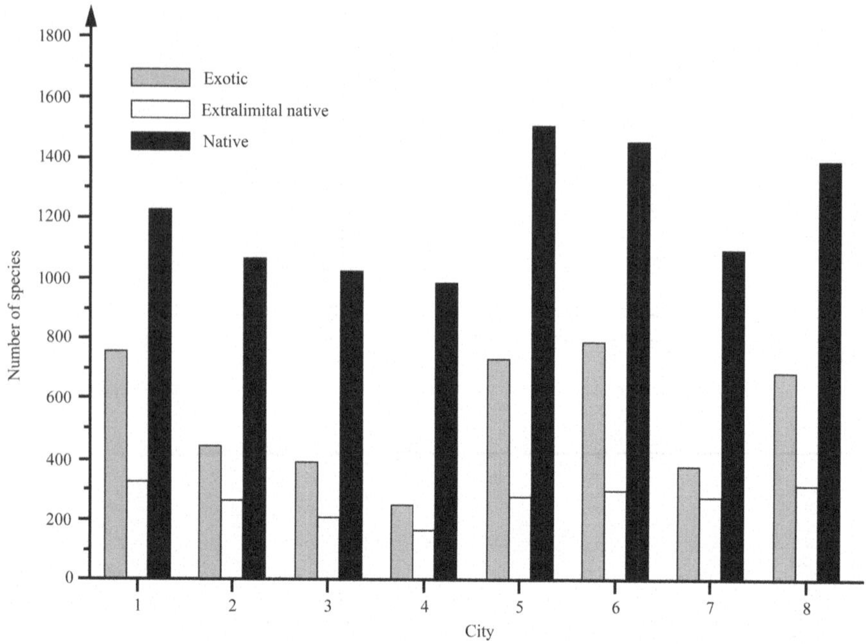

Fig. 6.4 Distribution of the number of plant species according to their origin in eight urban floras in cities of the United States (La Sorte and McKinney 2006). The urban areas are (1) Boston, (2) Chicago, (3) Detroit, (4) Minneapolis, (5) New York, (6) Philadelphia, (7) Saint Louis, and (8) Washington DC. The categories of origin are the following: (a) exotic, species introduced into the country; (b) native extralimit, species native to the country but that have expanded to a certain city where they were not originally found; (c) native, groups the species whose original distribution corresponds to the country and the city

showed that the highest concentration corresponded to exotic species (97.1%; 23,449 grains/ m^3), while a percentage representation of 2.4% was of native species (570 grains/m^3; Table 6.3). These authors recognized the abundant presence of allergenic pollen throughout the year (Fig. 6.5), from ornamental species such as the Oriental plane (*Platanus orientalis*), two species of Acer (*Acer negundo* and *Acer pseudoplatanus*) and Eastern cottonwood (*Populus deltoides*), as well as various species of exotic herbs (*Plantago lanceolata* and *Plantago major*, respectively), Lambsquarter (*Chenopodium album*, *Chenopodium multifidum*, *Chenopodium ambrosioides*), and several grasses (*Dactylis glomerata*, *Cynodon dactylon*, and *Lolium* spp.) (see Table 6.3; Fig. 6.5). Interestingly, the additional problem associated with allergenic pollen is that it can interact with other components of global change synergistically, increasing public health risk (Belmonte and Vilà 2004). For example, it has been shown that the increase in CO_2 increases pollen production by increasing the number of flower spikes in species of the genus *Ambrosia* (Ziska and Caulfield 2000; Wan et al. 2002; Wayne et al. 2002).

Animals, both vertebrates and exotic invertebrates, can also have an impact on the health of the human population. Although not documented to date in Chile or

Table 6.3 Annual average and percentage composition of pollen in the atmosphere of the city of Santiago de Chile, between 1993 and 1996 (Rojas and Roure 2001)

Taxa	Annual average (grains/m^3)	%
Platanus	11,175	46.3
Poaceae	1,830	7.6
Acer	1,715	7.1
Cupressus	874	3.6
Chenopodiaceae	641	2.7
Urticaceae	582	2.4
Fraxinus	581	2.4
Morus	550	2.3
Plantago	525	2.2
Total Oleaceae	676	2.1
Total Fabaceae	412	1.7
Populus	341	1.4
Pinus	335	1.4
Ulmus	322	1.3
Arecaceae	313	1.3
Eucalyptus	292	1.2
Brassicaceae	262	1.1
Total Asteraceae	253	1.1
Apiaceae	246	1.0
Species not determined	807	3.3
	Subtotals	
Exotic species	23,449	97.1
Native species	570	2.9
TOTAL	24,146	100

Argentina, the beaver (*Castor canadensis*) can transmit to human beings a protozoan called *Giardia lamblia*. This microorganism causes giardiasis, a disease that affects the intestinal tract and can only be treated with antibiotics. The pigeon (*Columba livia*), widely introduced and distributed globally, has been associated with the transmission of at least three diseases to humans: histoplasmosis, cryptococcosis, and psittacosis (Krauss et al. 2003). Histoplasmosis is a disease caused by the fungus *Histoplasma capsulatum* which grows in pigeon droppings. When a person cleans the droppings, he or she can be infected by aspiration of airborne particles. On the other hand, cryptococcosis is associated with immunocompromised patients whose condition facilitates the invasion of the fungus *Cryptococcus neoformans* which causes this disease. Its contagion in humans is also done by the respiratory tract. Psittacosis (also known as ornithosis or parrot fever) is an infectious disease that mainly affects parrots and parakeets (birds of the Psittacidae family) but can also affect and be transmitted to other birds, such as pigeons (Krauss et al. 2003). This disease is caused by a bacteria (*Chlamydia psittaci*) whose transmission to humans is produced by aspirating residues of excrement in suspension.

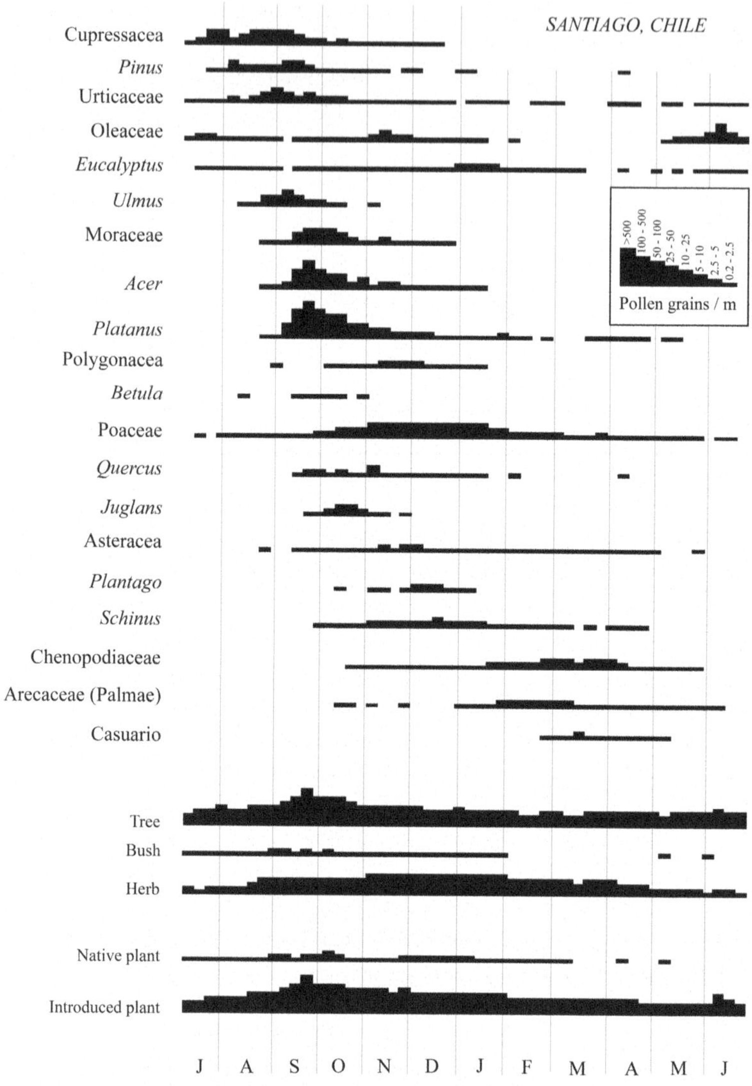

Fig. 6.5 Composition of atmospheric pollen in Santiago de Chile, representing its monthly variation throughout a year of study (Rojas and Roure 2001). Months are indicated correlatively with an initial letter, starting in July 1999 and ending in June 2000

Among the arthropods with pecking habits are numerous species that can affect health, as several can transmit toxins or parasites directly to humans. These typically include spiders, such as those of the genus *Loxosceles* and *Latrodectus*, introduced in Europe from America and Australia, respectively (Kobelt and Nentwig 2008). Also in Europe, more than half of the 47 introduced nematodes are human endoparasites that can be transmitted through meat consumption (Roques et al. 2008).

Associated with international traffic patterns and global climate change, the emergence of new epidemiological events in the human population has been observed (National Institutes of Health US 2007; Canals et al. 2014; Kuri-Morales et al. 2015). These diseases respond to changes in climatic conditions and international traffic that aid in the expansion of pathogens (National Institutes of Health US 2007). In addition, pathogens also acquire resistance. In fact, more than 30 new pathogenic microorganisms have reappeared and spread rapidly after an extended period when they were no longer considered public health problems (Canals et al. 2014; Kuri-Morales et al. 2015). Emerging diseases include Cryptosporidiosis, caused by a protozoan, and Lyne's disease, caused by a spirochete (Suárez and Berdasquera 2000). Reemerging diseases include dengue, yellow fever, cholera, diphtheria, and malaria (Suárez and Berdasquera 2000). These antecedents point to the importance of international traffic pathways as vectors of pathogens, so that future changes in the means of epidemiological control are foreseeable (Canals et al. 2014).

6.5 Impacts on the Economy

Exotic species have important economic effects (Pimentel et al. 2000). These effects involve increases or decreases in the production of human goods and services (Born et al. 2005; Colautti et al. 2006; Lovell et al. 2006). For example, the production of food for human consumption is based on the exploitation of a small number of species whose cultivation has become overcrowded (McNeely 2001). Species such as potatoes (*Solanum tuberosum*), rice (*Oryza sativa*), and corn (*Zea mays*) are the best known among plants, while among the animals are the cow (*Bos taurus*), the goat (*Capra hircus*), the horse (*Equus ferus caballus*), and the pig (*Sus scrofa domestica*). In contrast, other exotic species may act as pests or weeds, since their introduction reduces the efficiency and productivity of crops and pastures, alters the agroforestry infrastructure, reduces the aesthetic value of natural areas, and/or increases the cost associated with ecological restoration (McNeely 2001).

Several studies have evaluated the economic impact of biological invasions, although few of them are taxonomically or territorially comprehensive, focusing on those species with the greatest impact. Exceptions to this rule are the studies by Pimentel et al. (2000, 2005), who evaluated the economic impact of exotic species in the United States (Table 6.4). This study included a taxonomic diversity of 50,000 exotic species, including vascular plants, vertebrates, invertebrates, and microorganisms. According to these authors, the annual economic benefits derived from exotic species resulted in some US$ 800 billion, while the values associated with the economic cost totaled US$ 120 billion per year (see Table 6.4). Clearly, following a strictly economic criterion, the monetary gains associated with exotic species are eight times greater than the economic losses.

In a similar study conducted in South Africa, van Wilgen et al. (2001) estimated economic impacts in the order of $ 18 billion, attributable only to 180 species of

Table 6.4 Estimated annual costs associated to some exotic species introduced into the United States (Pimentel et al. 2000, 2005). Values are expressed in millions of dollars (US$). NA = not applicable

Taxonomic categories	Number of exotic species	Losses, damages	Control costs	Total annual costs (million $)
Plants	25,000			
Purple loosestrife		–	–	45
Aquatic weeds		10	100	110
Melaleuca tree		NA	3–6	3–6
Crop weeds		24,000	3,000	27,000
Weeds in pastures		1,000	5,000	6,000
Weeds in lawns, gardens, and golf courses		NA	1,500	1,500
Mammals	20			
Wild horses and donkeys		5	NA	5
Feral pigs		800	0.5	800.5
Mongooses		50	NA	50
Rats		19,000	NA	19,000
Cats		17,000	NA	17,000
Dogs		620	NA	620
Birds	97			
Pigeons		1,100	NA	1,100
Starlings		800	NA	800
Reptiles and amphibians	53			
Brown tree snake		1	11	12
Fishes	138	5,400	NA	5,400
Arthropods	4,500			
Fire ant		600	400	1,000
Formosan termite		1,000	NA	1,000
Green crab		44	NA	44
Gypsy moth		NA	11	11
Crop pests		13,900	500	14,400
Pests in lawns, gardens, and golf courses		NA	1500	1500
Forest pests		2,100	NA	2,100
Mollusks	88			
Zebra mussel		–	–	1,000
Asian clam		1,000	NA	1,000
Shipworm		205	NA	205
Microbes		20,000		
Crop plant pathogens		21,000	500	21,500
Plant pathogens in lawns, gardens		NA	2000	2,000
Forest plant pathogens		2,100	NA	2,100
Dutch elm disease		NA	100	100
Livestock diseases		14,000	NA	14,000
Human diseases		NA	7500	7,500
Total (million $)				120,105

exotic plants, weeds in natural environments or agroecosystems. Various species of aquatic plants that have spread over large areas in southern Africa are responsible for this effect, among them, the water hyacinth (*Eichhornia crassipes*), water lettuce (*Pistia stratiotes*), giant salvinia (*Salvinia molesta*), parrot feather (*Myriophyllum aquaticum*), and water fern (*Azolla filiculoides*). These species, in the absence of natural enemies and the presence of eutrophicated waters, form dense patches that modify aquatic ecosystems and restrict human use of water. Thus, because these species affect crop irrigation, they have also been associated with rural poverty in South Africa (Le Maitre et al. 2002).

Within the framework of the "Delivering Alien Invasive Species Inventories for Europe" (DAISIE) project, Vilà et al. (2010) assessed the impact of the more than 10,000 invasive species in Europe, focusing basically on recognizing those taxonomic groups for which there is information on their economic impact. This impact has been described, but not necessarily evaluated. Table 6.5 lists some species for which quantitative information on the costs associated with their presence is available. Of the total number of exotic species present in Europe, information only exists for 1347 taxa, i.e., 13%. Here, the two most diverse taxonomic groups were terrestrial plants and invertebrates, which generate the greatest economic and ecological impact (Vilà et al. 2010). The analysis of monetary cost items indicated that the resources involved in management, eradication, and environmental education programs concentrated the largest expenditures, followed by the economic impact on crop production (Table 6.5).

In South America, knowledge about the economic impact of exotic taxa is focused on a few species, and there is no regional or national information program. For example, Matthei (1995), referring to Chile's weeds, which are mostly exotic species, examined the cost of importing chemical agents that allow their control. This author documented that in 1991 alone, the import of herbicides amounted to CLP$ 21 million pesos, a value that does not include their commercialization and application in field conditions. This is only an indirect value, as it does not apply to the productive damage that weeds inflict on production. Clearly, this is a neglected area from the point of view of ecological economics research.

However, recent estimates have established the economic damage of the seven most damaging invasive species in Chile: American mink (*Neovison vison*), North American beaver (*Beaver canadensis*), wild boar (*Sus scrofa*), gorse (*Ulex europaeus*), blackberry (both *Rubus ulmifolius* and *Rubus constrictus*), European wasp (*Vespula germanica*), and European rabbit (*Oryctolagus cuniculus*). This study considered the accumulated costs of biological control of these species, the annual economic loss, and the economic impact over 20 years. Both *Oryctolagus cuniculus* and *Rubus* spp. are the species whose control has generated the highest economic costs (Table 6.6); at the same time, the species with the highest annual economic demands were *Sus scrofa* and *Vespula germanica* (Table 6.6). The projected economic impact over 20 years has been estimated at US$ 1991 billion, *Sus scrofa*, *Vespula germanica*, and *Neovison vison*, being the three major contributing species. It is important to note that various aspects were considered in estimating these costs, including forest biomass losses, grazing land depletion impacts, and negative

Table 6.5 Exotic species in Europe that generate high economic costs in various human activities (Vilà et al. 2010)

Species	Biome/taxa	Country	Spatial extension	Cost item	Period	Cost (Million € / year)	Reference
Carpobrotus spp.	Terrestrial plant	Spain	Localities	Control/eradication	2002–2007	0.58	Andreu et al. (2009)
Anoplophora chinensis	Terrestrial invertebrate	Italy	Country	Control	2004–2008	0.53	Van der Gaag (2007)
Cervus nippon	Terrestrial vertebrate	Scotland	Localities	Control		0.82	White and Harris (2002)
Myocastor coypus	Terrestrial vertebrate	Italy	Localities	Control/damages	1995–2000	2.85	Panzacchi et al. (2007)
Sciurus carolinensis	Terrestrial vertebrate	United Kingdom	Country	Control	1994–1995	0.46	White and Harris (2002)
Azolla filiculoides	Freshwater plant	Spain	Protected area	Control/eradication	2003	1.00	Andreu et al. (2009)
Eichhornia crassipes	Freshwater plant	Spain	River basin	Control/eradication	2005–2007	3.35	Andreu et al. (2009)
Oxyura jamaicensis	Freshwater vertebrate	United Kingdom	Country	Eradication	2007–2010	0.75	Scalera and Zaghi (2004)
Chrysochromulina polylepis	Marine alga	Norway	Country	Toxic bloom		8.18	Hopkins (2002)
Rhopilema nomadica	Marine invertebrate	Israel	Coast	Infrastructure damage	2001	0.04	Galil and Zenetos (2002)

The biome (taxa) occupied, the country where the problem has been documented, the item of cost inflicted, and the economic estimate in millions of euros are indicated for each of them impacts on carbon sequestration. In the case of *Oryctolagus cuniculus*, for example, the economic valuation was made by evaluating their impact on the productive sectors of fruit trees and forest plantations, as well as on the native species affected. In the case of *Sus scrofa*, its impact was estimated on forestry and agriculture; on European hazelnut, milk, and meat production; and on the consumption of araucaria seeds. For the North American mink (*Neovison vison*), its estimated impact was restricted to costs of its biological control in order to protect native biodiversity. The European wasp (*Vespula germanica*) affects the wine, honey, and fruit industry; meanwhile, for the species *Rubus ulmifolius* and *Rubus constrictus*, the assessment considered the impacts of the occupation of grasslands and agricultural and forestry crops. Finally, in the case of *Ulex europaeus*, its impact was assessed on cattle production and forest timber production, as well as the potential cost of fire control and public resources to investigate the species.

6.6 Impacts on Biodiversity

Invasions are obvious modifications to the distribution of biological richness on the planet, since by definition these taxa arrive at places where they were not previously found (Drake et al. 1989). Studying 23 avifaunas and 13 flower communities on oceanic islands, including Easter Island, Sax et al. (2002) documented that bird richness has been maintained, as the number of extinctions (recorded in a range between 0 and 34 species) has been offset by the number of invasions (in a range between 1 and 55 species) (Fig. 6.6). In the case of vascular plants, invasions (in a range between 25 and 2069 species) have far surpassed the number of extinctions (range between 0 and 71 species), resulting in a doubling of the specific richness of these islands (Table 6.7). Although Sax et al. (2002) did not investigate whether invasion events were correlated with extinction events, they concluded that the island communities studied have maintained or increased the number of species present.

Several authors argue that invasions can cause the extinction of species in host ecosystems, a situation that worries conservation biologists, since today most of communities have one or more naturalized species (Usher 1986). In a widely publicized study, Wilcove et al. (1998) compiled a catalogue of native species with various conservation statuses. For each of them, the causal factors that explained its conservation status were assigned; for this, the authors used information obtained from experts or from the available literature. The loss of habitat derived from anthropogenic activities was the main and most frequent factor, while the presence of exotic species took the second place. From this study, the idea that biological invasions are the second cause of species extinction has become widespread.

On the other hand, Rodríguez (2001), in an equivalent effort for animals listed in red books from four South American countries (vertebrates from Bolivia, mammals from Brazil, terrestrial vertebrates from Peru, and vertebrates and invertebrates from Venezuela), found a different pattern: habitat modification was the main

Table. 6.6 Three types of economic costs provoked by the seven most invasive exotic species in Chile

Costs/losses	*Castor canadensis*	*Oryctolagus cuniculus*	*Sus scrofa*	*Neovison vison*	*Vespula germanica*	*Rubus* spp.	*Ulex europaeus*	Total
Cumulative cost of biological control (US$)	64,539,600	1,465,754	62,321	608,271	66,766	145,085,398	54,257	211,882,367
Annual loss (US$)	733,094	3,249,337	38,278,724	9,526,620	21,532,771	10,580,010	4,034,771	87,935,327
Projection of loss (US$)	69,606,555	93,428,701	603,312,032	417,332,256	497,037,440	273,718,394	49,158,106	2,003,593,484

The species are North American beaver (Castor *canadensis*), European rabbit (*Oryctolagus cuniculus*), wild boar (*Sus scrofa*), American mink (*Neovison vison*), European wasp (*Vespula germanica*), blackberry (*Rubus* spp., brings together *Rubus ulmifolius* and *Rubus constrictus*), and gorse (*Ulex europaeus*). The economic costs were calculated based on three estimates: cumulative cost of biological control, annual loss, and projection of this loss in two more decades (US$)

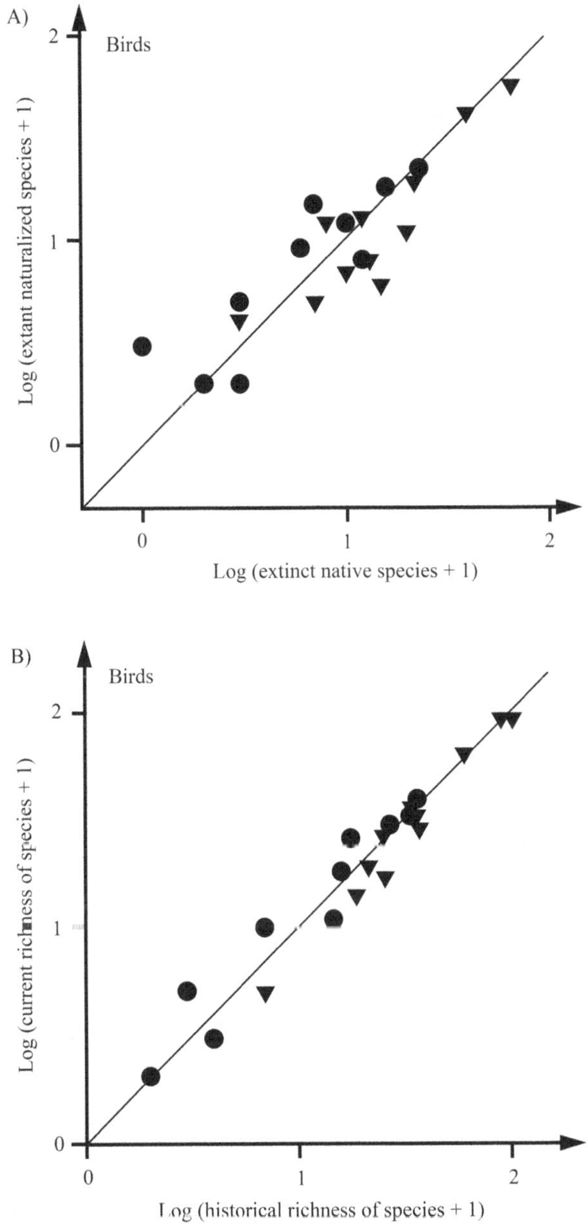

Fig. 6.6 Relationship between native and exotic bird species richness on oceanic islands (Sax et al. 2002). Data points represent islands that were inhabited (circles) or uninhabited (triangles) before European contact. (**a**) Richness of extinct native species versus the current number of naturalized species (regression: $y = 0.77x + 0.23$, $r^2 = 0.78$, $P < 0.001$); (**b**) historical bird richness versus the current number of naturalized species

Table 6.7 Diversity of vascular plants on oceanic islands, indicating numbers of current and extinct native species and naturalized taxa. All islands are currently inhabited by humans but have been classified into one of four categories, based on their history of human occupation prior to European contact. Class A islands have no evidence of human occupation or are unlikely to have been visited prior to the arrival of Europeans. Class B islands have evidence of human visit or are likely to have been visited, but there is no evidence of human settlement, and it is unlikely that persistent occupation has been established. Class C islands had human settlements but were abandoned before European contact. Class D islands had human settlements at the time of European contact (see Sax et al. 2002)

Island (or archipelago)	Class	Natives (extant)	Natives (extinct)	Naturalized species (extant)	Total no. of extant species
Christmas Island (Indian Ocean)	B	201	2	151	352
Cocos Island	B	61	1	53	114
Easter Island[a]	D	43	7	68	111
Hawaiian Islands[a]	D	1,223	71	1,090	2,313
Heron Island	B	27	...[b]	25	52
Lord Howe Island	A	219	2	202	421
Mangareva Island	D	85	...[b]	60	145
Nauru Island	D	50	1	85	135
New Zealand Islands[a]	D	2,065	3	2,069	4,134
Northern Line Islands	C	35	0	41	76
Norfolk Island	C	157	3	244	401
Pitcairn Island	C	40	4	40	80
Tristan da Cunna Island	A	70	0	54	124

[a]A fossil record exists
[b]Insufficient information was available to determine whether any species had become extinct

conservation risk for the species studied, followed by overexploitation of species (Table 6.8). Unlike the results obtained by Wilcove et al. (1998), in South America the presence of exotic species occupied the third place of risk, which involved only 6 of the 378 species studied (Table 6.8).

Although numerous studies suggest a causal relationship between the extinction of native flora and fauna as a consequence of the arrival of exotic species, this relationship is not easy to demonstrate. In fact, Gurevitch and Padilla (2004) indicated that the evidence linking biological invasions as a causal agent of extinction is weak, showing that most of it rests on anecdotal field observations, a problem also present in the influential work of Wilcove et al. (1998). Didham et al. (2005), based mainly on the results obtained by MacDougall and Turkington (2005), suggested that invasions could be associated with extinctions indirectly, and not directly as is usually considered. His argument was based on the fact that most exotic species are competitively inferior, so that their presence in the recipient communities would be associated with anthropogenic disturbances that cause their persistence. With this evidence, Didham et al. (2005) proposed that invasive species behave more like passengers and not as driver agents of the community diversity.

Table 6.8 Number and percentage of South American species threatened by the three main risk factors indicated (Rodríguez 2001). Because two or more causes may threaten the same species, the sum of the percentages does not necessarily result in 100%. For comparative purposes, the last column shows the information available for the United States from Wilcove et al. (1998)

Class	No.	Exotic species (%)	Habitat conversion (%)	Overexploitation (%)	Threatened by exotic species in the United States[a] (%)
Amphibia	20	30	100	45	27
Birds	170	3	87	43	69
Crustacea	4	75			
Gastropod	1	100			
Insecta	30	93	7		
Mammalia	101	4	87	65	27
Oligochacta	3	100	33		
Onychophora	1	100			
Osteichthyes	21	24	67	38	53
Reptilia	27	7	70	78	37
Total	378	6	86	48	49

[a]Wilcove et al. (1998)

This evidence shows that the role of biological invasions as a causal agent of extinction constitutes an arena of debate and controversy (Ricciardi 2004; Hulme et al. 2015; Thomas and Palmer 2015), since this effect is often masked by anthropic influence, which modifies the habitat favoring both the persistence of exotic species and the extinction or extirpation of native species (Kowarik 2003). What cannot be ignored, however, is that biological invasions can modify other functional and structural attributes of the ecosystems they invade. These aspects will be reviewed more closely in the next chapter.

References

Anderson CB, Rosemond AD (2007) Ecosystem engineering by invasive exotic beavers reduces in-stream diversity and enhances ecosystem function in Cape Horn, Chile. Oecologia 154:141–153

Anderson CB, Martínez-Pastur G, Lencinas MV, Wallem PK, Moorman MC, Rosemond AD (2009) Do introduced North American beavers *Castor canadensis* engineer different in southern South America? An overview with implications for restoration. Mammal Rev 39:33–52

Andreu J, Vilà M, Hulme PE (2009) An assessment of stake-holder perceptions and management of noxious alien plants in Spain. Environ Manag 43:1244–1255

Arismendi I, Szejnerc P, Larab A, González ME (2008) Impacto del castor en la estructura de bosques ribereños de *Nothofagus pumilio* en Tierra del Fuego, Chile. Bosque 29:146–154

Armesto JJ, Manuschevich D, Mora A, Smith-Ramírez C, Rozzi R, Abarzúa AM, Marquet PA (2010) From the Holocene to the Anthropocene: a historical framework for land cover change in southwestern South America in the past 15,000 years. Land Use Policy 27:148–160

Belmonte J, Vilà M (2004) Atmospheric invasion of non-native pollen in the Mediterranean region. Am J Bot 91:1243–1250

Born W, Rauschmayer F, Brauer I (2005) Economic evaluation of biological invasions: a survey. Ecol Econ 55:321–336

Camus P, Castro S, Jaksic F (2008) El conejo europeo en Chile: historia de una invasión biológica. Historia 41:305–339

Canals M, Torres-Contreras H, Castro SA, Jaksic FM (2014) Dinámica poblacional de la expansion de enfermedades infecciosas como modelo de invasiones biologicas. In: Jaksic FM, Castro SA (eds) Invasiones biológicas en Chile. Ediciones Universidad Católica de Chile, pp 459–485

Castro SA, Bozinovic F, Jaksic FM (2008) Ecological efficiency and legitimacy in seed dispersal of an endemic shrub (*Lithrea caustica*) by the European rabbit (*Oryctolagus cuniculus*) in Central Chile. J Arid Environ 72:1164–1173

Celesti-Grapow L, Pyšek P, Jarošík V, Blasi C (2006) Determinants of native and alien species richness in the urban flora of Rome. Divers Distrib 12:490–501

Colautti RI, Bailey SA, van Overdijk CDA, Admunsen K, MacIsaac HJ (2006) Characterised and projected costs of nonindigenous species in Canada. Biol Invasions 8:45–59

Coronato A, Escobar J, Mallea C, Roig C, Lizarralde M (2003) Características geomorfológicas de ríos de montaña colonizados por *Castor canadensis* en Tierra del Fuego, Argentina. Ecol Austral 13:15–26

Correa C, Bravo AP, AP Hendry AP. (2012) Reciprocal trophic niche shifts in native and invasive fish: salmonids and galaxiids in Patagonian lakes. Freshw Biol 57:1769–1781

Crooks JA (2002) Characterizing ecosystem-level consequences of biological invasion: the role of ecosystem engineers. Oikos 97:153–166

Davis MA (2009) Invasion biology. Oxford University Press, Oxford

Didham RK, Tylianakis JM, Hutchinson MA, Ewers RM, Gemmel NJ (2005) Are invasive species the drivers of ecological change? Trends Ecol Evol 20:470–474

Drake JR, Mooney HA, di Castri F, Groves RH, Kruger FJ, Rejmánek M, Williamson M (1989) Biological invasions: a global perspective. In: Scientific committee on problems of the environment (SCOPE). Wiley, Chichester

Dunham JB, Rieman BE, Peterson JT (2002) Patch-based models of species occurrence: lessons from salmonid fishes in streams. In: Scott JM, Heglund PJ, Samson F, Haufler J, Morrison M, Raphael M, Wall B (eds) Predicting species occurrences: issues of accuracy and scale. Island Press, Covelo, pp 327–334

Elgueta A, González J, Ruzzante D, Walde S, Habit E (2012) Trophic interference by *Salmo trutta* on *Aplochiton zebra* and *Aplochiton taeniatus* in Southern Patagonian lakes. J Fish Biol 82:430–443

Elton CS (1958) The ecology of invasions by animals and plants. University of Chicago Press, Chicago

Fanelli G, Tescarollo P, Testi A (2006) Ecological indicators applied to urban and suburban flora. Ecol Indic 6:444–457

Fuentes ER, Otaiza RD, Alliende MC, Poiani AJ (1984) Shrub clumps of the Chilean matorral vegetation: structure and possible maintenance mechanisms. Oecologia 42:405–411

Galil BS, Zenetos A (2002) A sea change: exotics in the eastern Mediterranean Sea. In: Leppäkoski E, Gollasch S, Olenin S (eds) Invasive aquatic species of Europe: distribution, impacts and management. Springer, Dordrecht, pp 325–336

Groves RH (2002) The impacts of alien plants in Australia. In: Pimentel D (ed) Biological invasions: economic and environmental costs of alien plant, animal and microbe species. CRC Press LLC, Boca Raton, pp 11–23

Gurevitch J, Padilla DK (2004) Are invasive species a major cause of extinctions? Trends Ecol Evol 19:470–474

Habit E, Dyer B, Vila I (2006) Estado de conocimiento de los peces dulceacuícolas de Chile. Gayana 70:100–113

Hopkins CCE (2002) Introduced marine organisms in Norwegian waters, including Svalbard. In: Leppäkoski E, Gollasch S, Olenin S (eds) Invasive aquatic species of Europe: distribution, impacts and management. Springer, pp 240–252

Hulme PE (2007) Biological invasions in Europe: drivers, pressures, states, impacts and responses. In: Hester R, Harrison RM (eds) Biodiversity under threat. Cambridge University Press, Cambridge, pp 56–80
Hulme PE, Pauchard A, Pyšek P, Vilà M, Alba C, Blackburn TM, Bullock JM, Chytrý M, Dawson W, Dunn AM, Essl F, Genovesi P, Maskell LC, Meyerson LA, Nuñez MA, Pergl J, Pescott OL, Pocock MJO, Richardson DM, Roy HE, Smart SM, Štajerová K, Stohlgren T, van Kleunen M, Winter M (2015) Challenging the view that invasive non-native plants are not a significant threat to the floristic diversity of Great Britain. Proc Natl Acad Sci U S A 112:2988–2989
Instituto Forestas, INFOR (2010) Estadísticas del sector forestal (en línea). Santiago, Chile. Disponible en http//www.infor.cl Consultado 20 julio. 2018
Iriarte JA, Lobos GA, Jaksic FM (2005) Invasive vertebrate species in Chile and their control and monitoring by governmental agencies. Rev Chil Hist Nat 78:143–154
Jaksic FM (1996) Ecología de los vertebrados en Chile. Ediciones Universidad Católica de Chile, Santiago
Jaksic FM (1998) Vertebrate invaders and their ecological impacts in Chile. Biodivers Conserv 7:1427–1445
Jaksic FM, Fuentes ER (1991) Ecology of a successful invader: the European rabbit in Central Chile. In: Grooves RH, di Castri F (eds) Biogeography of mediterranean invasions. University of Cambridge, Cambridge, pp 273–283
Jones CG, Lawton JH, Shachak M (1994) Organisms as ecosystem engineers. Oikos 69:373–386
Kitano S (2004) Ecological impacts of rainbow, brown, and brook trout in Japanese inland waters. Glob Environ Res 8:41–50
Kobelt M, Nentwig W (2008) Alien spider introductions to Europe supported by global trade. Divers Distrib 14:273–280
Kowarik I (2003) Human agency in biological invasions: secondary releases foster naturalisation and population expansion of alien plant species. Biol Invasions 5:293–312
Krauss H, Weber A, Appel M, Enders B, Graevenitz A, Isenberg HD, Schiefer HG, Slenczka W, Zahner H (2003) Zoonoses: infectious diseases transmissible from animals to humans. ASM Press, American Society for Microbiology, Washington, DC
Kuri-Morales PA, Guzmán-Morales E, de la Paz-Nicolau E, Salas-Fernández A (2015) Enfermedades emergentes y reemergentes. Gaceta de Médica Mexico 151:674–680
La Sorte FA, McKinney ML (2006) Compositional similarity and the distribution of geographical range size for native and non-native species in urban floras. Divers Distrib 12:679–686
Le Maitre DC, van Wilgen BW, Gelderblom CM, Bailey C, Chapman RA, Nel JA (2002) Invasive alien trees and water resources in South Africa: Case studies of the costs and benefits of management. For Ecol Manag 160:143–159
Lovell SJ, Stone SF, Fernandez L (2006) The economic impacts of aquatic invasive species: a review of the literature. Agricult Resour Econom Rev 35:178–194
Lowe S, Browne M, Boudjelas S, De Poorter M (2000) 100 of the world's worst invasive alien species: a selection from the global invasive species database. Published by The Invasive Species Specialist Group (ISSG) a specialist group of the Species Survival Commission (SSC) of the World Conservation Union (IUCN)
MacDougall AS, Turkington R (2005) Are invasive species the drivers or passengers of change in degraded ecosystems? Ecology 86:42–52
Matthei O (1995) Manual de las malezas que crecen en Chile. Alfabeta Impresores, Santiago, Chile
McNeely JA (2001) The great reshuffling: human dimensions of invasive alien species. IUCN, Gland/Cambridge
National Institutes of Health U.S. (2007) Understanding emerging and re-emerging infectious disease. Biological Sciences Curriculum Study
Panzacchi M, Bertolino S, Cocchi R, Genovesi P (2007) Population control of coypu *Myocastor coypus* in Italy compared to eradication in UK: a cost-benefit analysis. Wildl Biol 13:159–171

Parker IM, Simberloff D, Lonsdale WM, Goodell K, Wonham M, Kareiva PM, Williamson MH, von Holle B, Moyle PB, Byers JE, Goldwasser L (1999) Impact: toward a framework for understanding the ecological effects of invaders. Biol Invasions 1:3–19
Parkes JP, Robley A, Forsyth DM, Choquenot D (2006) Adaptive management experiments in vertebrate pest control in New Zealand and Australia. Wildl Soc Bull 34:229–236
Pavez EF, González CA, Jiménez JE (1992) Diet shifts of black-chested eagles (*Geranoaetus melanoleucus*) from native prey to European rabbits. J Raptor Res 26:27–32
Pavez EF, Lobos GA, Jaksic FM (2010) Cambios de largo plazo en el paisaje y los ensambles de micromamíferos y rapaces en Chile central. Rev Chil Hist Nat 83:99–111
Penaluna B, Arismendi I (2009) Peces nativos en aguas continentales del Sur de Chile. Imprenta América Ltda
Pimentel D, Lach L, Zuniga R, Morrison D (2000) Environmental and economic cost of nonindigenous species in the United States. Bioscience 50:53–65
Pimentel D, Zúñiga R, Morrison D (2005) Update on the environmental and economic costs associated with alien-invasive species in the United States. Ecol Econ 52:273–288
Revenga C, Campbell I, Abell R, de Villiers P, Bryer M (2005) Prospects for monitoring freshwater ecosystems towards the 2010 targets. Philos Trans R Soc B 360:397–413
Ricciardi A (2004) Assessing species invasions as a cause of extinction. Trends Ecol Evol 19:619
Ricotta C, La Sorte FA, Pyšek P, Rapson GL, Celesti-Grapow L, Thompson K (2009) Phyloecology of urban alien floras. J Ecol 97:1243–1251
Rodríguez JP (2001) Exotic species introductions into South America: an underestimated threat? Biodivers Conserv 10:1981–1994
Rojas G, Roure J (2001) Atmospheric pollen in Santiago, Chile. Grana 40:126–132
Roques A, Rabitsch W, Lopez-Vaamonde C, Nentwig W, Kenis M (2008) Alien terrestrial invertebrates of Europe. In: Drake JA (ed) Handbook of alien species in Europe. Invading nature. Springer Series in Invasion Ecology, Amsterdam, pp 63–79
Sax DF, Gaines SD, Brown JH (2002) Species invasions exceed extinctions on islands worldwide: a comparative study of plants and birds. Am Nat 160:766–783
Scalera R, Zaghi D (2004) Alien species and nature conservation in the EU: the role of the LIFE program. European Commission, Brussels
Simberloff D (2005) Non-native species do threaten the natural environment. J Agric Environ Ethics 18:595–607
Simonetti JA (1986) Human-induced dietary shift in *Dusicyon culpaeus*. Mammalia 50:406–408
Skewes RO, Gonzalez F, Rubilar L, Quezada OM, Olave IT, Vargas V, Avila AC (1999) Investigación, aprovechamiento y control del castor, islas Tierra del Fuego y Navarino. Servicio de Gobierno Regional XII Región. Magallanes y Antártica Chilena, Punta Arenas
Soto D, Arismendi I, González J, Sanzana J, Jara F, Jara C, Guzmán E, Lara A (2007) Southern Chile, trout and salmon country: invasion patterns and threats for native species. Rev Chil Hist Nat 79:97–117
Suárez CL, Berdasquera D (2000) Enfermedades emergentes y reemergentes: factores causales y vigilancia. Revista Cubana de Medicina General Integral 16:593–597
Thomas CD, Palmer G (2015) Non native plants add to the British flora without negative consequences for native diversity. Proc Natl Acad Sci U S A 112:4387–4392
Usher MB (1986) Invasibility and wildlife conservation: invasive species on nature reserves. Philos Trans R Soc Lond B 314:695–709
van der Gaag DJ (2007) Report workshop management of *Anoplophora*, Wageningen. www.minlnv.nl/cdlpub/servlet/CDLServlet?p_file_id=22662
van Wilgen BW, Richardson DM, Le Maitre DC, Marais C, Magadlela D (2001) The economic consequences of alien plant invasions: examples of impacts and approaches to sustainable management in South Africa. Environ Dev Sustain 3:145–168
Vila I, Habit E (2015) Current situation of the fish fauna in the Mediterranean region of Andean river systems in Chile. Fishes Mediterranean Environ 2015:002, 19p

Vilà M, Basnou C, Pyšek P, Josefsson M, Genovesi P, Gollasch S, Nentwig W, Olenin S, Roques A, Roy D, Hulme PE, DAISIE Partners (2010) How well do we understand the impacts of alien species on ecological services? A pan-European cross-taxa assessment. Front Ecol Environ 8:135–144

Wallem PK, Anderson CB, Martínez-Pastur G, Lencinas MV (2010) Community re-assembly by an exotic herbivore, *Castor canadensis*, in subantarctic forests, Chile and Argentina. Biol Invasions 12:325–335

Wan S, Yuan T, Bowdish S, Wallace L, Russell SD, Luo Y (2002) Response of an allergenic species, *Ambrosia psilostachya* (Asteraceae), to experimental warming and clipping: implications for public health. Am J Bot 89:1843–1846

Wayne P, Foster S, Connolly J, Bazzaz F, Epstein P (2002) Production of allergic pollen by ragweed (*Ambrosia artemisiifolia* L.) is increase in CO_2-enriched atmospheres. Anna Allerg Asthma Immunol 88:279–282

White PCL, Harris S (2002) Economic and environmental costs of alien vertebrate species in Britain. In: Pimentel D (ed) Biological invasions: economic and environmental costs of alien plant, animal and microbe species. CRC Press, Boca Raton, pp 113–149

Wilcove DS, Rothstein D, Dubow J, Phillips A, Losos E (1998) Quantifying threats to imperiled species in the United States. Bioscience 48:607–615

Williamson M (1996) Biological invasions. Chapman & Hall, London

Ziska LH, Caulfield FA (2000) Rising CO_2 and pollen production of common ragweed (*Ambrosia artemisiifolia*), a known allergy-inducing species: implications for public health. Aust J Plant Physiol 27:893–898

Chapter 7
Ecological Impacts

7.1 Preamble

As we have seen in the previous chapter, the presence of invasive species is often associated with the extinction of native species (Wilcove et al. 1998; Ricciardi 2004). Although the generality of this fact has been questioned by some authors (Didham et al. 2005; Davis 2009), there is no doubt that once arrived and established, invasive species can affect to a greater or lesser extent the structure and functioning of the ecosystems they invade (Vitousek et al. 1987; Chesson 2000; Crooks 2002).

In this chapter, we will review and analyze some of these ecological effects, not necessarily linked to the extinction of native taxa. In this section, we will go through different types of interactions and levels of organization, covering a wide range of cases ranging from organismic to biogeographic scales. The aim is to highlight the multiplicity of effects and nuances in which invasive species are involved. Due to the nature of this exercise, it is not always discernible when the effects referred to belong to one or another level of organization, so our presentation should rather be considered a referential systematization.

7.2 Ecological Impact on an Organismic Scale

The presence of organisms belonging to exotic species may modify various attributes at the individual level in the native biota (Parker et al. 1999). It is not difficult to imagine, for example, that the introduction of a predator can modify the behavior of its potential prey or, conversely, the introduction of prey species could modify the behavior of native predators. In the latter case, perhaps an interesting example is the change in European rabbit consumption patterns (*Oryctolagus cuniculus*) by native predators in central Chile (Jaksic 1996). Indeed, until recently in these

F. M. Jaksic, S. A. Castro, *Biological Invasions in the South American Anthropocene*, https://doi.org/10.1007/978-3-030-56379-0_7

communities, the culpeo fox (*Lycalopex culpaeus*) and eagle (*Geranoaetus melanoleucus*) consumed a marginal proportion of European rabbit. This fact was traditionally explained by the inefficiency of these native predators in capturing a new and elusive prey (Jaksic and Yáñez 1980, 1983; Jaksic and Ostfeld 1983; Jaksic and Fuentes 1991). However, more recent data indicate that the proportion of rabbits as food items has increased in these two predators' diets (Pavez et al. 2010), suggesting that these carnivores have learned to capture rabbits more efficiently.

The above example is also useful to illustrate how a species introduced under less predation pressure can also modify its behavior. In Spain, *Oryctolagus cuniculus* is under high predation pressure from its numerous natural enemies; apparently, this is why its use of habitat in shrub formations is restricted to habitats protected by tree or shrub cover (Jaksic and Fuentes 1980). Thus, in this type of habitat, the rabbit can escape from its predators. In contrast, in the scrub of central Chile, where predation pressure does not seem to be as high as in Spain, the pattern of space use is different, as this lagomorph tends to use open spaces with direct access to the herbaceous matrix it consumes (Fig. 7.1; Jaksic and Soriguer 1981; Jaksic and Fuentes 1991).

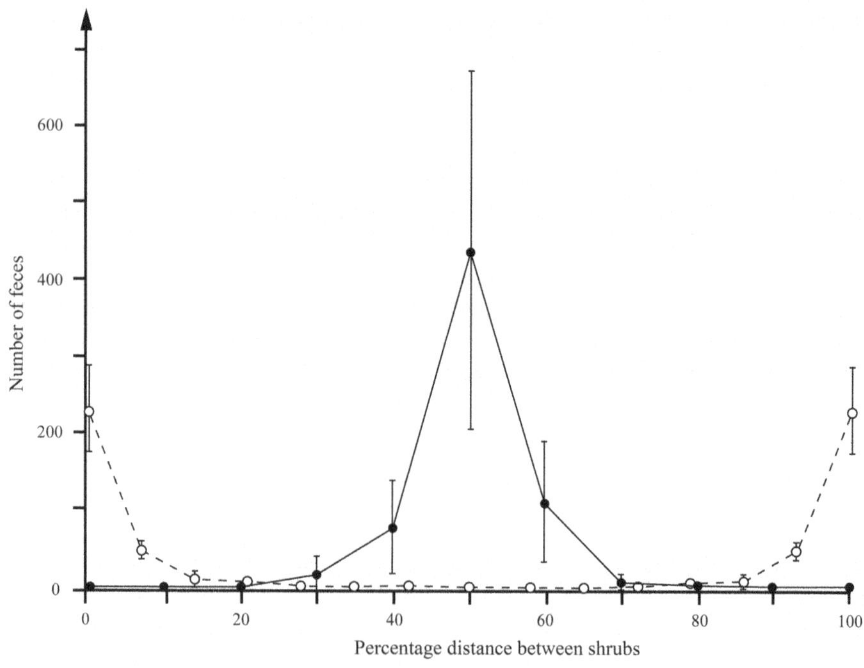

Fig. 7.1 Spatial distribution of feces of *Oryctolagus cuniculus* (continuous line) and of *Octodon degus* (segmented line), found between neighboring shrubs in the scrub of central Chile. The data show a differential use of space by *Oryctolagus cuniculus* and *Octodon degus*. The distance between sampled shrubs ranged from 4 to 13 m. The vertical bars are equivalent to two standard errors; N = 10 for each point sampled (Jaksic and Soriguer 1981)

Invasive species can not only affect hunting behavior and use of space but also behaviors associated with reproduction. Oviposition in insects, for example, is often associated with certain plant species that act as hosts (Wiklund 1975; Thompson 1988), as is the case with the monarch butterfly (*Danaus plexippus*) that oviposits exclusively in the milkweed, *Asclepias syriaca*. This behavioral pattern was affected by the introduction of the plant *Vincetoxicum rossicum*, since currently the monarch butterfly oviposits on this exotic plant (Mattila and Otis 2003).

When reproductive behavior is modified at the fecundity level, this can clearly affect the survival patterns of offspring and their genetic identity (Lee 2002; Hails and Timms-Wilson 2007; Largiadèr 2008). In this latter aspect, the hybridization between native and introduced organisms constitutes an arena of growing interest and concern (Largiadèr 2008; Miura 2007). On the one hand, it is estimated that introgression of genetic material into invasive species may result in a stimulus for invasiveness (Rejmánek et al. 2005; Hails and Timms-Wilson 2007; Largiadèr 2008). This is because initially exotic species are made up of a small number of organisms that carry a small sample of the original genetic diversity; under these conditions, exotic species will have little expectation of expanding population and geographically. Thus, the arrival of new genes or alleles introgressively would undoubtedly contribute to increasing the genetic diversity needed to cope with the new environment.

When the hybridization or introgression between exotic and native species has been consigned, the greatest interest is focused on the study of native organisms, because reproductive processes and genetic mechanisms contribute to erode their genetic diversity (Hails and Timms-Wilson 2007). In plants, hybridization is a relatively frequent phenomenon (Petit 2004), which tends to occur preferentially between related taxa. For example, genders *Poa* and *Tamarix* are recognized for their ability to hybridize, including native and exotic species (Gaskin and Kazmer 2008), while introgressive hybridization has been documented in cordgrass species (*Spartina* spp.), where the best known case is that of the California cordgrass (*Spartina foliosa*), which hybrids by introgression with the smooth cordgrass (*Spartina alterniflora*) in the coastal marshes of San Francisco, California (Daehler and Strong 1997).

Ornamental plants are often a source of new genetic information that can be incorporated into wild populations, such as the domestic radish (*Raphanus sativus*) when crossed with the wild radish (*Raphanus raphanistrum*). Their hybrids proliferate and abound in cultivation areas and ruderal habitats of the European Mediterranean region (Stanton 1987). The incorporation of genes from transgenic cultivated plants into wild plants can also have negative ecological and economic consequences, especially if these genes confer advantages on wild plants in terms of reproductive capacity or resistance to natural enemies (Kalin-Arroyo et al. 2000; Pauchard et al. 2011).

The hybridization process between exotic and native species is not unique to plants but has also been documented in vertebrates, although involving a small number of species (Stebbins 1959; Arnold 1997). In Spain, for example, the hybridization between the white-headed duck (*Oxyura leucocephala*) and the ruddy duck

(*Oxyura jamaicensis*) has proved to be a particularly complex phenomenon for the conservation of native taxa (Martí 1993). The white-headed duck is a species with a wide geographical distribution, as it is present between the Iberian Peninsula and Mongolia. Its population decline has been widely recognized, estimating that from a world population of hundreds of thousands of individuals it has declined to about 8,000 or 14,000 individuals in recent years (Torres et al. 1994). This population reduction is generated by the anthropogenic modification of freshwater habitats; however, the detection of hybrids with the introduced species *Oxyura jamaicensis* (Fig. 7.2) has raised concerns about the loss of genetic diversity and integrity of the *Oxyura leucocephala* (Muñoz-Fuentes et al. 2007).

The introduction of fish of the same species, but of different variety, race, or subspecies, may lead to genetic changes in local populations if the fish specimens reproduce with each other. This hybridization can lead to the homogenization of genetic diversity, as has occurred with trout restocking initiatives (*Salmo trutta*) in the Iberian Peninsula (Machordom et al. 1999). In addition, the populations of *Salmo trutta* also hybridize with salmon *Salmo* salar, representing a threat for this latter species because of its small populations (Verspoor 1988).

A number of invasive species have genetic or reproductive mechanisms that allow them to quickly fix certain genotypes. In plants, for example, these mechanisms include the acquisition of new sets of chromosomes by polyploidy,

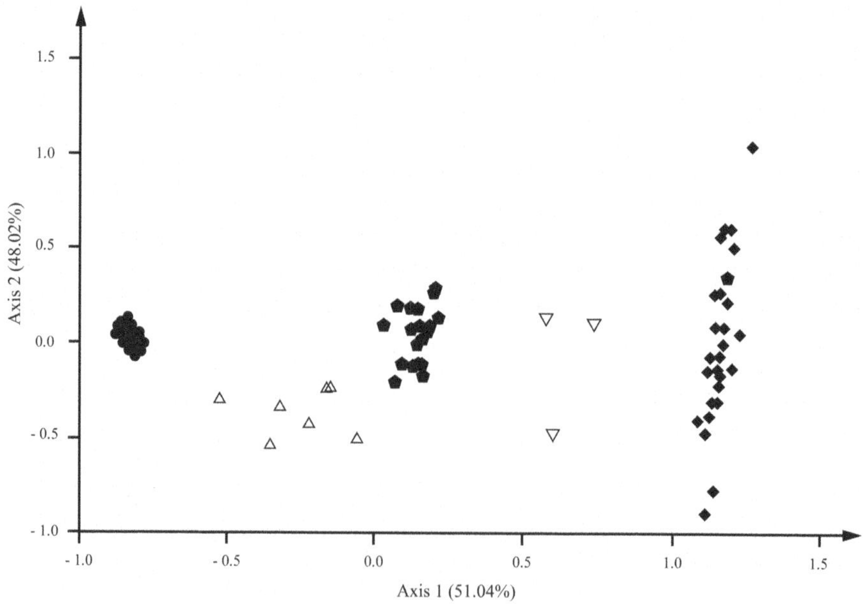

Fig. 7.2 Two-dimensional factorial correspondence analysis of multilocus genotypes based on microsatellite and intron data for specimens of *Oxyura* (Muñoz-Fuentes et al. 2007). The birds were morphologically identified as white-headed ducks (black circles), ruddy ducks (black diamonds), or hybrids (triangles, pentagons, and inverted triangles). A subsequent analysis of the hybrid genotypes corroborated their classification as hybrids

agamospermia or asexual seed production, and vegetative reproduction. The crossing between *Fallopia japonica* var. *compacta* and *Fallopia japonica* var. *japonica* originates the fertile allopolyploid hybrid *Fallopia* × *bohemica*, which exhibits an extraordinary rhizomatous growth that reduces the establishment and growth of native species in the community (Bailey and Child 1996). On the other hand, the common cordgrass (*Spartina anglica*), considered one of the world's most aggressive invasive taxa by the International Union for Conservation of Nature (IUCN; Nehring and Hesse 2008), is an allotetraploid that propagates vegetatively. Genetic analysis of *Spartina anglica* shows an extraordinary genetic uniformity of the individuals; its origin giving it hybrid vigor, *Spartina anglica*, originated in England, has been planted in several European countries to control coastal erosion. The invasion and expansion of its populations have led to the exclusion of numerous native plant species and the reduction of favorable habitats for various waterbirds (Daehler and Strong 1997).

7.3 Impact on Population and Community Scale

Interspecific interactions in the population and/or community context constitute the best documented field in the study of the impact of exotic species (Parker et al. 1999). Interactions such as competition, mutualism, predation, and parasitism (Davis and Pelsor 2001; Richardson et al. 2000; Sax et al. 2005) can cause modifications in population abundances and dynamics, species distribution, as well as in functional and structural attributes of communities (Krebs 2001). Since these two hierarchical levels are closely related to each other, we will discuss them together for each type of interaction.

7.4 Competitive Displacement

Interspecific competition is the type of interaction most addressed in the ecology of biological invasions (Lockwood et al. 2007). Exotic species can compete for resources and displace native species through exploitation or interference (Holway 1999). Exploitation competition occurs when the consumption of resources by a certain species affects the availability to the rest of the competitors (Petren and Case 1996). This seems to be the case with the competition between exotic rainbow trout (*Oncorhynchus mykiss*) and native silverside (*Basilichthys australis*) in Chile, where trout has been widely introduced and shares similar trophic habits with the native silverside. Using a system of experimental exclusions in field conditions, Pardo et al. (2009) found that the growth of this salmonid negatively affected the silverside, which resulted in lower body growth of the latter species (Fig. 7.3).

Some invasive species may become dominant in some habitat types because they have greater tolerance to resource scarcity (Tilman 2004). As a general rule, the

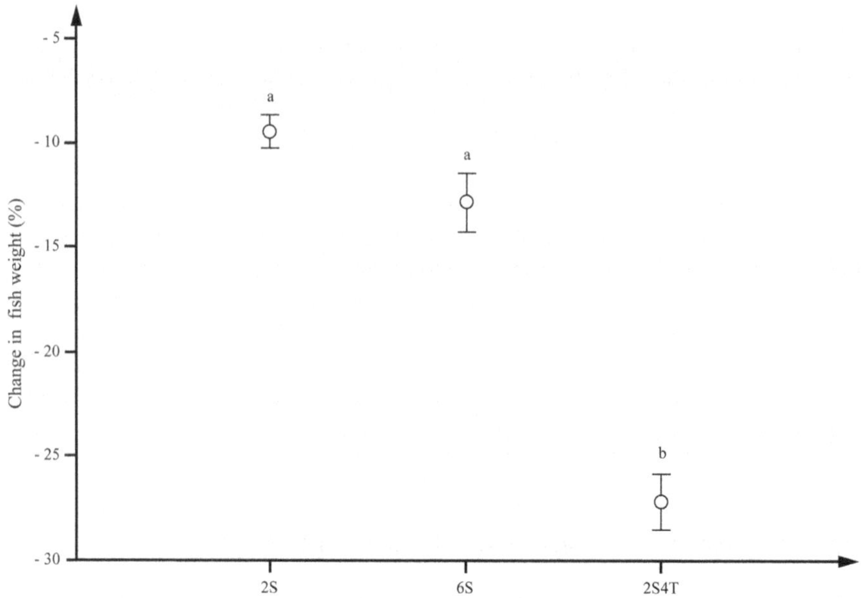

Fig. 7.3 Effect of the presence of rainbow trout (*Oncorhynchus mykiss*) on native Chilean silverside (*Basilichthys australis*) (Pardo et al. 2009). The graph shows the percentage temporal change in total body weight of *Basilichthys australis* in experimental treatments in the absence and presence of trout (2S: two silverside; 6S: six silverside and 2S4T: two silverside and four trout). The circles show the mean and the bars the standard deviation; the same letters indicate a non-significant effect for a level of $\alpha = 0.05$

species that is capable of exploiting a limiting resource at the lowest level of its availability will be the species that will maintain the highest levels of abundance (Tilman 1988). In this sense, some exotic species have become successful in the colonization of anthropized environments, highly disturbed and with scarce resources, competitively displacing native species (Kowarik 2008).

Competition for interference occurs when one of the competitors prevents access to a limiting resource from the part of another competitor (Case and Gilpin 1974). In animals, for example, this type of competition is usually manifested as territorial behavior, as is the case with the Argentine ant (*Linepithema humile*) (Holway et al. 2002). Part of its colonization success in the northern hemisphere is based on the formation of supercolonies where members exhibit low levels of aggressiveness towards their conspecifics but high towards other species (Holway et al. 2002). The above contrasts with what happens *Linepithema humile* in its native range of distribution, where it forms multiple colonies with high intra-specific aggressiveness (Holway and Suarez 1999).

Invasive plants can also compete by exhibiting interference mechanisms. Indeed, overgrowth and shade, as well as the emission of allelopathic compounds, are manifestations of this type of competition in plants. Overgrowth and shade seem to be the cause of the dominance of the burr chervil (*Anthriscus caucalis*) in habitats

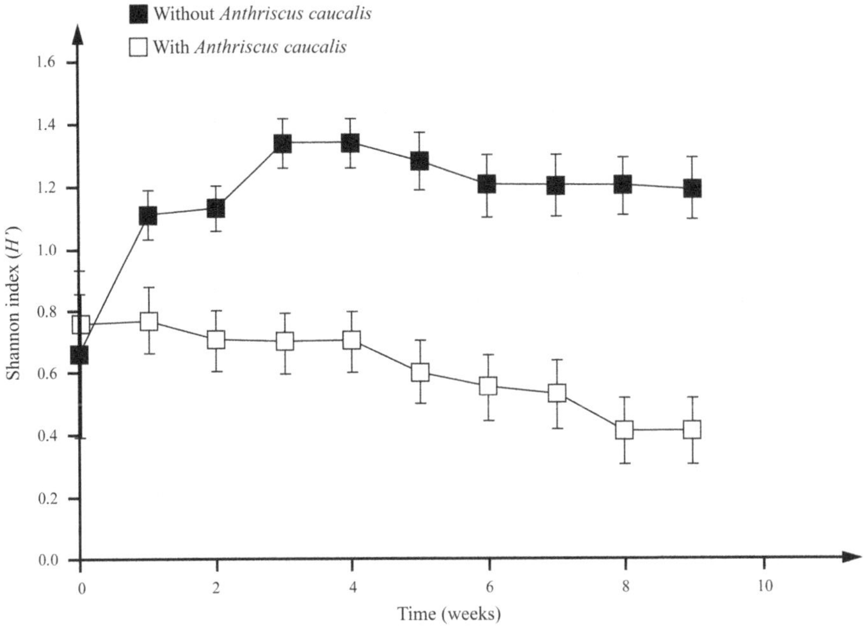

Fig. 7.4 Herb diversity (Shannon index, *H'*; mean ± SD) recorded in experimental quadrats of 0.5 × 0.5 m in the central Chilean shrubland (Castro et al. 2009). The curves correspond to two different treatments: without and with *Anthriscus caucalis* present inside the quadrants

bordering shrub fragments (Castro et al. 2009). This annual herb, originally from Europe, was first documented in Chile in 1899 (Reiche 1899). It is currently distributed in the central portion of the territory (Montenegro et al. 1991), associated with mesic-humid sites, especially on the perimeter edge of fragments of secondary sclerophyllous vegetation. The experimental exclusion of *Anthriscus caucalis* in these microhabitats leads to increased diversity (Fig. 7.4) and survival of native herbs. According to Castro et al. (2010), *Anthriscus caucalis* reaches spatial domination because it covers the ground forming a continuous and dense canopy, preventing other species from arriving and surviving under these conditions.

As mentioned above, the emission of allelopathic compounds can be an important competitive interference mechanism (Hierro and Callaway 2003). As a result of their metabolic processes, vascular plants release a number of chemical compounds, either into the soil or into the atmosphere (Callaway 2002). In some cases, these substances may be toxic to adjacent plants, a phenomenon known as allelopathy. When allelopathy is exerted by invasive plants, they can produce a competitive imbalance in their favor, impacting on the native plant community (Hierro and Callaway 2003).

Gómez-González et al. (2009) evaluated the competitive effect of the invasive species *Centaurea solstitialis* on two species native to Chile, *Baccharis linearis* and *Baccharis paniculata*. *Centaurea solstitialis* is an annual/biannual species whose native range corresponds to the Eurasian region. From here it has expanded to various locations around the world (Hierro et al. 2005; Fuentes et al. 2007; Rodríguez et al. 2018), including Chile (Matthei 1995). In Chile, *Centaurea solstitialis* is distributed in the Mediterranean region where it is considered an agricultural weed (Matthei 1995). In this study, competition was evaluated in experiments that combined different stages of the life cycle of these plants, including seed-seed, seed-plant, and plant-plant treatments. The germination of *Centaurea solstitialis* was abundant and accelerated, in comparison with both *Baccharis species*. As a plant or seed, *Centaurea solstitialis* did not affect the germination of the two native species; however, the presence of established plants of *Centaurea solstitialis* significantly decreased the plant biomass of both *Baccharis* species (Gómez-González et al. 2009). Gómez-González et al. (2009) suggested that under field conditions in the Chilean scrub, the early and explosive emergence of *Centaurea solstitialis* would constitute an important competitive dominance mechanism.

Competition between native and exotic plants can be determined not only by spatial interference but also by access to resources such as light and soil nutrients and even by the attraction of seed dispersers or pollinators (Brown et al. 2002). In the latter case, there is evidence showing that introduced plants can have a direct or indirect effect on the fertility of native plants by altering the feeding behavior of native pollinators and floral visits. In an 11-year survey (1995–2006), Morales (2007) showed changes in the frequency of visits in a population of the endemic liuto herb (*Alstroemeria aurea*). Thus, since the invasion of *Bombus ruderatus*, the frequency of visits of the native bumblebee *Bombus dahlbomii* declined significantly over time (Fig. 7.5).

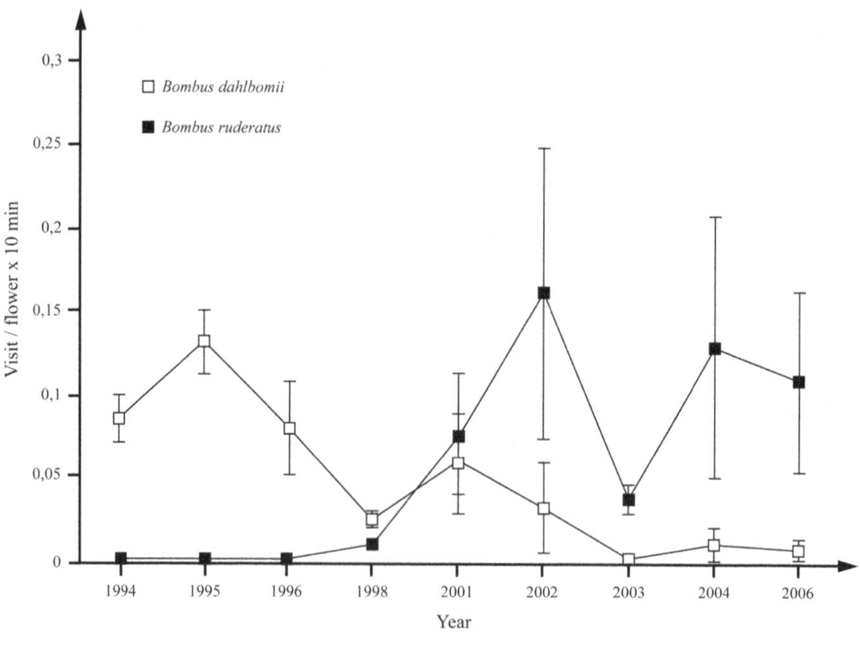

Fig. 7.5 Temporary variation in the frequency of visits from *Bombus dahlbomii* (native species) and *Bombus ruderatus* (non-native species) to flowers of *Alstroemeria aurea*, in southwestern Argentina since the invasion of the *Bombus ruderatus* (Morales 2007)

Interestingly, Muñoz and Cavieres (2008) documented the effect on the visiting frequency and seed production by two high Andean native Asteraceae, *Hypochaeris thrincioides*, and *Perezia carthamoides* in the presence of the exotic Asteraceae *Taraxacum officinale*. Their observations and experiments were conducted in the Andes of central Chile, at an altitude of 2800 m elevation, during the flowering season. Their results indicated that the three Asteraceae broadly shared the diversity of potential pollinators, a necessary condition for evidence of competition for this service; however, the visit indicators and the reproductive output associated to these visits allowed establishing that the presence of *Taraxacum officinale* generated significant negative effects to the other two species in high density conditions (Fig. 7.6).

7.5 Invasion by Facilitation

Traditionally, positive interactions between exotic and native species or between exotic species themselves have received less attention (Simberloff and Von Holle 1999; Richardson et al. 2000), which does not imply that they are any less important. Among the most documented types of mutualism are frugivory and seed dispersal interactions and those related to pollination services. All these imply a reciprocal positive effect between plants and animals (Sax et al. 2005). The plant obtains the service of propagule dispersion or pollination, while the animals obtain a variable nutritional reward.

Recent studies have revealed the importance that this type of interaction can have in various ecosystems. Cosyns et al. (2005a, b) showed that domestic livestock, including horses (*Equus ferus caballus*), cows (*Bos taurus*), goats (*Capra hircus*), and sheep (*Ovis orientalis aries*), has a great influence on the dispersal of seeds for numerous plant species. These herbivores usually consume seeds that they deposit by defecation in places where germination is possible. Other naturalized herbivores, such as rabbits and hares, can also exert seed dispersal in the ecosystems into which they have been introduced. Castro et al. (2008) documented the effect on the germination of seeds of a native shrub, litre (*Lithraea caustica*) after the passage through the digestive tract of the European rabbit (*Oryctolagus cuniculus*), noting that germination increased by 74.9% compared to seeds that did not pass through the digestive tract (Fig. 7.7). These authors also evaluated the effect of coprophagy – a well-documented behavior of the European rabbit – on the germination of litre seeds, by offering litre seeds that had already been defecated back to the rabbits (Fig. 7.7). This time the germination was 63.2%, still higher than the germination of seeds under control treatment (Castro et al. 2008).

Invasion by non-native pollinators can significantly affect native pollinators and the plants they visit, both native and introduced (Goulson 2003; Winter et al. 2006). In this line of work, the bees (Apoidea) have attracted attention, because at least five possible effects are attributed to them: (a) competition with native floral visitors for the floral resource, (b) competition with native organisms for nesting sites, (c)

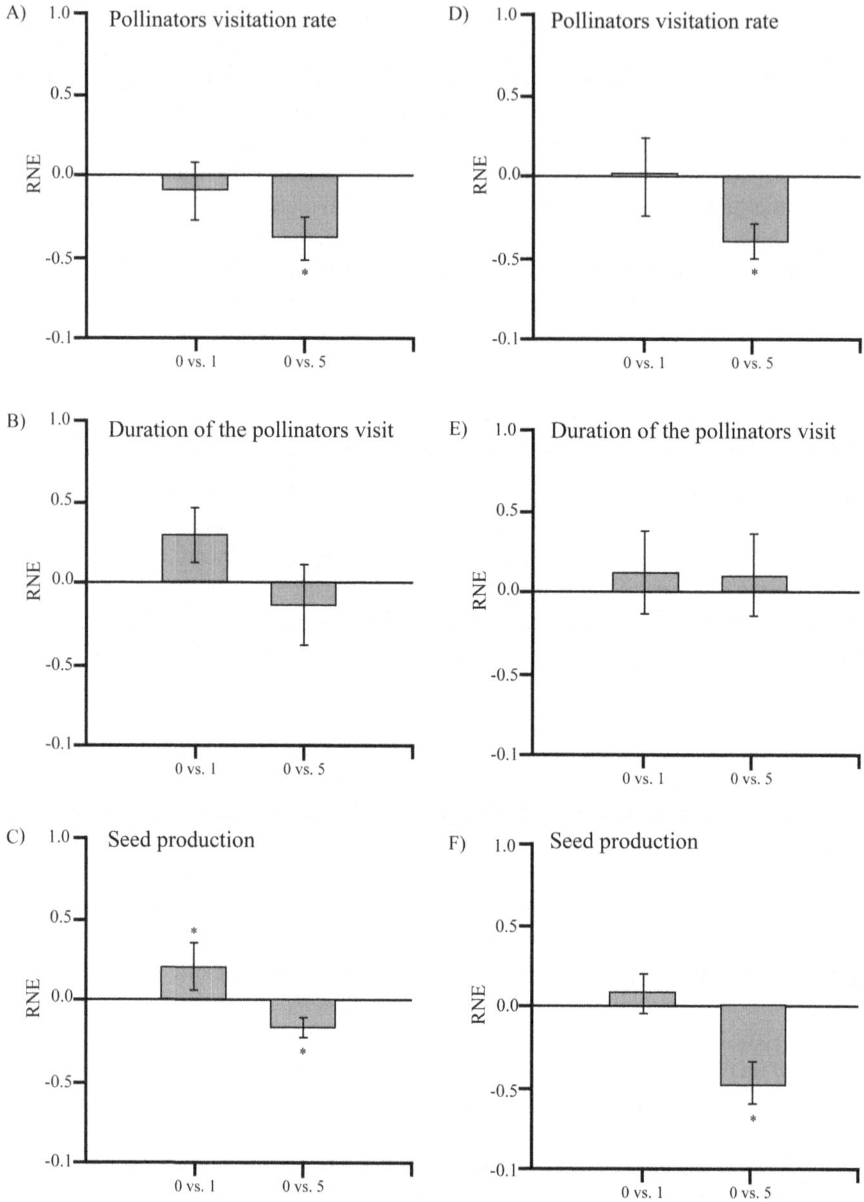

Fig. 7.6 Effect of the presence-absence of one *Taraxacum officinale* individual (0 vs 1, on the x axis) and five plants of *Taraxacum officinale* (0 vs 5, on the *x axis*) on (**a**) the visit rate of pollinators, (**b**) duration of the pollinator visit, and (**c**) seed production of *Hypochaeris thrincioides* (Muñoz and Cavieres 2008). Graphs (**d**), (**e**), and (**f**) represent the same effect but evaluated on the plant *Perezia carthamoides*. Error bars are ±2 SE, and the asterisks represent significant differences ($P < 0.05$). RNE: relative neighbor effect index

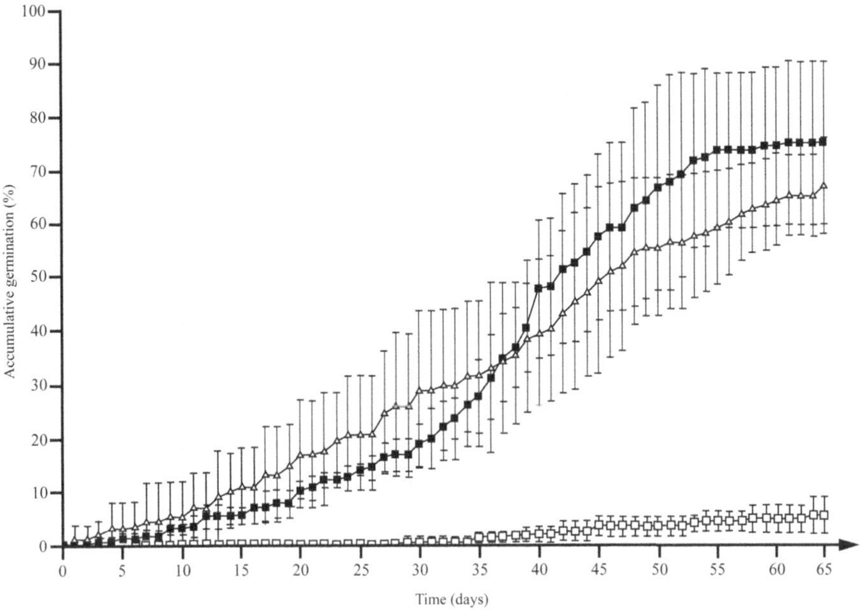

Fig. 7.7 Accumulative germination (mean ± SE) of *Lithraea caustica* seeds (Castro et al. 2008). Black squares, seeds passed once throughout rabbit digestive tract; white triangles, seeds passed twice; white squares, seeds not ingested by rabbits

transmission of pathogens to native organisms, (d) changes in native plant seed production, and (e) weed pollination (Goulson 2003).

According to Klein et al. (2007), 35% of global food production depends on the pollination of crops by animals, a service mostly provided by the honeybee (*Apis mellifera*), and by different species of wild bumblebees (e.g., *Bombus* spp.). It is estimated that among this type of organisms, bumblebees are pollinators that exhibit greater pollination efficiency in certain crops (Buchmann 1983; Fussell and Corbet 1991; Willmer et al. 1994; Velthuis and van Doorn 2006). In a recent review, Morales (2007) gathered evidence about the effects exerted by the introduction of some species of *Bombus* as pollinators. This author verified that only 5 of the nearly 239 known species of the genus have been used commercially (Table 7.1). At present, the commercial beehives of *Bombus* are found in some 36 countries, where they contribute to pollinate more than 25 crops (Velthuis and van Doorn 2006). This trade, together with the release into the wild of fertilized queens, has been the source of invasion of the five species of *Bombus* in some regions of South America, Asia, and Oceania (Table 7.1).

Population interactions of mutual benefit may promote the invasion of exotic species, for example, in extreme environments where invasion is facilitated by the resident community (Choler et al. 2001; Callaway et al. 2002). In high-Andean

Table 7.1 Bumblebee species (*Bombus*) introduced outside its natural range (Morales 2007), indicating their region of origin, type of introduction, country where introduced, their current establishment status, and countries other than those of introduction that have been invaded as a byproduct

Species	Origin	Type of introduction	Country of introduction	Establishment	Invaded country
Bombus terrestris	Europe	Release of queens	New Zealand	Yes	Tasmania Argentina
			Chile	Yes	
			Japan	Yes	
Bombus ruderatus	Western Europe	Release of queens	New Zealand	Yes	Argentina
	Northern Africa		Chile	Yes	
Bombus impatiens	Northeast USA	Import of colonies	Canada	Not confirmed	
			Mexico	Not confirmed	
			Guatemala	No data	
Bombus hortorum and *Bombus subterraneus*	Europe	Release of queens	New Zealand	Yes	

environments (above 3000 m elevation), Cavieres et al. (2005) showed that the presence of the cushion plants *Azorella monantha* and *Laretia acaulis* favors the survival of the exotic *Taraxacum officinale* in this type of environment (Fig. 7.8). This is because the native cushions modify the microclimatic and edaphic conditions in such a way that they facilitate the establishment of *Taraxacum officinale* (Cavieres et al. 2007; Badano et al. 2015). In turn, the abundance that *Taraxacum officinale* can achieve in these environments depends on the species that harbors it, as it reaches higher levels of abundance when the patches are of *Azorella monantha* rather than *Laretia acaulis* (Cavieres et al. 2008; Badano et al. 2015).

These facilitative interactions expected in stressful environments do not necessarily occur in other types of harsh environments. In the semiarid scrub of central Chile, characterized by high temperature and low availability of humidity to the flora, Becerra and Bustamante (2011) evaluated the effect of shade produced by litre (*Lithraea caustica*) on two invasive exotic species present in the region, *Pinus radiata* and *Eucalyptus globosus*. These authors expected the presence of *Lithraea caustica*, and its shadow was a mitigating factor of xeric stress. However, the germination and recruitment parameters of both exotic species were not improved, with no evidence of a possible facilitation effect of the native species on the invasion success of *Pinus radiata* and *Eucalyptus globosus*.

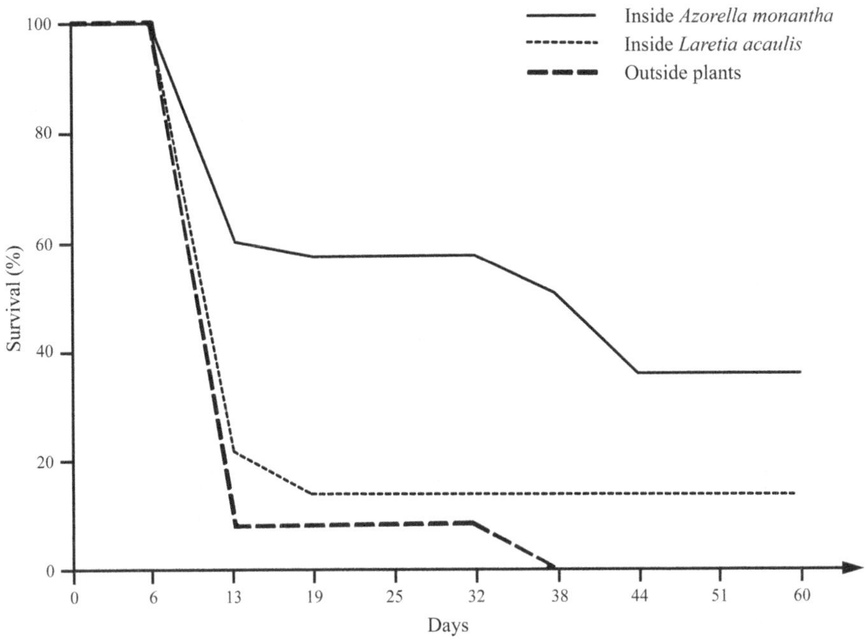

Fig. 7.8 Survival (%) of *Taraxacum officinale* afforded the shelter of *Azorella monantha*, *Laretia acaulis*, and on bare ground, in sites located at 3200 m above sea level in the Andes of central Chile (Cavieres et al. 2005)

On the other hand, Simberloff and Von Holle (1999) distinguished that when facilitating interactions occur between two exotic species, invasiveness can be enhanced simultaneously for the populations involved, leading to the phenomenon of "invasive meltdown" (Simberloff 2006). Among the studied examples of mutualistic interaction between invasive populations is the documented mutual facilitation between populations of the yellow crazy ant (*Anoplolepis gracilipes*) and some species of homoptera (*Coccus celatus*) on Christmas Island (O'Dowd et al. 2003).

7.6 Antagonistic Interactions

Consumption relationships between predators and prey, parasites and hosts, and herbivores and plants are often considered high-impact interactions when one of the taxa involved is native. Perhaps one of the most cited cases is the introduction of Nile perch (*Lates niloticus*) in Lake Victoria in Africa, whose predation on native fish has caused the extinction or threat to more than 200 species of native cichlids (Pringle 2005).

In Europe, the introduction of the American mink (*Neovison vison*) has had an important impact on other native animals as a consequence of predation. On the shores of continental waters, American mink preys on ground nesting birds (Rallidae, Anatidae, Laridae), which can cause local extinction (Vidal-Figueroa and Delibes 1987). In Chile and Argentina, the American mink was introduced in the 1930s, in the southern portion of both territories (Jaksic et al. 2002). A recent study has shown that its diet is mainly based on the consumption of vertebrates, such as mammals, birds, and fish (Fig. 7.9), although it is also capable of consuming invertebrates (Schüttler et al. 2009). The voracious and generalist character of their trophic habits, added to the rugged landscape of the Patagonian and archipelagic territory of Tierra del Fuego, which impedes easy human access for its control, guarantees the expansion of exotic species such as mink (Rozzi and Sherriffs 2003). In addition, this mustelid has no natural enemies in the region, so it probably behaves as a top predator in riparian communities (Ibarra et al. 2009).

Like carnivores, introduced herbivores can also modify the distribution and abundance patterns of the plants they consume, which are ultimately analogous to their prey. The yellow crazy ant (*Anoplolepis gracilipes*), by excluding land crabs (e.g., *Gecarcoidea natalis*) that inhabit the forest floor and feed on seeds, leaves, fruits, and flowers, generates a drastic effect on the composition of the understory on Christmas Island, increasing the density and richness of seedlings (Table 7.2; O'Dowd et al. 2003). In Europe, the coypu (*Myocastor coypus*) can cause local

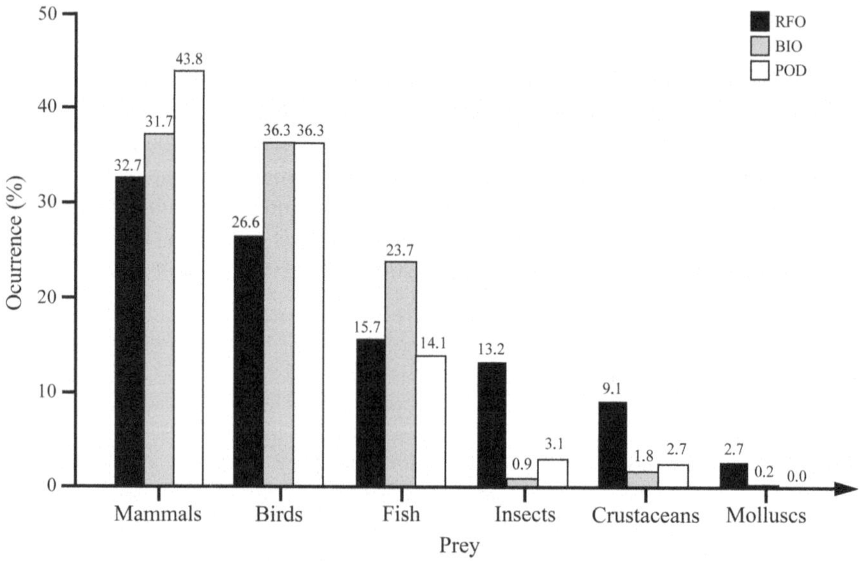

Fig. 7.9 American mink prey (*Neovison vison*) observed in fecal remains on Isla Navarino, Chile (Schüttler et al. 2009). The percentages are based on data from 512 fecal samples collected in three semiaquatic habitats (marine, riparian, and lacustrine) between 2005 and 2007. *RFO* relative frequency of occurrence of each prey category, *BIO* percentage of biomass consumed, *POD* percentage of occurrence of each prey category as the dominant item

Table 7.2 Impact on forest soil after an invasion of the exotic ant *Anoplolepis gracilipes*, after excluding the herbivorous crab *Gecarcoidea natalis* (O'Dowd et al. 2003). Different variables (first column) are compared in invaded and non-invaded sites. F and P are statistics derived from ANOVA of random blocks for each variable; all block effects were non-significant ($P > 0.05$)

Variables associated with forest floor	Invaded site (mean ± 1 SD)	Not invaded site (mean ± 1 SD)	$F_{1,5}$	*p*-value
(a) *Anoplolepis* activity index	6.96 ± 1	0.06 ± 0.06	45.4	0.001
(b) Land crabs				
Crab burrows/80m^2 [a]	2.3 ± 1	95.7 ± 24	44.8	0.001
Dead crabs/80m^2 [a]	51.8 ± 17	0.0 ± 2	18.8	0.007
(c) Litter cover (%)	87.0 ± 3	43.0 ± 13	21.4	0.006
(d) Seedlings				
Number of seedlings/80 m^2 [a]	1375.8 ± 166	44.7 ± 16.7	123.6	<0.001
Number of spp./80 m^2 [a]	22.2 ± 2	6.3 ± 1.3	34.0	0.002

[a]Sum of 5 sites of 4 × 4 m

extinction of plants such as water lilies (*Nymphaea* spp.), reeds (*Phragmites* spp.), and bulrush (*Typha* spp.), due to overgrazing (Prigioni et al. 2005). Domestic herbivores, such as goats (Coblenz 1978), wild pigs that modify soil properties (Graves 1984; Sierra 2001), and the European rabbit (Holmgren et al. 2000), have a reputation for affecting the composition and plant dynamics of the communities they invade (Holmgren 2002).

Invasive species can also affect native species by transmission of their own parasites (Prenter et al. 2004). Perhaps one of the most emblematic cases is the introduction of phylloxera (*Daktulosphaira vitifoliae*), associated with the American vine strains (*Vitis labrusca* and *Vitis riparia*). This hemipteran attacks the root of the European vine (*Vitis vinifera*) and caused extensive mortality that devastated European vineyards at the end of the nineteenth century. Today European vine cultivation is only possible by grafting on American vines (Porten and Huber 2003). In Chile, aphids are known plant parasites that can consume photosynthetic products directly from leaves (Niemeyer 1992). Curiously, this taxon constitutes one of the groups with the highest representation of exotic species in the country, since of a total of 135 reported species, 77% correspond to exotic species (Fuentes-Contreras et al. 1997).

7.7 Modifying Ecosystems

Various ecosystem processes can be modified by the arrival of exotic species (Vitousek 1990). These include the transfer of matter and energy, such as hydrological and nutrient cycles, primary productivity, modification of disturbance regimes, and biogeochemical, geomorphological, and microclimatic processes (Mack et al. 2000). All these changes constitute more or less perceptible effects at a global level

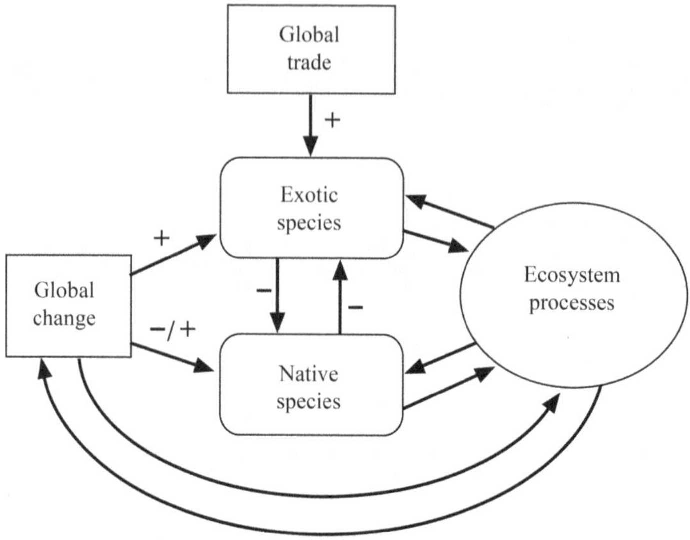

Fig. 7.10 Conceptual model of interactions between exotic species, native species, ecosystem processes, global trade, and global change (adapted from Dukes and Mooney 1999). Arrows indicate direction of interactions, and symbols (+, −) indicate whether a given influence is generally interpreted as positive or negative. Global transport has brought invasive species to new regions, where some of these species suppress native species. Together, native and exotic species modulate ecosystem processes. In this schematic, nitrogen deposition and habitat fragmentation are considered elements of global change, which in turn affect both the processes and the balance between native and exotic species

(Mooney et al. 1987; Vitousek 1990; Mooney and Hobbs 2000), one more reason to note the importance of biological invasions in the planetary context. Figure 7.10 presents a conceptual model initially proposed by Dukes and Mooney (1999), in which relevant constraints between exotic and native species and ecosystem processes are recognized as factors of global change. From this model, it is not only possible to determine that biological invasions constitute an important consequence of global change but also that they behave as a causal agent or promoter of it (Dukes and Mooney 1999; Sala et al. 2000; Chapin III et al. 2000). Among the first studies that documented impacts of exotic species on biogeochemical processes is the work of Vitousek et al. (1987), who studied the introduction of the faya (*Morella faya,* initially = *Myrica faya*) in the Hawaiian Islands and its effect on soil nitrification. The faya is a nitrogen-fixing species, whose presence ended up quadrupling the entry of nitrogen by fixation in these islands, historically limited by this nutrient (Vitousek et al. 1987). In South American ecosystems, nitrogen fixation by introduced species has been widely recognized, but poorly studied (Figueroa et al. 2004). Naturalized species such as *Acacia dealbata*, *Teline monspessulana*, and *Ulex europaeus* are able to fix nitrogen, increasing its availability, especially in soils that are disturbed and nutrient-poor (Scherer-Lorenzen et al. 2007).

Di Tomaso (1998) studied the impact of different species of tamarisks (*Tamarix* spp.) in the water cycle on the banks of streams and rivers in the southwestern portion of the United States. The invasion of *Tamarix* spp. can be quite aggressive. Some species already occupy an area greater than 4700 km^2 of flat, riparian and wetlands in the United States (Stenquist 2000). Tamarisks are native to the Eurasian and African regions and were introduced to North America in the early nineteenth century (Di Tomaso 1998). According to ecophysiological studies (Whitcraft 2007), this shrub is not capable of efficiently regulating its evapotranspiration rate, so its roots make an excessive consumption of water, causing an increase in aridity and salinity of riparian areas (Di Tomaso 1998).

Obviously, these ecosystem effects are not restricted to terrestrial environments alone. In aquatic ecosystems, the zebra mussel (*Dreissena polymorpha*) has been able to alter the availability of nitrogen, phosphorus, and carbon for other species living in Lake Huron, United States. It has been described that because of its filtering habit, this mussel is capable of sifting the water column (Gardner et al. 1995), causing a decrease in plankton and suspended solids (Arnott and Vanni 1996; James et al. 1997). As a result, the abundance of harmful cyanobacteria increases due to changes in the nitrogen-phosphorus ratio (Raikow et al. 2004). Such changes indirectly impact a wide variety of ecological processes, which ultimately affect the composition and distribution of native populations (Higgins et al. 2008).

From the above examples, it is important to note that the presence of certain exotic species may contribute to the increase (or sometimes decrease) rate of a pre-existing ecosystem process. This is because the arrival of a new species may be redundant with the functional role of existing species (Olden et al. 2004; González et al. 2010). In contrast, other species possess biological characteristics that give them novel functionalities for the ecosystems they invade (Crooks 2002). This has occurred, for example, in the case of the arrival in central Chile of some annual Eurasian herbs (Arroyo et al. 2000). Several of these species facilitate the occurrence of fires because their tissues have flammable characteristics (Figueroa et al. 2004, 2009; Pauchard et al. 2011). These characteristics emerged as an evolutionary response to fires that naturally occur in the range of origin (Figueroa et al. 2009, 2014). Once naturalized, the use of fire as a land management agent allowed the expansion of these species throughout central Chile, where this type of disturbance was never natural (Figueroa et al. 2014).

When the effect imprinted by the exotic species is capable of profoundly modifying the structure of the habitat, invasive species may be considered ecosystem engineers (Jones et al. 1994). Engineering species are those that, due to their abundance or biomass, have a gravitational influence on the generation of habitats for the rest of the community (Jones et al. 1997). As controversial as this concept has been (Wright and Jones 2006), many exotic species have been syndicated as ecosystem engineers because of their relevant role in host ecosystems (Badano et al. 2007; Menge et al. 2008). The case of the beaver (*Castor canadensis*) is perhaps one of the most easily understood because of its notoriety. In the Patagonian ecosystems, it builds burrows and dikes in watercourses, which create large dams. Anderson and

Table 7.3 Ecological variables of benthic macroinvertebrate communities in forested sites, beaver lagoon, and downstream sites, compared by ANOVA. Values are annual means (±SD) based on seasonal samples for each habitat in four rivers. *H′*: Shannon diversity index. AFDM is ash-free dry mass (Anderson and Rosemond 2007). The letters A, B, and C denote significant differences for each category represented in the different treatments

Category	Forest	Beaver pond	Downstream	*F*	*df*	*P*
Richness (no. taxa/m^2)	15.3 ± 1.7 A	10.0 ± 0.9 B	15.8 ± 1.2 A	5.7	2,6	0.04
Diversity (H′)	2.0 ± 0.1 A	1.4 ± 0.1 C	1.9 ± 0.1 B	1,540.4	2,6	< 0.001
Abundance (individuals/m^2).	2,611.5 ± 504.9	14,350 ± 10,381	5,086.2 ± 1,029.8	0.99	2,6	0.42
Biomass (mg AFDM/m^2)	257.9 ± 89.4 A	864.1 ± 179.8 B	443.3 ± 68.2 AB	7.2	2,6	0.03

Rosemond (2007) found that introduced beavers impact the properties of aquatic ecosystems, transforming lotic systems into lentic ones, thus affecting nutrient flow and macroinvertebrate composition. Estimates of macroinvertebrate richness, biodiversity, and biomass showed significant differences depending on the location of the dam (see Table 7.3).

On the other hand, services provided by native species acting as ecosystem engineers may encourage invasion. Badano et al. (2007), through careful experimental design, showed *Azorella monantha* which, a cushion plant in high Andean environments, is capable of exerting a nurse effect on two invasive species under field conditions: the dandelion (*Taraxacum officinale*) and the mouse ear (*Cerastium arvense*). This effect was observed at different elevation levels (3200, 3400 and 3600 m elevation), showing a greater magnitude at higher altitudes (Fig. 7.11).

Another dimension of ecosystems that has been studied in some detail is the effect of exotic species on disturbance dynamics. In this area, one of the most studied aspects is the alteration of the fire regime by exotic plants (Brooks et al. 2004). Fire is a common type of disturbance in many ecosystems, leading to the removal of vegetation cover (Bond and Keeley 2005; Pausas and Keeley 2009). In an early review, D'Antonio and Vitousek (1992) gathered evidence about the role of European grasses as fire facilitators in various regions of North and South America. In Argentina and Chile, numerous species appear to promote the frequency and intensity of fires (Figueroa et al. 2004). It has been documented that the growth habits, life cycles, and flammability of the vegetative tissues of these species make them prone to incineration. This would contribute to the gestation and/or spread of fire, affecting the distribution of native vegetation, to the extent that these exotic taxa are overrepresented. In this way, a positive feedback dynamic is established between the fire and the pyrogenic species (Fig. 7.12). Curiously, the fires in this region have not been part of its most recent evolutionary history (Figueroa et al. 2004), so their occurrence constitutes a true novelty in the structuring and functionality of the ecosystem. However, few studies have deepened this aspect of invasive pastures in Chile (Figueroa et al. 2004, 2009). There is evidence that fire stimulates the germination of exotic species and reduces the germination of native species. In

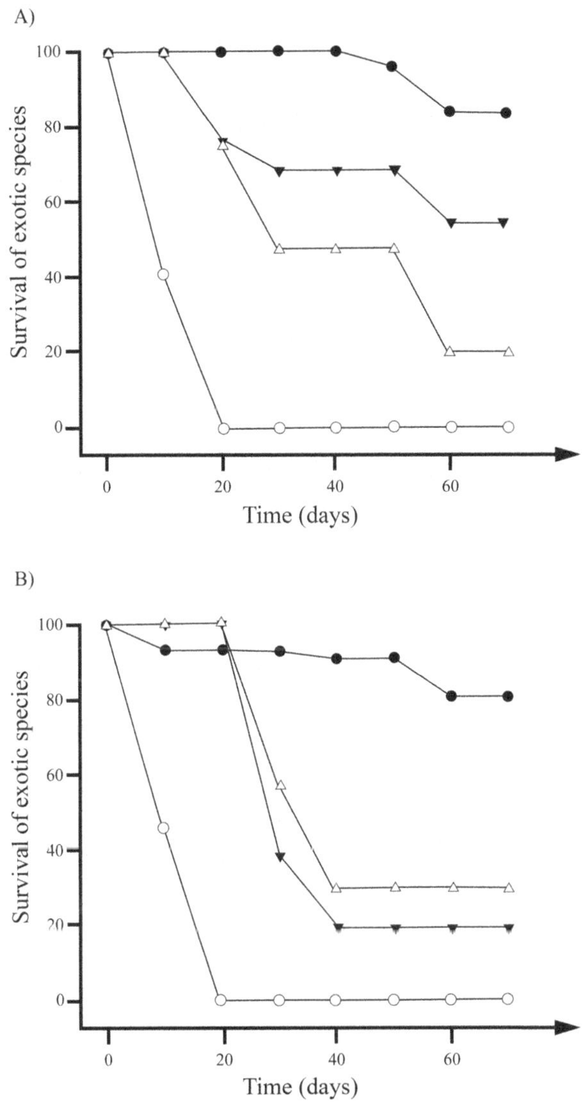

Fig. 7.11 Survival (%) of *Taraxacum officinale* (**a**) and *Cerastium arvense* (**b**) transplanted, sheltered by *Azorella monantha* (black symbols) and in open areas (white symbols), at higher elevation (circles) and at lower elevation (triangles) (Badano et al. 2007)

central Chile, for example, Gómez and Cavieres (2009) have shown that the germination of exotic species is significantly increased in the face of high intensity fires. García et al. (2010) showed that the germination of *Teline monspessulana* increased from 11% to 46% after the experimental application of temperatures that simulated fire. As previously indicated, this could explain, at least in part, the successful naturalization and subsequent expansion of exotic species in the region.

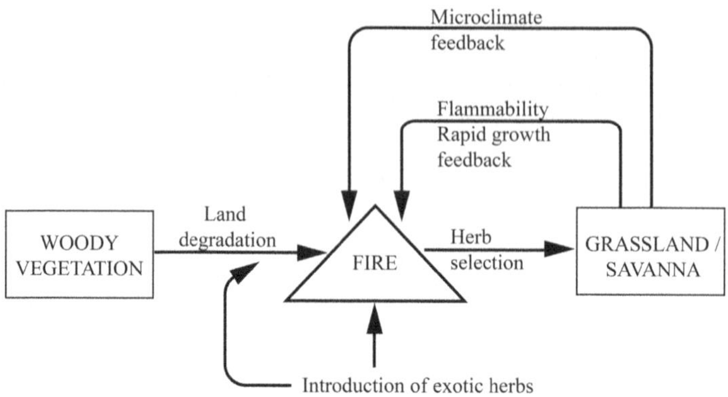

Fig. 7.12 Conceptual diagram linking deforestation and the fire cycle with the invasion of exotic grasses in the USA (D'Antonio and Vitousek 1992). In some cases, the invasion itself is sufficient to initiate positive feedback of grass fires; at the same time, this factor interacts with man-made changes in land use

7.8 Changing Biogeography to Homogeneity

It is widely known that the current biogeographic regions of the planet are made up of biota and characteristics of each region (Sax et al. 2005). This geographical ordering has taken millions of years to configure and derives from the geological history of the planet that has led to differentiation in the composition of species between regions (Brown and Lomolino 1998). The appearance of the ancient sea of Tethys, for example, served as a barrier of separation from the supercontinent Pangea in two: Gondwana and Laurasia (Brown and Lomolino 1998). Thus, the flora and fauna of these two continents took different evolutionary paths from that moment on. However, this did not prevent that with time and spontaneously, some species had the ability to be exchanged between these regions, as happened, for example, with the appearance of the isthmus of Panama (Webb 1991). This isthmus served as a bi-directional bridge of dispersion and colonization (i.e., invasion) between South America and North America (Fig. 1.2).

In the context of a longstanding geological and evolutionary history of isolation, Mooney and Hobbs (2000) visualized that invasions constituted transgressions to the biological history of the planet. This transgression essentially is that a certain species that originated in a certain region is inoculated in a second region with which it does not share a recent evolutionary history. Although this idea was not new, since it had been previously suggested by other researchers (De Candolle 1855; Elton 1958), Mooney and Hobbs (2000) conceived that, if the rate of exchange of species between regions of the globe were not controlled, this phenomenon would lead to the reconfiguration of the old Pangea, the New Pangea they called it in attention to the fact that the geographical barriers that have historically separated the different regions of the planet are disappearing thanks to human transport. Clearly, what may represent an insurmountable barrier for some organisms is not so for others. What is important here is that human-assisted transport tends to dissipate the functional importance of some of these barriers, allowing new species to arrive in a given region. In this same line, Gordon Orians (according to Rosenzweig 2001) proposed the occurrence of a new stage in the history of the planet what he called "Homogecene", referring to the fact that the biological composition of each region is in a process of homogenization.

Biotic homogenization constitutes one of the most worrying processes at the biogeographic level, and little by little, it has begun to take greater gravitation in the conservation agenda at this scale (Olden 2006; Rooney et al. 2007). McKinney and Lockwood (1999) were the first to propose a conceptual definition for biotic homogenization. On the basis of a metaphor, "there are few winning species that replace many losers" these authors managed to install the idea that, at a global level, biodiversity is being reduced and simplified; it is reduced as a consequence of the extinction of species and simplified because few species (sometimes native, sometimes exotic) extend their distributional range and replace those extinct. By extension, the global biota would be homogenizing (McKinney and Lockwood 1999). Rahel (2000) was one of the first researchers to evidence the occurrence of biotic homogenization in a quantitative manner. To do this, he analyzed the composition of assemblages of freshwater fish present in hydrographic basins in different states of the United States. An important advance of this work was to analogize the concept of homogenization to that of compositional similarity, which provided a quantitative way of measuring it. Thus, if two regions share a greater number of species, this means that these should be more homogeneous in their composition or, similarly, exhibit a higher level of compositional similarity. Rahel (2000) managed to establish the composition of freshwater fish assemblages present in the 48 states at two points in time. The first moment corresponded to the most probable composition of native fish prior to the introduction of exotic species and to the anthropic impact on the river basins of the United States, while the second moment represented the current situation, in which not only has the introduction of exotic species occurred in diverse basins but also the extinction or local extirpation of native species. Applying the Jaccard similarity index (J) at this time and then calculating the current difference from the original by the index (ΔJ), it was possible to observe that the variation in similarity has increased towards the present. In other words, the current similarity of the composition of freshwater fish is greater than the previous or historical similarity, a revealing fact of the biotic homogenization process.

The homogenization studies acquired greater impulse after the contributions of McKinney and Lockwood (1999) and Olden and Poff (2003, 2004), who not only provided formal definitions for the process but also developed conceptual tools, recognizing both the mechanisms associated with biotic homogenization and the possible compositional scenarios. Thus, biotic homogenization has taken part of the conservation agenda at a biogeographic scale (Olden 2006), highlighting also the diverse edges that this phenomenon can take under different ecological, spatial, and temporal scales. It is now accepted that the introduction of new species into a region not only contributes to changing taxonomic composition and similarity relationships but may also affect other components of diversity, such as functional and phylogenetic diversity (Olden and Rooney 2006).

Compositional homogenization depends on two mechanisms. As we have seen, one of them is the introduction of new species, while the other is extinction or local extirpation (Olden and Poff 2003). The extinction of native species can also increase compositional similarity because this process contributes to the loss of species unique to each region, which maintained the compositional difference at first. Therefore, with their disappearance, the similarity increases.

The balance of invasions and extinctions in space and time can lead to biotic homogenization but also to different compositional scenarios. Indeed, biotic differentiation, seen as the alternative process to homogenization, implies a decrease in compositional similarity after invasion and extinction events (Olden and Poff 2003). This could happen because the areas receive a different set of exotic species (Fig. 2.7), while the native species they may have in common become extinct. On the other hand, a third compositional scenario is the biotic tracking, which would occur when similarity levels are not modified (at least significantly) despite the occurrence of invasions and extinctions (Castro et al. 2010); this scenario may be a product of the fact that changes in species richness are not yet large enough to modify similarity relationships or because invasions and extinctions occur in a balanced manner (Olden and Poff 2003).

Because biotic homogenization studies have been conducted profusely in northern hemisphere regions (e.g., Olden 2006; Olden and Rooney 2006), few antecedents are available to establish their global character or their occurrence at other latitudes. In order to fill this gap, it was necessary, on the one hand, to have evidence of regions located in the southern hemisphere and, on the other, to extend the geographical coverage, spanning diverse areas of the planet. In this sense, a set of evidence appeared recently after analyzing the floristic and faunistic composition of diverse areas or regions of South America. Let us look at some of this evidence.

Oceanic islands such as Easter Island, the Juan Fernández archipelago, and Desventuradas archipelago are islands that politically belong to Chile; they are located in the Pacific Ocean, between 3000 and 600 km from the South American mainland coast. They emerged as volcanoes from the seabed, and from that moment, they began to receive propagules of different plant species from different regions of the planet. Archeological evidence indicates that only Easter Island was populated by Polynesians from 400 AD, who took advantage of introducing some species of plants and animals useful for their survival (Vargas et al. 2006). However, after their discovery by Europeans (during the sixteenth and seventh centuries), these islands began to be visited more frequently, first by navigators who introduced species such as rabbits and goats, and then gave rise to a resident human population. The Desventuradas archipelago, the only islands that do not have a permanent human population, is visited by fishermen, sailors, and scientific explorers, often throughout the year. The European influence, as well as the Chilean territorial colonization, has contributed to introduce new species of plants, now established or naturalized, and to extinguish native species, among them the woody taxa (i.e., *Sophora*) that were used as a source of wood and firewood. The patterns of taxonomic similarity between the islands, considering the original flora (the most probable prior to the European discovery), and the current flora, which has a large number of exotic species, were analyzed by Castro et al. (2008). According to these authors, naturalized vascular plant species now almost double the number of native species. After comparing the current similarity differential with the original, the islands have shown a 2% increase in compositional similarity, a meager value considering the diversity of resident exotic species that exceeds 200 taxa (Castro et al. 2008). In a second instance, these authors made these comparisons this time including species of ornamental plants (not naturalized) that in rigor constitute an important contingent of the exotic flora introduced in the islands (Castro and Jaksic 2008). In this opportunity, the results showed that the floristic similarity was increased between the islands, which allows to conclude that the nonnaturalized flora also contributes to the floristic homogenization. These authors warned that despite the potential ecological and functional importance of ornamental plants (Corbet et al. 2001; Reichards and White 2001a, b; Gaston et al. 2005, 2006; Smith et al. 2006), they are usually not included in floristic homogenization studies (Castro and Jaksic 2008).

The phylogenetic analysis has offered a complementary perspective to the homogenization studies of these islands. Carvallo and Castro (2017) analyzed the changes in the phylogenetic diversity of angiosperm flora in oceanic islands of Chile both in their pre- and post-colonization condition. In spite of the extinction of 18 species of angiosperms, these have not significantly altered the α-phylogenetic diversity of the islands; neither have the 756 exotic species introduced to the islands. In terms of the β-spatial diversity, results showed that the contribution of exotic angiosperms was greater than native angiosperms. Because exotic angiosperms belong to a set of phylogenetically related species, Carvallo and Castro (2017) suggested that the introduction of species might be associated with a functional effect not yet evaluated on these islands.

Floristic changes tending to homogenization have also been studied in other localities of South America. For example, Lobo et al. (2011) analyzed whether the composition of fragmented forests of the northeastern coast of Brazil is in the process of taxonomic homogenization as a result of the colonization of species considered pioneers. Lobo and his team compared the composition of 12 tropical forest sub-regions strongly disturbed by anthropogenic activities. These authors recorded the presence of 5122 trees at two times: pre-1980, a time span extending from 1902 to 1980, and post-1980, a time span extending from 1981 to 2006. The results showed that the levels of floristic similarity were 28% higher in the post-1980 period than in the previous period, denoting a process of floristic homogenization. The interesting thing about this case is that this homogenization process was determined by a contingent of 24 species of native trees (i.e., non-exotic), all of short life, low stature, and pioneering successional characteristics (Lobo et al. 2011).

The addition of one or more species to a given region, by extension, will imply the modification of the composition of the region's biodiversity. In many cases, when the introduction of species occurs in communities with a high degree of isolation, such as in the case of islands, basins, rivers, or lakes, this not only implies the arrival of a new species but also that of a new evolutionary lineage that could theoretically modify the relationships of evolutionary diversity or phylogenetics (Olden et al. 2004; Lockwood et al. 2007). Carvallo et al. (2013) analyzed phylogenetic diversity in 67 communities in the Mediterranean region of central Chile. These authors used five different descriptors to characterize diversity in native and exotic plant species. Thus, they documented that the native flora, represented by 204 species, possessed a greater phylogenetic diversity than the exotic plants represented by 179 taxa. In both exotic and native flora, the families Asteraceae, Brassicaceae, Fabaceae, Papaveraceae, and Poaceae were the most diverse and widely distributed. This suggests that the presence of environmental filters would make it more favorable to the establishment of members of these lineages in the Mediterranean region of central Chile (Carvallo et al. 2013).

Other examples can be obtained from floristic analyses of continental Chile (Castro and Jaksic 2008). Patterns of modification in floristic diversity beyond the species level have been poorly documented and analyzed. Examination of the taxonomic membership patterns of naturalized vascular flora in continental Chile has shown that, for a sample of 419 species, 41% belongs to a genus represented in the native flora, 56% belongs to a family represented in the native flora, and only 3% corresponds to taxa not represented as a genus or family. These values indicate that the representation of most biological lineages are not entirely new in continental Chile, as they were already represented in native flora (Castro and Jaksic 2008; Carvallo et al. 2014).

Assemblages of freshwater fish show a special propensity for biotic homogenization as a consequence of their original isolation (Rahel 2002). According to several studies, in this type of communities, biotic homogenization is mainly driven by the processes of invasion of species in concomitance with a minor importance of the extirpation or extinction of native species (Lockwood et al. 2001; Rahel 2002; McKinney 2005a, b; Baiser et al. 2014). In basins distributed throughout continental Chile, Vargas et al. (2015) analyzed the composition of freshwater fish in search of evidence for biotic homogenization. According to these authors, Chile's ichthyofauna consists of 44 native species, most of which are registered under some category of conservation concern (72%; Vargas et al. 2015). In addition, about 82% of these species are endemic to the territory, which is enhanced by the presence of relict species that come from an extensive history of biogeographic isolation from the rest of South America (Dyer 2000; Habit et al. 2006). Vargas et al. (2015) documented that the addition of 27 exotic fish species to Chile's watersheds has increased the average value of watershed similarity from 40.3% to 41.5%, i.e., by 1.2%. These authors concluded that the basins of the territory would be in the process of homogenization of their ichthyofauna. This fact is supported by previous studies (Castro et al. 2014), which have shown that the increase in compositional similarity constitutes an incipient trend in these communities. Indeed, both studies highlight that the process of compositional change in these basins is mainly determined by the occurrence of invasions and, to a lesser extent, by the extinction of native species, as has been evidenced in other regions of the world (Olden 2006). In view of the high level of endemism of the native ichthyofauna associated with its conservation state, it is possible that biotic homogenization will worsen in the basins of continental Chile.

Indeed, due to increasing habitat modification due to climate change and anthropogenic effect, as well as the ecological impact of exotic species, an increase in the extinction of native species is expected (Castro et al. 2014; Habit and Cussac 2016), a phenomenon called "extinction debt" (Kauussaari et al. 2009); equivalently, invasive species are expected to continue to extend their distributions on a geographic scale, a phenomenon called "invasion debt" (Essl et al. 2011).

Despite these previous results, what seems to be out of controversy is that the freshwater ichthyofauna of the Mediterranean regions of the planet are in global homogenization. In effect, Marr et al. (2010) found that the basins located in central Chile are homogenizing together with the other four Mediterranean regions of the world: California (United States), southwestern Australia, Iberian Peninsula (which includes Spain and Portugal), and the southwestern Cape of South Africa. Although the levels of endemism recorded for some of these regions exceed 70% (as in the South African Cape region, Australia, and Chile), in all of them, the richness of exotic species exceeds that of native species. A fraction of exotic fish species has resulted in an 8% increase in compositional similarity between these regions, showing that the process of homogenization occurs concomitantly between these regions despite the distances that separate them.

Studies to evaluate the occurrence and importance of biotic homogenization on a biogeographic scale are an important need (Olden 2006). Among the first studies that addressed biotic homogenization analysis under a wide geographic scale is Spear and Chown (2008). These authors compared the representation of species belonging to the orders Artiodactyla and Perissodactyla present in 41 countries of the globe, among them South Africa, a Mediterranean-type ecosystem. The variation in the compositional similarity of ungulates between these countries increased by 2%, due mainly to the increase in invasions and, to a lesser extent, to the increase in extinctions. Similarly, Castro et al. (2010) analyzed the compositional and floristic similarity changes in 12 oceanic island systems located in the Pacific and Atlantic (Fig. 7.13) and recorded that these islands and archipelagos originally contained 5958 species of native vascular plants, a value that increased to 7749 species after being colonized by Europeans. Thus, 2679 species were translocated to the islands and only 142 went extinct, implying an average increase in similarity from 1.6% to 6.3%. From these results, it is estimated that floristic homogenization is the predominant biotic change trend experienced by these insular systems (Fig. 7.14), despite the distance and geographic location that separates them. On the other hand, Cassey et al. (2007) analyzed the patterns of change in the compositional similarity of the avifauna present in oceanic islands of the Pacific and Atlantic oceans. The number of islands analyzed in this study is considerable (N = 152), comprising a total bird species richness of 1491 taxa. These authors documented that the avifauna in islands located in the Pacific Ocean is in a process of biotic differentiation, since its current similarity is less than the islands originally had. Conversely, the Atlantic islands are in the process of homogenizing their avifauna, as evidenced by the significant increase in levels of compositional similarity. However, for the global analysis, Cassey et al. (2007) found an increase in the similarity of the avifauna studied, with values varying between 0% and 3%.

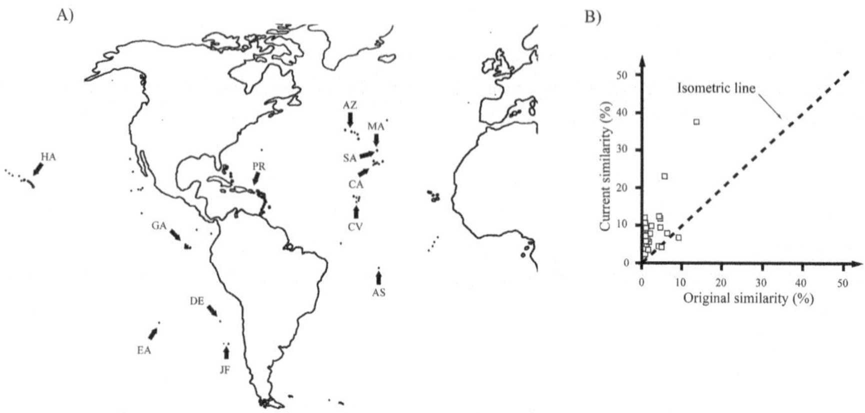

Fig. 7.13 Floristic homogenization between different oceanic archipelagos of the globe (Castro et al. 2010). (**a**) Geographical location of the islands and archipelagos studied. Ascension (AS), Azores (AZ), Canary (CA), Cape Verde (CV), Desventuradas (DE), Easter Island (EA), Galápagos (GA), Hawaii (HA), Juan Fernández (JF), Madeira (MA), Puerto Rico (PR), and Savage Islands (SA). (**b**) Contrast of taxonomic similarity values measured in pre- and post-European conditions in the 12 oceanic archipelagos. Similarity was calculated using the Jaccard similarity index and expressed as a percentage. The *x* axis indicates the original similarity values, while on the *y* axis, the current similarity values are displayed. The diagonal line represents the ideal position of the isometric values between the two axes. The point cloud is located above the diagonal, a graphic way of highlighting the floristic homogenization

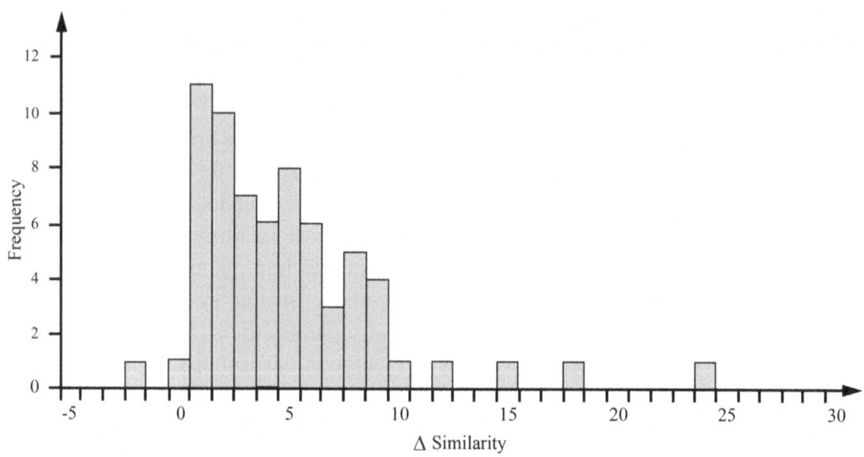

Fig. 7.14 Frequency distribution of similarity differentials (Δ)*J*) among 12 archipelagos (Castro et al. 2010). Δ*J* was calculated for each archipelago pair as the difference between the Jaccard similarity index calculated for the current flora (Ja) and prior (Jo) to the European colonization. These values denote floristic homogenization for the archipelagos studied, as there is a greater frequency of positive values (Δ J > 0)

Although the studies mentioned above have been carried out on different types of organisms, they have in common that the relocation of species constitutes the main mechanism that modifies the patterns of compositional similarity of regions on a global scale and whose numerical importance surpasses extinctions by several orders of magnitude.

Although similarity values appear to differ between different taxonomic groups (vascular plants, fish, mammals, and birds), evidence to date shows that biotic homogenization (i.e., increased compositional similarity) appears to be a trend of biotic change on a planetary scale. In view of this fact, a central challenge seems to be to address how biotic homogenization modifies other components of biodiversity beyond compositional similarity (Olden 2006; Olden and Rooney 2006).

References

Anderson CB, Rosemond AD (2007) Ecosystem engineering by invasive exotic beavers reduces in-stream diversity and enhances ecosystem function in Cape Horn, Chile. Oecologia 154:141–153

Arnold ML (1997) Natural hybridization and evolution. Oxford University Press, New York

Arnott DL, Vanni MJ (1996) Nitrogen and phosphorus recycling by the zebra mussel (*Dreissena polymorpha*) in the western basin of Lake Erie. Can J Fish Aquat Sci 53:646–659

Arroyo MTK, Marticorena C, Matthei O, Cavieres L (2000) Plant invasions in Chile: present patterns and future predictions. In: Mooney HA, Hobbs RJ (eds) Invasive species in a changing world. Island Press, Washington, DC, pp 385–421

Badano EI, Villarroel E, Bustamante RO, Marquet P, Cavieres LA (2007) Ecosystem engineering facilitates invasions by exotic plants in high-Andean ecosystems. J Ecol 95:682–688

Badano EI, Bustamante RO, Villarroel E, Marquet PA, Cavieres LA (2015) Facilitation by nurse plants regulates community invasibility in harsh environments. J Veg Sci 26:756–767

Bailey JP, Child LE (1996) A survey of the distribution of *Fallopia* × *bohemica* (Chrtek & Crtková) J Bailey (Polygonaceae) in the British Isles. Watsonia 21:187–198

Baiser B, Olden JD, Record S, Lockwood JL, McKinney ML (2014) Pattern and process of biotic homogenization in the New Pangaea. Proc R Soc Lond B 279:4772–4777

Becerra PI, Bustamante RO (2011) Effect of a native tree on seedling establishment of two exotic invasive species in a semiarid ecosystem. Biol Invasions 13:2763–2773

Bond WJ, Keeley JE (2005) Fire as a global 'herbivore': the ecology and evolution of flammable ecosystems. Trends Ecol Evol 20:387–394

Brooks ML, D'Antonio CM, Richardson DM, Grace JB, Keeley JE (2004) Effects of invasive plants on fire regimes. Bioscience 54:677–688

Brown JH, Lomolino MV (1998) Biogeography. Sinauer, Sunderland

Brown BJ, Mitchell RJ, Graham SA (2002) Competition for pollination between an invasive species (*Purple loosestrife*) and a native congener. Ecology 83:2328–2336

Buchmann SL (1983) Buzz pollination in angiosperms. In: Dones, Little RJ (eds) Handbook of experimental pollination biology. Scientific and Academic Editions, New York, pp 73–113

Callaway RM (2002) The detection of neighbors by plants. Trends Ecol Evol 17:104–105

Callaway RM, Brooker RW, Choler P, Kikvidze Z, Lortie CJ, Michalet R, Paolini L, Pugnaire FI, Newingham B, Aschehoug ET, Armas C, Kikodze D, Cook BJ (2002) Positive interactions among alpine plants increase with stress. Nature 417:844–848

Carvallo GO, Castro SA (2017) Invasions but not extinctions change phylogenetic diversity of angiosperm assemblage on southeastern Pacific Oceanic islands. PLoS One 12:e0182105

Carvallo GO, Teillier S, Castro SA, Figueroa JA (2014) The phylogenetic properties of native- and exotic-dominated plant communities. Austral Ecol 39:304–312

Case TJ, Gilpin ME (1974) Interference competition and niche theroy. Proc Natl Acad Sci U S A 71:3073–3077

Cassey P, Lockwood JL, Blackburn TM, Olden JD (2007) Spatial scale and evolutionary history determine the degree of taxonomic homogenization across island bird assemblages. Divers Distrib 13:458–466

Castro SA, Jaksic FM (2008) How general are global trends in biotic homogenization? Floristic tracking in Chile, South America. Glob Ecol Biogeogr 17:524–531

Castro SA, Bozinovic F, Jaksic FM (2008) Ecological efficiency and legitimacy in seed dispersal of an endemic shrub (*Lithrea caustica*) by the European rabbit (*Oryctolagus cuniculus*) in Central Chile. J Arid Environ 72:1164–1173

Castro SA, Badano E, Guzman D, Cavieres L (2009) Biological invasion of a refuge habitat: *Anthriscus caucalis* (Apiaceae) decreases diversity, evenness, and survival of native herbs in the Chilean matorral. Biol Invasions 12:1295–1303

Castro SA, Daehler CC, Silva L, Torres-Santana CW, Reyes-Betancort JA, Atkinson R, Jaramillo P, Guezou A, Jaksic FM (2010) Floristic homogenization as a teleconnected trend in oceanic islands. Divers Distrib 16:902–910

Castro SA, Escobedo VM, Aranda J, Carvallo GO (2014) Evaluating Darwin's naturalization hypothesis in experimental plant assemblages: phylogenetic relationships do not determine colonization success. PLoS One 9:e105535

Cavieres LA, Quiroz CL, Molina-Montenegro MA, Muñoz AA, Pauchard A (2005) Nurse effect of the native cushion plant *Azorella monantha* on the invasive non-native *Taraxacum officinale* in the high-Andes of central Chile. Perspect Plant Ecol Evol System 7:217–226

Cavieres LA, Badano EI, Sierra-Almeida A, Molina-Montenegro MA (2007) Microclimatic modifications of cushion plants and their consequences for seedling survival of native and non-native herbaceous species in the high Andes of central Chile. Arct Antarct Alp Res 39:229–236

Cavieres LA, Quiroz CL, Molina-Montenegro MA (2008) Facilitation of the non-native *Taraxacum officinale* by native nurse cushion species in the high-Andes of central Chile: are there differences between nurses? Funct Ecol 22:148–156

Chapin FS III, Zavaleta ES, Eviner VT, Naylor RL, Vitousek PM, Reynolds HL, Hooper DU, Lavorel S, Sala OE, Hobbie SE, Mack MC, Díaz S (2000) Consequences of changing biodiversity. Nature 405:234–242

Chesson P (2000) Mechanisms of maintenance of species diversity. Annu Rev Ecol Syst 31:343–366

Choler P, Michalet R, Callaway RM (2001) Facilitation and competition on gradients in alpine plant communities. Ecology 82:3295–3308

Coblenz BE (1978) The effects of feral goats (*Capra hircus*) on island ecosystems. Biol Conserv 13:279–286

Corbet SA, Bee J, Dasmahaoatra K, Gale S, Gorringe E, La Ferla B, Moorhouse T, Trevail A, Van Bergen Y, Yorontsova M (2001) Native or exotic? Double or single? Evaluating plants for pollinator-friendly gardens. Ann Bot 87:219–232

Cosyns E, Delporte A, Lens L, Hoffmann M (2005a) Germination success of temperate grassland species after passage through ungulate and rabbit guts. J Ecol 9:353–361

Cosyns E, Claerbout S, Lamoot I, Hoffmann M (2005b) Endozoochorous seed dispersal by cattle and horse in a spatially heterogeneous landscape. Plant Ecol 178:149–162

Crooks JA (2002) Characterizing ecosystem-level consequences of biological invasion: the role of ecosystem engineers. Oikos 97:153–166

D'Antonio CM, Vitousek PM (1992) Biological invasions by exotic grasses, the grass/fire cycle, and global change. Annu Rev Ecol Syst 23:63–87

Daehler CC, Strong DR (1997) Reduced herbivore resistence in introduced smooth cordgrass (*Spartina alterniflora*) after a century of herbivore-free growth. Oecologia 110:99–108

Davis MA (2009) Invasion biology. Oxford University Press, Oxford

Davis MA, Pelsor M (2001) Experimental support for a resource-based mechanistic model of invasibility. Ecol Lett 4:421–428

De Candolle ALP (1855) Géographie botanique raissoné. V. Masson, Paris

Di Tomaso JM (1998) Impact, biology, and ecology of saltcedar (*Tamarix* spp.) in the southwestern United States. Weed Technol 12:326–336

Didham RK, Tylianakis JM, Hutchinson MA, Ewers RM, Gemmel NJ (2005) Are invasive species the drivers of ecological change? Trends Ecol Evol 20:470–474

Dukes JS, Mooney HA (1999) Does global change increase the success of biological invaders? Trends Ecol Evol 14:135–139

Dyer B (2000) Systematic review and biogeography of the freshwater fishes of Chile. Estudios Oceanológicos 19:77–98

Elton CS (1958) The ecology of invasions by animals and plants. University of Chicago Press, Chicago

Essl F, Stefan D, Rabitsch W, Hulme PE, Hülber K, Jarošík V, Kleinbauer I, Krausmann F, Kühn I, Nentwig W, Vilà M, Genovesi P, Gherardi F, Desprez-Loustau ML, Roques A, Pyšek P (2011) Socioeconomic legacy yields an invasion debt. Proc Natl Acad Sci U S A 108:203–207

Figueroa JA, Castro SA, Marquet PM, Jaksic FM (2004) Exotic plant invasions to the mediterranean region of Chile: causes, history and impacts. Rev Chil Hist Nat 77:465–483

Figueroa JA, Cavieres LA, Gómez-González S, Molina-Montenegro MA, Jaksic FM (2009) Do heat and smoke increase emergence of exotic and native plants in the matorral of central Chile? Acta Oecol 35:335–340

Figueroa JA, Castro SA, Abarca L, Contreras TE, Jaksic FM (2014) El fuego facilita las invasiones biológicas de plantas en comunidades mediterráneas, Invasiones biológicas en Chile. In: Jaksic FM, Castro SA (eds) . Ediciones Universidad Católica de Chile, pp 343–355

Fuentes N, Ugarte E, Kühn I, Klotz S (2007) Alien plants in Chile: inferring invasion periods from herbarium records. Biol Invasions 10:649–657

Fuentes-Contreras E, Muñoz R, Niemeyer HM (1997) Diversidad de áfidos (Hemiptera: Aphidoidea) en Chile. Rev Chil Hist Nat 70:531–542

Fussell M, Corbet SA (1991) Forage for bumble bees and honeybees in farmland: a case-study. J Apic Res 30:87–97

García RA, Pauchard A, Cavieres LA, Peña E, Rodríguez MF (2010) Fire promotes *Teline monspessulana* (Fabaceae) invasion by increasing its germination. Rev Chil Hist Nat 83:443–452

Gardner SM, Cabido MR, Valladares GR, Diaz S (1995) The influence of habitat structure on arthropod diversity in Argentine semiarid Chaco forest. J Veg Sci 6:349–356

Gaskin JF, Kazmer DJ (2008) Introgression between invasive saltcedars (*Tamarix chinensis* and *T. ramosissima*) in the USA. Biol Invasions 11:1121–1130

Gómez S, Cavieres LA (2009) Litter burning does not equally affect seedling emergence of native and alien species of the Mediterranean-type Chilean matorral. Int J Wildland Fire 18:213–221

Gómez-González S, Lohengrin LA, Cavieres A, Torres P, Torres-Díaz C (2009) Competitive effects of the alien invasive *Centaurea solstitialis* on two Chilean *Baccharis* species at different life-cycle stages. Gayana Botánica 66:71–83

González AL, Kominoski JS, Danger M, Ishida S, Iwai N, Rubach A (2010) Can ecological stoichiometry help explain patterns of biological invasions? Oikos 119:779–790

Goulson D (2003) Effects of introduced bees on native ecosystems. Annu Rev Ecol Evol Syst 34:1–26

Graves HB (1984) Behavior and ecology of wild and feral swine (*Sus scrofa*). J Anim Sci 58:482–492

Habit E, Cussac V (2016) Conservation of the freshwater fauna of Patagonia: An alert to the urgent need for integrative management and sustainable development. J Fish Biol 89:369–371

Habit E, Dyer B, Vila I (2006) Estado de conocimiento de los peces dulceacuícolas de Chile. Gayana 70:100–113

Hails R, Timms-Wilson T (2007) Genetically modified organisms as invasive species? In: Caldwell MM, Heldmaier G, Jackson RB, Lange OL, Mooney HA, Schulze ED, Sommer U (eds) Biological invasions: analysis and synthesis. Springer, Berlin/Heidelberg/New York, pp 293–310

Hierro JL, Callaway RM (2003) Allelopathy and exotic plant invasion. Plant Soil 256:29–39
Hierro JL, Maron JL, Callaway RM (2005) A biogeographic approach to plant invasions: the importance of studying exotics in their introduced and native range. J Ecol 93:5–15
Higgins TM, McCutchan JH, Lewis WM (2008) Nitrogen ebullition in a Colorado plains river. Biogeochemistry 89:367–377
Holmgren M (2002) Exotic herbivores as drivers of plant invasion and switch to ecosystem alternatives states. Biol Invasions 4:25–33
Holmgren M, Avilés R, Sierralta L, Segura A, Fuentes ER (2000) Why have European herbs so successfully invaded the Chilean matorral? Effects of herbivory, soil nutrients, and fire. J Arid Environ 44:197–211
Holway DA (1999) Competitive mechanisms underlying the displacement of native ants by the invasive Argentine ant. Ecology 80:238–251
Holway DA, Suarez AV (1999) Animal behavior: an essential component of invasion biology. Trends Ecol Evol 14:328–330
Holway DA, Suaarez AV, Case TJ (2002) Role of abiotic factors in governing susceptibility to invasion: a test with argentine ants. Ecology 83:1610–1619
Ibarra JT, Fasola L, MacDonald DW, Rozzi R, Bonacic C (2009) Invasive American mink (*Mustela vison*) in wetlands of the Cape Horn biosphere reserve, southern Chile: what are they eating? Oryx 43:87–90
Jaksic FM (1996) Ecología de los vertebrados en Chile. Ediciones Universidad Católica de Chile, Santiago
Jaksic FM, Fuentes ER (1980) Why are native herbs in the Chilean matorral more abundant beneath bushes: microclimate or grazing? J Ecol 68:665–669
Jaksic FM, Fuentes ER (1991) Ecology of a successful invader: the European rabbit in Central Chile. In: Grooves RH, di Castri F (eds) Biogeography of mediterranean invasions. University of Cambridge, Cambridge, pp 273–283
Jaksic FM, Ostfeld RS (1983) Numerical and behavioral estimates of predation upon rabbits in mediterranean-type shrublands: a paradoxical case. Rev Chil Hist Nat 56:39–49
Jaksic FM, Soriguer RC (1981) Predation upon the European rabbit (*Oryctolagus cuniculus*) in mediterranean habitats of Chile and Spain: a comparative analysis. J Anim Ecol 50:269–281
Jaksic FM, Yáñez JL (1980) ¿Quién controla las poblaciones de conejos introducidos? Medio Ambiente 4:41–44
Jaksic FM, Yáñez JL (1983) Rabbit and fox introductions in Tierra del Fuego: history and assessment of the attempts at biological control of the rabbit infestation. Biol Conserv 26:367–374
Jaksic FM, Iriarte JA, Jiménez JE, Martínez DR (2002) Invaders without frontiers: cross-border invasions of exotic mammals. Biol Invasions 4:157–173
James WF, Barko JW, Eakin HL (1997) Nutrient regeneration by the zebra mussel (*Dreissena polymorpha*). J Freshw Ecol 12:209–216
Jones CG, Lawton JH, Shachak M (1994) Organisms as ecosystem engineers. Oikos 69:373–386
Jones CG, Lawton JH, Shachak M (1997) Positive and negative effects of organisms as physical ecosystem engineers. Ecology 78:1946–1957
Kauussaari M, Bommarco R, Heikkinen RK, Helm A, Krauss J, Lindborg R, Ockinger E, Pärtel M, Pino J, Rodà F, Stefanescu C, Teder T, Zobel M, Steffan-Dewenter I (2009) Extinction debt: a challenge for biodiversity conservation. Trends Ecol Evol 24:564–571
Klein AM, Vaissière BE, Cane JH, Steffan-Dewenter I, Cunningham SA, Kremen C, Tscharntke T (2007) Importance of pollinators in changing landscapes for world crops. Proc R Soc Lond B 274:303–313
Kowark I (2008) On the role of alien species in urban flora and vegetation. Urban Ecol 4:321–338
Krebs CJ (2001) Ecology: the experimental analysis of distribution and abundance. Benjamin Cummings, Menlo Park
Largiadèr CR (2008) Hybridization and introgression between native and alien species. In: Nentwig W (ed) Biological invasions. ecological studies (Analysis and synthesis). Springer, Berlin/Heidelberg, pp 275–292

Lee CE (2002) Evolutionary genetics of invasive species. Trends Ecol Evol 17:386–391
Lobo D, Leao T, Melo FPL, Santos AMM, Tabarelli M (2011) Forest fragmentation drives Atlantic forest of northeastern Brazil to biotic homogenization. Divers Distrib 17:287–296
Lockwood JL, Hoopes M, Marchetti M (2007) Invasion ecology. Blackwell Publishing, Massachusetts
Machordom A, García-Marín JL, Sanz N, Almodóvar A, Pla C (1999) Allozyme diversity in brown trout (*Salmo trutta*) from Central Spain: genetic consequences of restocking. Freshw Biol 41:707–717
Mack RN, Simberloff D, Lonsdale WM, Evans H, Clout M, Bazzaz FA (2000) Biotic invasions, causes, epidemiology, global consequences, and control. Ecol Appl 10:689–710
Marr SM, Marchetti MP, Olden JD, García-Berthou E, Morgan DL, Arismendi I, Day JA, Griffiths CL, Skelton PH (2010) Freshwater fish introductions in mediterranean-climate regions: are there commonalities in the conservation problem? Divers Distrib 16:606–619
Martí R (1993) The spreading of the Ruddy duck *Oxyura jamaicensis* in Europe and its effect on conservation of the White-headed duck *Oxyura leucocephala* in Spain. SEO/BirdLife Spain, Madrid
Matthei O (1995) Manual de las malezas que crecen en Chile. Alfabeta Impresores, Santiago, Chile
Mattila HR, Otis GW (2003) A comparison of the host preference of monarch butterflies (*Danaus plexippus*) for milkweed (*Asclepias syriaca*) over dog-strangler vine (*Vincetoxicum rossicum*). Entomol Exp Appl 107:193–199
McKinney ML, Lockwood JL (1999) Biotic homogenization: a few winners replacing many losers in the next mass extinction. Trends Ecol Evol 14:450–453
Menge BA, Chan F, Lubchenco J (2008) Response of a rocky intertidal ecosystem engineer and community dominant to climate change. Ecol Lett 11:151–162
Miura O (2007) Molecular genetic approaches to elucidate ecological and evolutionary issues of the biological invasions. Ecol Res 22:876–883
Montenegro G, Teillier S, Arce P, Poblete V (1991) Introductions of plants and animals into the mediterrean-type climate area of Chile. In: Groves RH, di Castri F (eds) Biogeography of mediterranean invasions. Cambridge University Press, Cambridge, pp 103–109
Mooney HA, Hobbs HA (2000) Invasive species in a changing world. Island Press, California
Mooney HA, Vitousek PM, Matson PA (1987) Exchange of materials between terrestrial ecosystems and the atmosphere. Science 238:926–932
Morales CL (2007) Introducción de abejorros (*Bombus*) no nativos: causas, consecuencias ecológicas y perspectivas. Ecol Austral 17:51–65
Muñoz AA, Cavieres LA (2008) The presence of a showy invasive plant disrupts pollinator service and reproductive output in native alpine species only at high densities. J Ecol 96:459–467
Muñoz-Fuentes V, Vilà C, Green AJ, Negro JJ, Sorenson MD (2007) Hybridization between white-headed ducks and introduced ruddy ducks in Spain. Mol Ecol 16:629–638
Nehring S, Hesse KJ (2008) Invasive alien plants in marine protected areas: the *Spartina anglica* affair in the European Wadden Sea. Biol Invasions 10:937–950
Niemeyer HN (1992) Aspectos ecológicos y moleculares de la interacción entre áfidos y sus plantas hospederas. Rev Chil Hist Nat 65:103–114
O'Dowd DJ, Green PT, Lake PS (2003) Invasional 'meltdown' on an oceanic island. Ecol Lett 6:812–817
Olden JD (2006) Biotic homogenization: a new research agenda for conservation biogeography. J Biogeogr 33:2027–2039
Olden JD, Poff NL (2003) Toward a mechanistic understanding and prediction of biotic homogenization. Am Nat 163:442–460
Olden JD, Poff NL (2004) Ecological processes driving homogenization: testing a mechanistic model using fish faunas. Ecology 85:1867–1875
Olden JD, Rooney TP (2006) On defining and quantifying biotic homogenization. Glob Ecol Biogeogr 15:113–120
Olden JD, Poff NL, Douglas M, Douglas ME, Fausch KD (2004) Ecological and evolutionary consequences of biotic homogenization. Trends Ecol Evol 19:18–24

Pardo R, Vila I, Capella JJ (2009) Competitive interaction between introduced rainbow trout and native silverside in a Chilean stream. Environ Biol Fish 86:353–359
Parker IM, Simberloff D, Lonsdale WM, Goodell K, Wonham M, Kareiva PM, Williamson MH, von Holle B, Moyle PB, Byers JE, Goldwasser L (1999) Impact: toward a framework for understanding the ecological effects of invaders. Biol Invasions 1:3–19
Pauchard A, García RA, Langdon B, Fuentes N (2011) Invasiones de plantas introducidas en Chile y su impacto en la biodiversidad: historia, estado actual y desafíos para su gestión. In: Figueroa E (ed) Conservación de la biodiversidad en las Américas: lecciones y recomendaciones de política. Editorial de la Universidad de Chile, Santiago, pp 147–182
Pausas JG, Keeley JE (2009) A burning story: the role of fire in the history of life. Bioscience 59:593–601
Pavez EF, Lobos GA, Jaksic FM (2010) Cambios de largo plazo en el paisaje y los ensambles de micromamíferos y rapaces en Chile central. Rev Chil Hist Nat 83:99–111
Petit RJ (2004) Biological invasions at the gene level. Divers Distrib 10:159–165
Petren K, Case TJ (1996) An experimental demonstration of exploitation competition in an ongoing invasion. Ecology 77:118–132
Porten M, Huber L (2003) An assessment method for the quantification of *Daktulosphaira vitifoliae* (Fitch) (hem. Phylloxeridae) populations in the field. J Appl Entomol 127:157–162
Prenter J, MacNeil C, Dick JTA, Dunn AM (2004) Roles of parasites in animal invasions. Trends Ecol Evol 19:385–390
Prigioni C, Balestrieri A, Remonti L (2005) Food habits of the coypu, *Myocastor coypus*, and its impact on aquatic vegetation in a freshwater habitat of NW Italy. Folia Zool 54:269–277
Pringle RM (2005) The origins of the Nile perch in Lake Victoria. Bioscience 55:780–787
Rahel FJ (2000) Homogenization of fish faunas across the United States. Science 288:854–856
Rahel FJ (2002) Homogenization of freshwater faunas. Annu Rev Ecol Syst 33:291–315
Raikow DF, Sarnelle O, Wilson AE, Hamilton SK (2004) Dominance of the noxious cyanobacterium *Microcystis aeruginosa* in low-nutrient lakes is associated with exotic zebra mussels. Limnol Oceanogr 49:482–487
Reiche K (1899) Estudios críticos sobre la flora de Chile. Anales de la Universidad de Chile 104:767–847
Rejmánek M, Richardson DM, Higgins SI, Pitcairn MJ, Grotkopp E (2005) Ecology of invasive plants: state of the art. In: Mooney HA, Mack RN, McNeely JA, Neville L, Schei PJ, Waage J (eds) Invasive alien species: a new synthesis. Island Press, Washington, DC, pp 104–161
Ricciardi A (2004) Assessing species invasions as a cause of extinction. Trends Ecol Evol 19:619
Richardson DM, Allsopp N, D'Antonio CM, Milton SJ, Rejmánek M (2000) Plant invasions: the role of mutualisms. Biol Rev 75:65–93
Rooney TP, Olden JD, Leach MK, Rogers DA (2007) Biotic homogenization and conservation priorization. Biol Conserv 134:447–450
Rosenzweig ML (2001) The four questions: what does the introduction of exotic species do to diversity? Evol Ecol Res 3:361–367
Rozzi R, Sherriffs M (2003) El visón (*Mustela vison*, Schereber), un nuevo mamífero exótico para la Isla Navarino. Anales del Instituto de la Patagonia, Serie Ciencias Naturales 31:97–104
Sala OE, Chapin FS, Armesto JJ, Berlow E, Bloomfield J, Dirzo R, Huber-Sanwald E, Huenneke LF, Jackson RB, Kinzig A, Leemans R, Lodge DM, Mooney HA, Oesterheld M, Poff NL, Sykes MT, Walker BH, Walker M, Wall DH (2000) Global biodiversity scenarios for the year 2100. Science 287:1770–1774
Sax DF, Stachowicz JJ, Gaines SD (2005) Species invasions: insights into ecology, evolution and biogeography. Sinauer Associates Incorporated, Sunderland
Scherer-Lorenzen M, Venterink HO, Buschmann H (2007) Nitrogen enrichment and plant invasions: the importance of nitrogen-fixing plants and anthropogenic eutrophication. Biol Invasions 193:163–180
Schüttler E, Klenke R, McGehee S, Rozzi R, Jax K (2009) Vulnerability of ground-nesting waterbirds to predation by invasive American mink in the Cape Horn Biosphere Reserve, Chile. Biol Conserv 142:1450–1460

Sierra C (2001) El cerdo cimarrón (*Sus scrofa*, Suidae) en la Isla del Coco, Costa Rica: escarbaduras, alteraciones al suelo y erosión. Revista de Biología Tropical (Costa Rica) 49:1158–1170

Simberloff D (2006) Invasional meltdown 6 years later: important phenomenon, unfortunate metaphor, or both? Ecol Lett 9:912–919

Simberloff D, von Holle B (1999) Positive interactions of nonindigenous species: Invasional meltdown? Biol Invasions 1:21–32

Smith RM, Thompson K, Hodgson JG, Warren PH, Gaton KJ (2006) Urban domestic gardens (IX): composition and richness of the vascular plant flora, and implications for native biodiversity. Biol Conserv 129:312–322

Spear D, Chown SL (2008) Taxonomic homogenization in ungulates: patterns and mechanisms at local and global scales. J Biogeogr 35:1962–1975

Stanton ML (1987) Reproductive biology of petal color variants in wild populations of *Raphnus sativus*: I. Pollinator response to color morphs. Am J Bot 74:178–187

Stebbins GL (1959) The role of hybridization in evolution. Proc Am Philos Soc 103:231–251

Stenquist S (2000) Saltcedar integrated weed management and the endangered species act. In: Spencer N (ed) Proceedings of the X international symposium on biological control of weeds. Montana State University, Bozeman, pp 487–504

Thompson JN (1988) Evolutionary ecology of the relationship between oviposition preference and performance of offspring in phytophagous insects. Entomol Exp Appl 47:3–14

Tilman D (1988) Plant strategies and the dynamics and structure of plant communities. Princeton University Press, Princeton

Tilman D (2004) Niche tradeoffs, neutrality, and community structure: a stochastic theory of resource competition, invasion, and community assembly. Proc Natl Acad Sci U S A 101:10854–10861

Torres JA, Moreno B, Alcalá-Zamora A (1994) La Malvasía Canela (*Oxyura jamaicensis*) en España y su relación con la Malvasía Cabeciblanca (*Oxyura leucocephala*). Oxyura 7:5–27

Vargas PV, Arismendi I, Gomez-Uchida D (2015) Evaluating taxonomic homogenization of freshwater fish assemblages in Chile. Rev Chil Hist Nat 88:16–26

Velthuis HHW, van Doorn A (2006) A century of advances in bumblebee domestication and the economic and environmental aspects of its commercialization for pollination. Apidologie 37:421–451

Verspoor E (1988) Widespread hybridization between native Atlantic salmon, *Salmo salar*, and introduced brown trout, *S. trutta*, in eastern Newfoundland. J Fish Biol 32:327–334

Vidal-Figueroa T, Delibes M (1987) Primeros datos sobre el visón Americano (*Mustela vison*) en el suroeste de Galicia y noroeste de Portugal. Ecología 1:145–152

Vitousek PM (1990) Biological invasions and ecosystem processes: towards an integration of population biology and ecosystem studies. Oikos 57:7–13

Vitousek PM, Walder LR, Whiteaker LD, Mueller-Dombois D, Matson PA (1987) Biological invasion by *Myrica faya*, alters ecosystem development in Hawaii. Science 238:802–804

Webb SD (1991) Ecogeography and the Great American interchange. Paleobiology 17:266–280

Whitcraft CR (2007) Wetland plant influence on sediment ecosystem structure and trophic function. Ph. D. Thesis, University of California, San Diego

Wiklund C (1975) The evolutionary relationship between adult oviposition preferences and larval host plant range in *Papilio machaon* L. Oecologia 18:156–197

Wilcove DS, Rothstein D, Dubow J, Phillips A, Losos E (1998) Quantifying threats to imperiled species in the United States. Bioscience 48:607–615

Willmer PG, Bataw AAM, Highes JP (1994) The superiority of bumblebees to honeybees as pollinators: insect visits to raspberry flowers. Ecol Entomol 9:271–284

Winter K, Adams L, Thorp R, Inouye D, Day L, Ascher J, Buchmann S (2006) Importation of non-native bumble bees into North America: potential consequences of using *Bombus terrestris* and other non-native bumblebees for greenhouse crop pollination in Canada, Mexico, and the United States. White Paper of the North American Pollinator Protection Campaign

Wright JP, Jones CG (2006) The concept of organisms as ecosystem engineers ten years on: Progress, limitations, and challenges. Bioscience 56:203–209

Part II
Case Studies

Chapter 8
European Rabbit (*Oryctolagus cuniculus* L.) in Chile: The Human Dimension Behind a Biological Invasion

Pablo Camus, Sergio A. Castro, and Fabián M. Jaksic

8.1 History of the European Rabbit in Chile

Globalization has caused modifications not only in cultural and economic patterns around the globe but also in the biological heritage of the countries involved (Vitousek et al. 1996). The increment in commercial connection and transport between remote regions of the planet has favored an unprecedented exchange of flora, fauna, and microorganisms (Mack et al. 2000). In many occasions, the introduction of species is affected intentionally, as in those with economic and productive importance. However, in other situations, the introductions are involuntary and, in many cases, undesired, as is the case with weeds or pests (Mack et al. 2000).

The invading species have awoken a renewed and growing interest in our society. On one side, they admittedly affect the natural (biological) heritage of the recipient

P. Camus (✉)
Instituto de Historia, Facultad de Historia, Geografía y Ciencia Política, Santiago, Chile

Center of Applied Ecology & Sustainability (CAPES), Pontificia Universidad Católica de Chile, Santiago, Chile
e-mail: pcamusg@uc.cl

S. A. Castro
Departamento de Biología, Facultad de Química y Biología, Universidad de Santiago de Chile, Santiago, Chile

Center of Applied Ecology & Sustainability (CAPES), Pontificia Universidad Católica de Chile, Santiago, Chile
e-mail: sergio.castro@usach.cl

F. M. Jaksic
Departamento de Ecología, Pontificia Universidad Católica de Chile, Santiago, Chile

Center of Applied Ecology & Sustainability (CAPES), Pontificia Universidad Católica de Chile, Santiago, Chile
e-mail: fjaksic@bio.puc.cl

F. M. Jaksic, S. A. Castro, *Biological Invasions in the South American Anthropocene*, https://doi.org/10.1007/978-3-030-56379-0_8

geopolitical units, since they can alter the existence of native species, and in some cases cause their extinction. On the other, the invading species negatively affect the productive areas, as they can behave as weeds or pests to crops (Pimentel et al. 1999). Finally, the arrival of new species can have effects on human health, introducing pathogens or deleterious substances.

Though the biological dimension of these invasions is beginning to be understood from the point of view of ecological theory, its human dimension is clearly less studied (McNeely 2001). This is particularly true if it is considered that global connectivity (i.e., globalization) is sustained by relationships between specific geopolitical units (i.e., countries) and that these have been profoundly modified by humans in the last 500 years (McNeely 2001). Thus, the study of the human dimension of biological invasions constitutes an aspect that helps to understand how species introductions have been supported historically, socially, and politically and what lessons can be learned from these experiences to apply in the future.

In continental Chile, the introduction of alien plant species began early on (Figueroa et al. 2004; Fuentes et al. 2008). The commercial exchange between the native peoples probably contributed to spread of numerous species and crops (quinoa, pepper, camelids, etc.). However, the Spanish colonization changed this dynamic, by introducing species from Europe, with very remote evolutionary and biogeographical links to local biota (Castro et al. 2005). With the founding of the Republic, there was an active stimulation of commercial liberalization and internal stockbreeding, forestry, and farming, and it was possible to introduce new breeds and crops (Castro et al. 2005). Similarly, the introduction of vertebrate species (cows, goats, pigs, sheep, and horses) was also early (Gay 1973). However, the species that ended up naturalizing and exerting a negative impact (including the European rabbit) correspond to species introduced more recently, from the nineteenth century onwards (Jaksic 1998). As a result, currently Chile has at least 1,110 alien species (PNUD 2017) with a largely unknown introduction history, in terms of the social and biological context of their introduction in the country.

The European rabbit (*Oryctolagus cuniculus*) has an important place among the invading fauna of Chile (Jaksic 1998). As an introduced species – apparently with commercial purposes – the breeding of European rabbits in central Chile led to their escape or liberation in natural habitats. Nowadays the European rabbit is distributed in central Chile, Juan Fernández archipelago, Tierra del Fuego, and part of Chilean-Argentinean Patagonia (Jaksic and Fuentes 1988). As a wild species, the European rabbit spread its geographical distribution in Chile, reaching high densities, to the point of causing economical damage to forestry and farming. In this chapter we studied the available historical antecedents that describe the introduction and spread of European rabbits in Chile. Specifically, we intend to cover two topics: on the one hand, to reconstruct the history of this invasion, focusing on the available documentary evidence, and on the other hand, to analyze the political and social context that triggered the introduction and naturalization of one of the most damaging species for Chilean forestry and farming. With these antecedents, we will try to gather some relevant lessons about the introduction of species in Chile and the need for a virtuous dialogue between scientific research and policy making, which have accompanied and will accompany the multilateral connection of our country in the face of globalization. We warn the reader that we built this chapter based on methodological and analytical protocols typical of history, using statements, testimonies, or reports that different social actors exposed in their time.

8.2 The European Rabbit as Alien Species in the World

The European rabbit is a native species of the Iberian Peninsula (Lees and Bell 2008). Its distribution was originally restricted by glaciations determining its diversification into two subspecies: *Oryctolagus cuniculus cuniculus* distributed in the northeastern region of the peninsula and southeast France and *Oryctolagus cuniculus algirus* occupying southwestern Spain and Portugal (Biju-Duval et al. 1997). During the fifteenth and sixteenth centuries onwards, *Oryctolagus cuniculus* was introduced to different continents, including oceanic islands, as eventual food for travelers and castaways (Flux and Fullagar 1992; Flux 1994; Thompson and Kong 1994). In several of these areas, *Oryctolagus cuniculus* naturalized and became an unwanted species (Thompson and Kong 1994; Mack et al. 2000).

In Australia, for example, the European rabbit was introduced in 1859, and speedily the population spread throughout the continent (Williams 1995). To stop its vertiginous population growth, foxes as natural predators were introduced from England, and a large wire fence of 11,000 kilometers was placed all along the country, both with no result (Williams 1995). Then, European rabbit hunting was encouraged for commercial purposes, and after a decade Australia exported 700 million rabbit skins and 160 million frozen rabbits. Nevertheless, all these measures of population control were vain. In 1950, Australian authorities started injecting the myxoma virus to European rabbits, using mosquitoes as transmitting agents of the disease (Voigt 1987). In almost every country where the European rabbit has been introduced, it has become naturalized and turned to pest species (Thompson and King 2004). So, *Oryctolagus cuniculus* has become one of the iconic examples of an invasive species.

8.3 Origins of the European Rabbit in Central Chile

The historical evidence regarding the introduction of the European rabbit in Chile is scarce and fragmentary (Camus et al. 2008). The French naturalist Lataste (1892) pointed out that the first specimens of *Oryctolagus cuniculus* were introduced to Chile in 1884. A small colony was supposedly freed on an island in Cauquenes (see Fig. 8.1) and spread across central Chile. Despite Lataste's version being accepted for a long time, recent findings provide an alternative explanation (Camus et al. 2008).

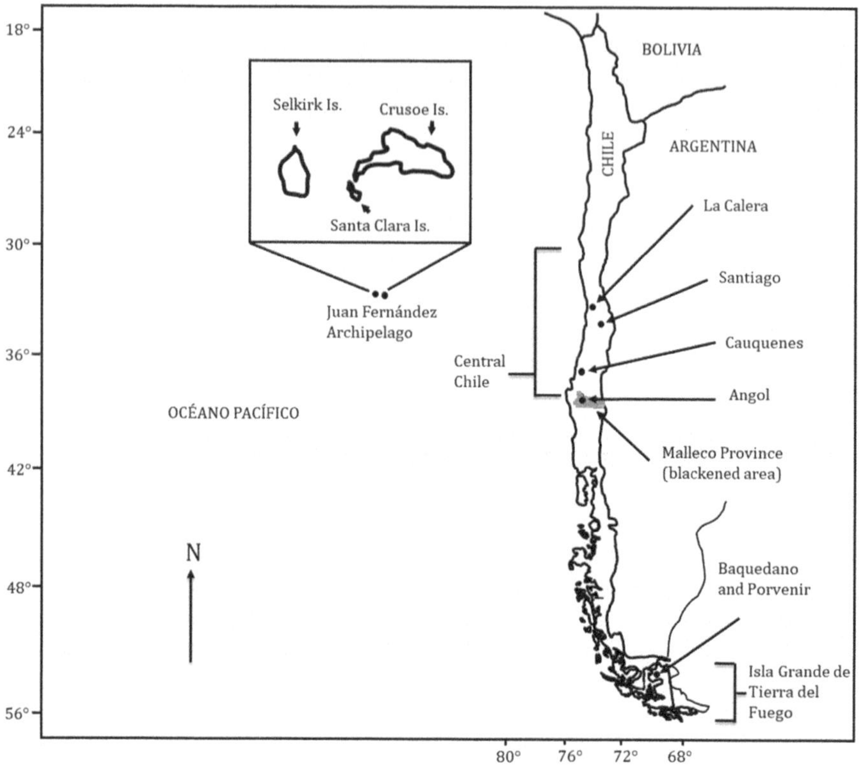

Fig. 8.1 Referential locations for the European rabbit (*Oryctolagus cuniculus*) invasion in Chile. See text

Molina (1788–1795), more than a century before Lataste, indicates indirectly the presence the European rabbits in central Chile. This author, when describing the guinea pig (*Cavia* sp.), wrote that despite all its morphological similarity with rabbits, it "flees from their company and these animals have never been seen in relation or together". Additionally, this author states that both guinea pigs and rabbits "are very afraid of cats and moles, who are their enemies and predators" (Molina 1987). At the same time, when referring to vizcachas (*Lagidium viscacia*), a Chilean endemic rodent, Molina (1788–1795) points out that "those peoples [peasants] prefer the meat of this animal that is white and very tender, to that of rabbits and hares". These historical records reveal that towards the middle of the eighteenth century, the European rabbits had already been introduced in central Chile (Camus et al. 2008).

Later, the French naturalist Claudio Gay wrote that in central Chile the European rabbit are "of a grey hue mixed with amber in wild state, with reddish blond in the nape of the neck; its throat and stomach are whitish. The ears almost as long as the head. The tail not as long as the thigh and brown op top; but in domesticity the colors vary a lot" (Gay 1847), evidencing the presence of naturalized rabbit populations. Additionally, Gay indicated that the rabbit is "unfortunately little abundant in relation to the numerous services its meat offers as food and its hairs as felt for hats" (Gay 1973), evidence that denotes a reduced abundance for the species. Consequently, Gay (1847) pointed out that in Chile "it would doubtless be useful to try to propagate them in the wild, especially in the large regions next to the mountain ranges where the lands are not being farmed yet, because they would offer a much tastier and healthier meat than that of domestic rabbits, a large quantity of furs, that the art of hat industry employs so frequently and to so much advantage". In these lines, written in the middle of the nineteenth century, naturalization with a reduced population abundance for *Oryctolagus cuniculus* was being documented. Towards the end of the nineteenth century, it began to appear the first reports indicating that the European rabbit constitute a potential pest in Chile. Miers (1900) remembers a fact that occurred before 1849, when a friend told him that "in his estate of La Calera he had a rabbit farm of about four blocks, with many rabbits. After inquiring about it, I deduce that after his death, his heirs did not take care of the rabbits, and it is likely, according to the explanation I was given, that as they were locked up inside walls of lime and bricks, receiving no food, most of them perished. Some must have escaped. Because somebody told me a few days ago that one can see rabbits in territories next to Calera" (see Fig. 8.1).

So far, the analysis of historical evidence shows that European rabbits were introduced in central Chile before 1884, probably during the eighteenth century if not before. Finalizing the nineteenth century, populations of *Oryctolagus cuniculus* were present both under captivity and as naturalized in central Chile, although showing small abundance in wild conditions.

8.4 Controverted Role of the European Rabbit in Central Chile

Finalizing the nineteenth century, a controversy began about the economic benefits and impacts ensuing the breeding of European rabbits (Camus et al. 2008). Two contrasting positions were established. On one hand, *Oryctolagus cuniculus* was considered as a highly profitable species that would provide meat and fur abundantly; on the other hand, it was considered as an unwanted species that could cause damage to agriculture, just as had happened in other regions of the world.

In 1870, for example, it was published "Breeding of Rabbits" by the Boletín de la Sociedad Nacional de Agricultura (Bulletin of the Nacional Society of Agriculture), highlighting the "secondary role" of the European rabbit in Chilean farms (Anonymous 1870). In this line, Santos-Tornero (1875) also highlighted the benefits of European rabbit breeding in central Chile, although warning that "rabbits can be damaging when in freedom". In that year (1875), the Boletín de la Sociedad Nacional de Agricultura published an article about the breeding of rabbits "to manifest the great utility farmers in other countries obtain from this field that is highly neglected in the Republic" (Anonymous 1875). Similarly, Le Feuvre (1885) stated that the breeding of European rabbits was convenient "especially for the poor of the countryside; both because it allows them to vary their diet and because it can be done in a very economic matter, having children – who normally have very little to do – pick up the herbs and weeds that these animals feed on". Le Feuvre was the director of the most important center of agricultural experimentation in the country, the Escuela Práctica de Agricultura de Chillán (Practical School of Agriculture of Chillán), and he informed that in his establishment rabbits "have multiplied enough to form a good basis for their exploitation on a bigger scale" (Le Feuvre 1900).

In contrast with the previous author, Philippi (1885) wrote "rabbits have not become wild for the reason of happiness, though some Frenchmen have tried to place them in one or another farmyard to have the pleasure to hunt them and to eat their meat, highly esteemed in France. The large damage they do was soon to be seen and they have been exterminated before they could become wild. The Englishmen, great rabbit hunters, have transported them to Australia, where they have multiplied in some places in such excess that it became a pest; in such a way that the settlers do not know how to get rid of these animals, that threaten to devastate the whole plantation". Other authors feared the effects of propagation of rabbits in Chile, due to their high fertility. Miers (1900), for example, wrote: "Last year I made the following observations regarding a pair of rabbits I had: the male was white and the female black. She bore eight little ones on October 26th; seven on November 27th, and nine on January 2nd. So, in 68 days one pair multiplied to 27. If we calculate that the first pair procreate during 7 months of the year; and that their offspring only from 6 months onwards, one can establish that in a year we could count on 570. This estimation is not fantastic, it is in fact still very limited, when one considers that the natural state of pure freedom and enough food is always more favorable for procreation than the artificial state of confinement". Similarly, Castillo (1912) wrote: "in Chile, especially in the central zone, which is similar to the original habitat of the rabbit, this rodent has found a suitable ground to undermine and an abundance of living fences to protect itself from the persecution it is subject to for the damages it causes. It finds itself so well here, so much to its taste that it multiplies prodigiously invading the farming lands with astonishing speed, for which it relies on the valuable resource of the blackberry bush, an invading plant that serves as an inviolable den, only violable by fire". Likewise, Castillo also pointed out that: "like many other organic beings adjusted to places outside of their country of origin, rabbits have found conditions in other regions of the earth that are so suitable to prosper that they have acquired a degree of rusticity completely unforeseen by their importers" (Castillo 1912). In the following years, the role of a deleterious species was attributed to the European rabbit, considering this species as a "pest" (Anonymous 1914).

Despite the recognized impact of *Oryctolagus cuniculus*, the economic benefits were considered as an opportunity for Chilean economic development, especially by the proprietors of smaller agricultural lands. From the 1920s onwards, it is possible to notice in the available documentation that there was a campaign to introduce the breeding of rabbits as an alternative for economic development. In 1921, a manual for rabbit breeding was published (Echeverría 1921); in 1928, the

Sociedad Nacional de Agricultura (National Society of Agriculture) published 'The breeding of the rabbit"; also in 1928 it was published "Advice for the production of skins" (Anonymous 1928).

Summarizing, in the last decades of the nineteenth century and first decades of the twentieth century, a debate began about the benefits or harms of the acclimatization and establishment of rabbit breeding, especially as an economic opportunity for the small landowners. *Oryctolagus cuniculus* was valued for its quick multiplication and easy breeding and the benefits obtained by the exploitation of skins and meat. On the other hand, other authors recognized the potential impact of the European rabbit in central Chile, shunning its commercial exploitation (e.g., Philippi 1885; Miers 1900; Castillo 1912). Nevertheless, together with the development of rabbit breeding promoted by government institutions like the Departamento de Tierras y Colonización and guild organizations like the Sociedad Nacional de Agricultura, at the end of the 1920s, the existence of a rabbit overpopulation was more and more evident in Chile.

8.5 European Rabbits as Pests in Central Chile

Towards the end of the 1920s, several legislative actions were set in motion, aimed at implementing measures for population control of *Oryctolagus cuniculus* (Camus et al. 2008). They were fundamentally based on promoting rabbit hunting and/or discouraging hunting of potential predators, such as carnivorous mammals and raptorial birds (see Congreso Nacional de Chile 1928; Camacho 1929; Anonymous 1930). The hunting evidenced an ostensible growth; Reed (1934), for example, wrote: "in the last 20 years the wild rabbit has multiplied extraordinarily in the countryside of Chile, and is already reaching the south [...] agriculture laments the large damages this rodent causes to crops [...] I ascribe an extraordinary power to the acclimatization of wild rabbits in the fields of Chile, from the point of view of our people [...] previously, farmers did not eat meat [...] now, with the multiplication of rabbits, meat is part of the farmers' daily diet and also of that of city dwellers". The statistics on rabbit and fox skins exported by Chile show that the number of fox furs sent abroad effectively decreased, which is evidence of a smaller hunting pressure, a result of the restrictions imposed by the hunting law of 1929. Also, the number of exported rabbit skins increased notably, at least until the 1960s, with a maximum of 479,031 skins exported from 1950 to 1954. These numbers indicate that there is no direct correlation between the presumed increment in the fox population and a reduction in the rabbit population, because despite the decrease in the fox skin exports, the export of rabbit skins increased. According to the antecedents available towards 1960, this would respond to two factors. On one hand, the mentioned boosting of rabbit breeding for that period, and on the other, the distributional outspread of the European rabbit in central Chile (Camus et al. 2008).

By 1934 the *Oryctolagus cuniculus* pest was far from being controlled. The historical evidence shows that requests to lift the hunting ban on foxes were rejected because the rabbits were still considered an enemy of the forestry and agricultural sectors. In fact, in this year a group of furriers from Santiago formally requested the opening of the fox hunt in the northern region of the country, arguing that this would simultaneously encourage the hunting of rabbits and hares (Anonymous 1934). A similar initiative was promoted by another group of furriers, this time, from south of Santiago, appealing to the need to favor furriers and hunters who saw their economic sustenance depleted (Anonymous 1934). In spite of the sectoral pressure, the authorities maintained the condition of closed hunting of carnivorous mammals.

Afterwards, the European rabbit spread successfully from Chile to Argentina, between 1945 and 1950 (Bonino and Gader 1987). Bonino and Gader (1987) state that these invading European rabbits "came almost certainly from Chile, since at the same latitude on the Chilean side there were populations of this species, and in that area there are many passages across the mountain range of an altitude that is no impediment for the advance of rabbits". On the other hand, Greer (1989) registered mammals observed in the province of Malleco (see Fig. 8.1) between 1960 and 1962 and estimated that the rabbit was found "all along the province of Malleco, except in the high parts of the Andes". These observations show that the rabbit continued to expand its distribution to the southern part of central Chile.

The legislative framework that implemented the ban of carnivorous mammals and raptorial birds rested on the presumption that these native predator species consumed European rabbits and, therefore, could control their populations. However, few antecedents were available to validate this hypothesis. Housse (1953) indicated that the culpeo fox (currently, *Lycalopex culpaeus* Mol.) fed on "small mammals that are incapable of resisting them: all the species of rats, hares, and rabbits of any age, viscacha rats, goats, lambs, pudus and sometimes domestic cats". Amazingly, it took almost 30 years for Jaksic and Yáñez (1980) to quantify for the first time the impact of rabbits in the diet of Chilean foxes, based on an analysis of droppings and stomachs of carnivores and of regurgitations of birds of prey, and they concluded that rabbits constituted a minority fraction of the prey hunted by the most common predators in the country. As an explanation of this phenomenon, they stated that Chilean predators were not efficient in the hunting of *Oryctolagus cuniculus*, because they had not developed the "behavioral adaptations needed to hunt a recently introduced species, as is the rabbit. In any case, whatever the explanation, it is apparent that predators play no important role in the abundance of rabbits in Chile" (Jaksic and Yáñez 1980). With this, these authors annihilated many years of public policies on the matter and renewed the historical challenge: how do we control the rabbit pest? What to do to avoid damages to agriculture and forestry?

From a political and legislative point of view, we can only state that the policy failure was based on the fallacious analogical reasoning that attributed the same role as European foxes had in controlling the rabbit population of Spain, to the intrinsic capacity of Chilean foxes to control the pest of rabbits in Chile. So, the myth that Chilean foxes were voracious rabbit eaters was built on suppositions and not on a scientific basis, with quantified data about the diet of these predators (Table 8.1).

Table 8.1 Number of exported skins, 1910–1980. From Iriarte and Jaksic (1986)

Years	European rabbits	Foxes
1910–1914	0	96,524
1915–1919	0	14,418
1920–1924	0	19,176
1925–1929	453,775	53,367
1930–1934	68,315	24,059
1935–1939	75,989	21,827
1940–1944	151,739	25,904
1945–1949	159,655	9,692
1950–1954	479,031	2,404
1955–1959	255,621	2,845
1960–1964	273,719	17,893
1965–1969	120,270	14,704
1970–1974	37,000	11,100
1975–1979	51,645	1,746
1980–1984	14,600	0
Total	2,141,359	315,659

8.6 European Rabbits as Pests in Tierra del Fuego

A different focal point where the European rabbit was introduced in Chile occurred in Tierra del Fuego (Jaksic 1998). Arentsen (1954) points out that the first rabbits were freed towards 1874 on Isla Grande of Tierra del Fuego (see Fig. 8.1). Nevertheless, this small population did not get established due to the adverse environmental conditions. In 1913, a second attempt for the rabbit's introduction was made in Baquedano and a little later in the proximities of Porvenir (see Fig. 8.1), but due to the harsh winters, none of these attempts prospered. Supposedly, the pest of 1950 came from two pairs of rabbits of European origin freed in the surroundings of Porvenir around 1936 or perhaps 2 or 3 years before (Arentsen 1954).

Together with the rabbit pest in the plains of Tierra del Fuego, the commerce of this species grew from 1942 onwards (Table 8.2). Since 1947 onwards, the major outbreak of rabbits occurred (Claro 1950). Mulgrue (1948) wrote "the rabbit does not only consume the superficial vegetation used for grazing, but he also eats the heart of the plant, so it dies, turning green fields into wastelands. Besides, the rabbit [...] feeds only on the best grasses of each region, leaving weeds that end up extirpating the usable grass rabbits did not eat. On the other hand, rabbit droppings poison and kill the pasturelands, and transform the devastated lands in dry soil that, lacking trees to evaporate humidity, can give way to disastrous droughts. The lack of vegetation caused by these rodents and the inevitable drought that comes with it lead to the erosion of the territory, so that in a short time the lands full of rabbits become patches of loose sand. It is necessary to seriously draw attention on what this can signify in a region of strong winds, like Chilean Patagonia. The food needed

Table 8.2 Export of rabbit skins to the north of Chile from the province of Magallanes (Anonymous 1950b)

Year	Amount	Value ($)
1939	4,171	15,732
1940	4,318	13,564
1941	10,624	64,877
1942	51,151	422,074
1943	148,289	1,216,495
1944	120,045	1,464,811
1945	89,793	772,961
1946	185,337	2,564,440
1947	318,031	4,159,150
1948	433,340	3,504,286
1949	379,490	3,203,472

for one sheep is only enough for 16 rabbits. If we consider that a pair of rabbits, with a normal number of kindling – at least four per year – becomes millions after one year, we will see that the danger of not having enough grasslands for sheep can become a sad reality in a relatively short time".

Some stockbreeders of Magallanes stated that "if this malady continues, the pasture capacity of Tierra del Fuego will be over in five or ten years up to the roots, giving way to the erosion of the soil that will make a great desert of the whole province" (Chaparro 1950). Likewise, he pointed out that "the rabbit pest originates in an imbalance produced in nature by the uncontrolled hunting of fine furred animals [...] he [Mr. Carlos Strauss] said we have to move towards a protection of our fauna as a way to avoid the disappearance of the species that feed on rabbits and prevent a disproportionate increment of the species" (Chaparro 1950).

With these arguments, the Union of Small Stockbreeders of Magallanes agreed to "negotiate with official organisms the enactment of a decree to consider the closed season of Andean skunks, grey foxes, wild cats, Andean mountain cats and Pampas cats, species that have run out in the province and that could contribute to the extermination of rabbits" (Anonymous 1950a). From Santiago, the Ministry of Agriculture requested the Ministry of Economy to attend "a suggestion of the Local Committee of Magallanes to enact a decree to prohibit, for a period not shorter than five years, the hunting and commerce of skins, hairs and feathers of various animal species that are enemies of the rabbit" (Anonymous 1950a). The minister of Economy referred the letter to the Department of Fishing and Hunting. On June 6th 1950, a decree was requested to prohibit the hunting of all the species that were enemies of rabbits, aiming to "combat the pest of those animals, that are causing such serious damage to the livestock in the regions of Magallanes and Aysén (see Fig. 8.1) due to the biological imbalance caused by the uncontrolled persecution, especially of foxes and Andean mountain cats". At the same time, to combat the rabbit pest, it was recommended to use their frozen meat in the cold stores of that area, sending it to the central zone, which would considerably increase the hunting of that species. The decree also entrusted the representative of the Stockbreeders

Association of Magallanes to acquire foxes to free in Tierra del Fuego, providing the funds for that campaign (Anonymous 1950a).

In 1950, the minister of Economy considered the rabbit pest had caused important damages to the livestock of that area and that the hunting and extermination of some enemy species of the rabbit caused the increment in their numbers. Consequently, it was agreed to prohibit for 5 years "the hunting and commerce of skins, hairs and feathers or eggs of the following species considered as enemies of the rabbit, in the provinces of Aysén and Magallanes" (Anonymous 1950a). On August 7th 1952, the Ministry of Agriculture dictated Supreme Decree N° 811 declaring officially the European rabbit as a pest for agriculture in Magallanes (Arentsen 1954). This promoted the use of more aggressive tools to control their population growth. In fact, in 1953, the biological control of the European rabbit began by using myxoma virus (Arentsen 1954). A campaign was planned to inject the virus in 100,000 rabbits towards 1954. Then all the isolated focus that might remain would be exterminated with all kinds of elements, like fats, poisons, gasses, dogs, predators, fire guns, etc. (Arentsen 1954). Based on this proposal, in 1954, it began the campaign against the rabbit pest (Sabat 1955).

From 1955 onwards, the rabbit pest abated in Magallanes, and from then on, it ceased to be a problem for local stockbreeders (Camus et al. 2008). With the passing of time and from a scientific viewpoint, Jaksic and Yáñez (1983) stated that the introduction of the myxomatosis virus was the most effective agent in the biological control of rabbits in Tierra del Fuego.

8.7 European Rabbits as Pests in Juan Fernández Archipelago

The available records indicate that the European rabbit was introduced in the island of Robinson Crusoe in 1935 (Kahler 1953) or 1936 (Brescia 1979), with the aim to breed this species in captivity and use it as a dietary supplement for the local human population. Some specimens escaped, and their distribution spread out through the island and its neighboring islet, Santa Clara (Ojeda et al. 2003). Towards 1970, the abundance of European rabbits began to be considered worrisome (Torres and Aguayo 1971). In fact, in 1975, the perception was alarming when it was stated that the rabbit "is currently represented by many thousands of specimens in the island of Robinson Crusoe. Because of its continuum growth, it represents one of the most serious problems for the biological equilibrium of this island, because it largely contributes to the continuum decrease of the herbal stratum and the consequent increment of erosion. Besides, it contributes to the extermination of diverse native plant species and even endemic ones. The intense hunting by the islanders does not seem to cause a significant decrease of their population" (Mann 1975). In 1976, the Management Plan of Juan Fernández National Park indicated that the European rabbit "is forming large populations that contribute significantly to the detriment of the vegetation cover on the islands of Robinson Crusoe and Santa Clara and, in a smaller amount, on Alejandro Selkirk" (CONAF 1976).

By 1981, the population abundance of *Oryctolagus cuniculus* in Santa Clara and Robinson Crusoe islands (see Fig. 8.1) was considered "very large, approximately 7,000 specimens on average and fluctuates greatly from year to year, and represents important pests when it reaches their maximum population number. In the competition with livestock over the vegetation cover and because of the construction of tunnels that constantly collapse under the feet of livestock, they facilitate the erosion of the soil; their destructive action is boosted" (Mann 1981).

Progressively, the European rabbits started to compete for food with the sheep on the island of Santa Clara, so the intervention of the authorities was required. Control measures on rabbits were implemented between 1983 and 1985, allowing ecological evaluations that showed a recovery of the herbaceous vegetation in areas where the rabbits had diminished. The sampling of the rabbits also showed that there was an important decrease in their size and weight as a consequence of rabbit hunting because the larger specimens – the most reproductive individuals – were declining in the wild (Sáiz and Ojeda 1988). In spite of two additional initiatives for biological control of the rabbits (see Camus et al. 2008), this species still devastates the archipelago and causes serious damage to the ecosystems of the islands, with their various ranges of endemic species. Evidently, hunting has proven to be insufficient as a means to control the rabbits in the islands of Juan Fernández.

8.8 Closing Comments

The first references about the introduction of European rabbits in Chile come from the mid-eighteenth century. These observations are indirect records and were registered by Molina (1788–1795) to describe to Europeans some Chilean species like guinea pigs and vizcachas. In the nineteenth century, several authors refer to the existence of European rabbits as domestic animals in farms in central Chile. Some authors recommend the breeding of this animal for its economic benefits, while others warn about the dangers of this species could mean to agriculture and forestry. Up to now, it is not possible to determine if it was intentional or casual that rabbits became wild.

At the beginning of the twentieth century, many observers began to warn about the dangers of the "rabbit in the wild". It is possible to suppose that many rabbit hutches were not built under strict security norms and that this was especially noticeable in the second half of the 1920s. Since then, the presence of the rabbit as a pest is described, sometimes dramatically, in central Chile, Tierra del Fuego, and the Juan Fernández archipelago. Given this situation, the government responded formulating public policies that have tried to put an end to the rabbit pest. In the 1930s, the hunting of foxes was banned, which permitted the protection of those species and dismantling of the national furrier industry. However, decades later it was discovered that Chilean foxes only fed sporadically on introduced rabbits, since they are better adapted to prey on other native animals. This shows that this policy was based on mistaken suppositions and could not have had meaningful results. According to Camus et al. (2008), the forestry industry had to recur to powerful and dangerous poisons that have no antidote in case of human poisoning. So, the difficulties to control this invading species in central Chile persist.

Currently, there is evidence that the problem posed by the "rabbit in the wild" has not been definitely resolved in central Chile. In the case of Tierra del Fuego, after a widespread alarm that included the arrival of hunters, traps, poisons, and introduction of predators, the Chilean State successfully implemented the introduction of the myxoma virus, at that time widely acknowledged worldwide for its efficiency to control the excessive growth of rabbit populations. On Juan Fernández archipelago, after years of international financing and national efforts, rabbits remain a species difficult to control and impossible to eradicate.

In a broader framework, the historical background examined here shows a lack of dialogue between scientists and decision-makers. We highlight the need to establish an effective and virtuous dialogue that can help prevent or control future invasions.

References

Anonymous (1870) Cría del conejo. Boletín de la Sociedad Nacional de Agricultura 1:168

Anonymous (1875) Cría del conejo. Boletín de la Sociedad Nacional de Agricultura 6:346

Anonymous (1914) Cartilla práctica sobre las enfermedades de árboles y cultivos, causadas por insectos y animales: remedios. Boletín de la Sociedad Nacional de Agricultura 45:528

Anonymous (1928) Consejos para la producción de piel. Boletín de la Sociedad Nacional de Agricultura 60:381–382

Anonymous (1930) Animales dañinos: procedimientos para exterminar la plaga de los conejos. Boletín de la Sociedad Nacional de Agricultura 62:198

Anonymous (1931) Expediente Decreto Ley N° 2.436, August 11

Anonymous. 1950a. Expediente Decreto ley N° 757., July 13

Anonymous (1950b) Exterminación total de la plaga de conejos. Boletín Ganadero 1:3

Arentsen P (1954) Control biológico del conejo. Boletín Ganadero 4:23–25

Biju-Duval C, Ennafaa H, Dennebouy N, Monnerot M, Mignotte F, Soriguer R, El-Gaiced AE, El-Hili A, Mounolou JC (1997) Mitochondrial DNA evolution in lagomorphs: origin of systematic heteroplasmy and organization of diversity in European rabbits. J Mol Evol 33:92–102

Bonino N, Gader R (1987) Expansión del conejo silvestre europeo (*Oryctolagus cuniculus* L.) en la República Argentina y perspectivas futuras. Anales del Museo de Historia Natural de Valparaíso 18:157–162

Brescia M (1979) Mares de leyenda. Talleres Gráficos García, Santiago, Chile

Camacho C (1929) El conejo silvestre. Boletín del Departamento de Agricultura 1:3

Camus P, Castro S, Jaksic F (2008) El conejo europeo en Chile: historia de una invasión biológica. Historia (Chile) 41:305–339

Castillo L. 1912. El conejo en libertad. Boletín de Bosques, Pesca y Caza (Chile) 1: 253

Castro SA, Figueroa JA, Muñoz-Schick M, Jaksic FM (2005) Minimum residence time, biogeographical origin, and life cycle as determinants of the geographical extent of naturalized plants in continental Chile. Divers Distribut 11:183–191

Chaparro M 1950. Plaga de Conejos invade Tierra del Fuego. Boletín Ganadero 1:2

Claro D (1950) Instituciones ganaderas aúnan sus esfuerzos para extirpar la plaga de conejos en la región. Boletín Ganadero 1:24
CONAF (1976) Plan de Manejo Parque Nacional Juan Fernández. Valparaíso, Chile
Congreso Nacional de Chile (1928). Cámara de Diputados, Diario de Sesiones. Publicación Oficial de la Cámara de Diputados, 41ª sesión ordinaria September 10th, 1.246
Echeverría C (1921) Conejos y conejeras: Lecciones prácticas sobre los medios de manejar esta industria. Impreso por el autor, Santiago, 102 pp
Figueroa J, Castro SA, Marquet P, Jaksic FM (2004) Exotic plant invasion to the Mediterranean region of Chile: causes, history and impacts. Rev Chil Hist Nat 77:465–483
Flux JEC (1994) World distribution. In: Thompson HV, King CM (eds) The European rabbit: the history and biology of a successful colonizer. Oxford University Press, Oxford, pp 8–21
Flux JEC, Fullagar PJ (1992) World distribution of the rabbit, *Oryctolagus cuniculus*, on islands. Mammal Rev 22:151–205
Fuentes N, Ugarte E, Kühn I, Klotz S (2008) Alien plants in Chile: inferring invasion periods from herbarium records. Biol Invasions 10:649–657
Gay C (1847) Historia física y política de Chile. Zoología. Santiago, Museo de Historia Natural de Santiago, 126
Gay C (1973) Agricultura chilena. ICIRA editores, Santiago
Greer JK (1989) Mamíferos de la provincia de Malleco, Angol. Editorial Museo Dillman S. Bullock, Angol
Housse R (1953) Animales salvajes de Chile en su clasificación moderna. Ediciones de la Universidad de Chile, Santiago
Iriarte JA, Jaksic FM (1986) The fur trade in Chile: an overview of seventy years of export data (1910–1984). Biol Conserv 38:243–253
Jaksic FM (1998) Vertebrate invaders and their ecological impacts in Chile. Biodivers Conserv 7:1427–1445
Jaksic FM, Fuentes E (1988) El conejo español: ¿un convidado de piedra? In Ecología del paisaje en Chile central: estudios sobre sus espacios montañosos. In: Fuentes E, Prenafeta S (eds) . Ediciones Universidad Católica, Santiago
Jaksic FM, Yáñez J (1980) ¿Quién controla las poblaciones de conejos introducidos? Medio Ambiente 4:41–44
Jaksic FM, Yáñez J (1983) Rabbit and fox introductions in Tierra del Fuego: history and assessment of the attempts at biological control of the rabbit infestation. Biol Conserv 26:367–374
Kahler W (1953) La isla Juan Fernández. Revista En Viaje Ferrocarriles del Estado de Chile) 238: 15
Lataste F (1892) A propos de lapins domestiques. Actas de la Societé Scientifique du Chili 2:210–222
Le Feuvre R (1885) Lecciones teórico-prácticas de agricultura y zootecnia: dadas a los alumnos de la Escuela Normal de Preceptores. Imprenta Excelsior, Valparaíso
Le Feuvre R (1900) Memoria del Ministerio de Industria y Obras Públicas, Santiago
Lees AC, Bell DJ (2008) A conservation paradox for the 21st century: the European wild rabbit *Oryctolagus cuniculus*, an invasive alien and an endangered native species. Mammal Rev 38:304–320
Mack RN, Simberloff D, Lonsdale M, Evans H, Clout M, Bazzaz FA (2000) Biotic invasions: causes, epidemiology, global consequences, and control. Ecol Appl 10:689–710
Mann G (1975) Observaciones sobre el estado actual de algunos representantes de flora y fauna en el Parque Nacional Juan Fernández. Boletín del Museo Nacional de Historia Natural 34:211
Mann G (1981) Análisis del plan de manejo y situación actual del Parque Nacional Juan Fernández. Informe 34 CONAF, Valparaíso
McNeely JA (2001) The great reshuffling: how alien species help feed the global economy. IUCN Biodiversity Policy Coordination Division
Miers N (1900) Invasión de los conejos. Boletín de la Sociedad Nacional de Agricultura 21:453–454
Molina JI (1788–1795) Compendio de la historia civil del Reyno de Chile. Imprenta de Sancha, Madrid

Molina JI. 1987. Ensayo sobre la historia natural de Chile. Ediciones Maule, CIUDAD?? Chile
Mulgrue J (1948) Una advertencia a Chile. In La Prensa Austral, Punta Arenas, June 19 1948
Ojeda P, González H, Araya G (2003) Erradicación del conejo europeo (*Oryctolagus cuniculus* Linnaeus, 1758) desde la Isla Santa Clara, Archipiélago de Juan Fernández. Informe Técnico 48, CONAF, Valparaíso
Philippi R (1885) Sobre los animales introducidos en Chile desde su conquista por los españoles. Anales de la Universidad de Chile 68:323–324
Pimentel D, Lach L, Zuniga R, Morrison D (1999) Environmental and economic costs of nonindigenous species in the United States. Bioscience 50:53–65
Programa de las Naciones Unidas, PNUD (2017) Catálogo de las especies exóticas asilvestradas /naturalizadas en Chile. Laboratorio de Invasiones Biológicas (LIB) Universidad de Concepción, Proyecto GEF/MMA/PNUD Fortalecimiento de los Marcos Nacionales para la Gobernabilidad de las Especies Exóticas Invasoras: Proyecto Piloto en el Archipiélago de Juan Fernández. Santiago de Chile
Reed C (1934) Aprovechemos la carne y la piel del conejo silvestre y con esto contribuiremos al bienestar nacional. Imprenta y Litografía La Ilustración, Santiago
Sabat E (1955) La mixomatosis no es una panacea infalible. Boletín Ganadero 45:23
Sáiz F, Ojeda P (1988) *Oryctolagus cuniculus* L. in Juan Fernández: problema y control. Anales del Museo de Historia Natural de Valparaíso 19:91-XXX
Santos-Tornero S (1875) De la cría y propagación de animales domésticos. Librería del Mercurio, Valparaíso
Thompson HV, Kong CM (1994) *The European Rabbit*. Oxford University Press, Oxford
Thompson H, King C (2004) The European rabbit: the history and biology of a successful colonizer. Oxford Science Publications, Oxford
Torres D, Aguayo A (1971) Algunas observaciones sobre la fauna del archipiélago de Juan Fernández. Boletín de la Universidad de Chile 112:34-XX
Vitousek PM, D'Antonio CM, Loope LL, Westbrooks R (1996) Biological invasions as global environmental change. Am Sci 84:218–228
Voigt J (1987) La destrucción de equilibrio biológico. Editorial Alianza, Madrid
Williams K (1995) Managing vertebrate pests: rabbits. CSIRO. Division of Wildlife and Ecology. Publisher Service, Canberra, Australia

Chapter 9
Invasive European Wild Rabbits (*Oryctolagus cuniculus*) in Argentina: State of the Art and Prospects for Research

S. Yasmin Bobadilla, Ricardo A. Ojeda, and M. Fernanda Cuevas

9.1 The Paradoxical European Wild Rabbit

The European wild rabbit, *Oryctolagus cuniculus*, is simultaneously a threatened species within its native range and yet, a successful colonizer worldwide (Lees and Bell 2008). The European rabbit is native to the Iberian Peninsula, south of France and north of Africa. In the Iberian Peninsula, its populations have undergone a massive decline during the twentieth century as a result of introduced pathogens (including myxoma virus and rabbit calicivirus), overhunting, habitat loss, and changes in land use (Lees and Bell 2008). The rabbit is a keystone species in the Mediterranean ecosystem of the Iberian Peninsula. For example, it is a key food source for more than 30 carnivorous species and a primary prey item for many of them, including critically endangered species such as the Iberian lynx (*Lynx pardinus*) and the Spanish imperial eagle (*Aquila adalberti*) (Delibes-Mateos et al. 2008). It also is an important ecosystem engineer for having the potential to modulate availability of resources for themselves and other organisms by building extensive open burrow systems (Gálvez-Bravo et al. 2008).

Paradoxically, the European rabbit is considered one among the 100 most harmful invasive alien species in the world due to its high plasticity and its impact on biological diversity and human activities (Lowe et al. 2000). The European rabbit is a unique case at global level, because no other invasive species is categorized as endangered in the Iberian Peninsula and, at the same time, viewed as an exotic and destructive threat to be eradicated in other places. The apparent contradiction that the European rabbit presents makes it an excellent model to study basic processes in

S. Y. Bobadilla (✉) · R. A. Ojeda · M. F. Cuevas
Biodiversity Research Group (GIB), National Council for Science & Technology of Argentina (CONICET) and the Institute for Aridland Research (IADIZA), Mendoza, Argentina
e-mail: ybobadilla@mendoza-conicet.gob.ar; rojeda@mendoza-conicet.gob.ar; fcuevas@mendoza-conicet.gob.ar

F. M. Jaksic, S. A. Castro, *Biological Invasions in the South American Anthropocene*, https://doi.org/10.1007/978-3-030-56379-0_9

population biology (i.e., life history, demographic models, and so on), evolution (e.g., rapid adaptive evolution), and ecology between its native and invasive ranges.

9.2 The European Wild Rabbit as an Invasive Species

Current generalizations about successful invaders are based on ecological traits of species and characteristics of vulnerability of invaded communities (Mack et al. 2000). Most successful invasive mammals exhibit some of these intrinsic and extrinsic attributes (Novillo and Ojeda 2008). The European rabbit shows remarkable ecological adaptability which is based on its broad diet, environmental tolerance, rapid dispersal, and high reproductive rate (Thompson and King 1994). Other hypotheses proposed about invasion success include predator and competitive release, a similar climate in native and invaded ranges (climate matching), and/or environmental disturbances in invaded habitats (e.g., an increase in annuals and decrease in perennials which typically follows extensive anthropogenic disturbance) (Lees and Bell 2008) (Table 9.1).

Table 9.1 Intrinsic, extrinsic, and human-related factors of the European rabbit, examples and explanations

Factors	Examples	Explanation
Intrinsic		
	Broad diet	Able to feed on different items
	Environmental tolerance	Type of habitat is not a limiting factor
	High dispersal capacity	Advantage to colonize new environments
	High reproductive capacity	Advantage to fast population increase and colonization
Extrinsic		
	Enemy release	Invader lacks its natural enemies (predators, pathogens) in the new habitat
	Climatic matching	Sets of species seem to be limited by climate
	Disturbed environments	Any relatively discrete event in time that increase resource availability and changes in habitats (e.g., herbivory, land-use change)
Associated to human		
	High propagule pressure	Many introduction events (frequency) and/or many individuals per introduction (supply)
	Old date introductions	More time for acclimatization
	Pathway of introduction	Food-valued organism

Furthermore, there are human-related factors that provide important insight into the understanding of successful colonization, such as the historical relationship between human activities and the structure and function of contemporary ecosystems and landscapes (i.e., historical ecology of the invasion) (Delibes and Delibes-Mateos 2015). European rabbits are edible animals that were semi-domesticated some thousands of years ago and have been transported to many places for human use; however, they only persisted in a few localized areas. Thereby, rabbits have been liberated on more than 800 different islands worldwide where this species has consolidated its global distribution (Flux and Fullagar 1992). Delibes and Delibes-Mateos (2015) argue that a high propagule pressure might have been necessary for successful colonization of European rabbit in several continental parts of the world (Table 9.1).

At a global scale, the economic impact of the European rabbit is associated with the high costs involved in its control, the cost of production loss due to vegetation degradation from the high grazing pressure it applies, reduced crop yields, and reduced domestic stock production (Williams et al. 1995). Besides, being a serious vertebrate pest for agricultural production, the European rabbit affects habitat structure, species composition, and ecosystem processes (Vázquez 2002). Particularly, rabbits change the structure of plant communities through herbivory and the structure of vertebrate communities through competition or via their role as a prey species (Courchamp et al. 2003; Davey et al. 2006). For example, in Chile the European rabbit along with cattle have altered the structure and functioning of the sclerophyllous scrub, generating a replacement of this ecosystem by savannas dominated by the native *Acacia caven* shrub (Holmgren 2002). Also, its warren-building behavior increases soil erosion and changes in soil biogeochemical properties (Eldridge and Koen 2008; Eldridge and Myers 2001, Crooks 2002). Rabbits directly compete for food with native herbivorous mammals (Jaksic 1998), birds (Courchamp et al. 2003), and livestock (Jaksic et al. 2002). European rabbits could also indirectly affect native fauna by facilitating mesopredator release and hyperpredation (Courchamp et al. 2000; Lees and Bell 2008).

9.3 History of European Rabbit Invasion and Expansion in Argentina

The history of introductions of the European rabbit in Argentina has been documented by several authors. Delibes and Delibes-Mateos (2015) documented from a sailor's diary that in 1765 a ship named Purísima Concepción shipwrecked near the coast of Tierra del Fuego and the crew introduced rabbits in the island with the intention of being supplied with food. However, nothing is known about the success of this early introduction. In 1880, Thomas Bridges and his sons introduced rabbits in several islands in the Beagle Channel (Bridges 1949, in Jaksic and Yáñez 1983) in order to provide food to castaways and yaganes (natives to the island) (Navas

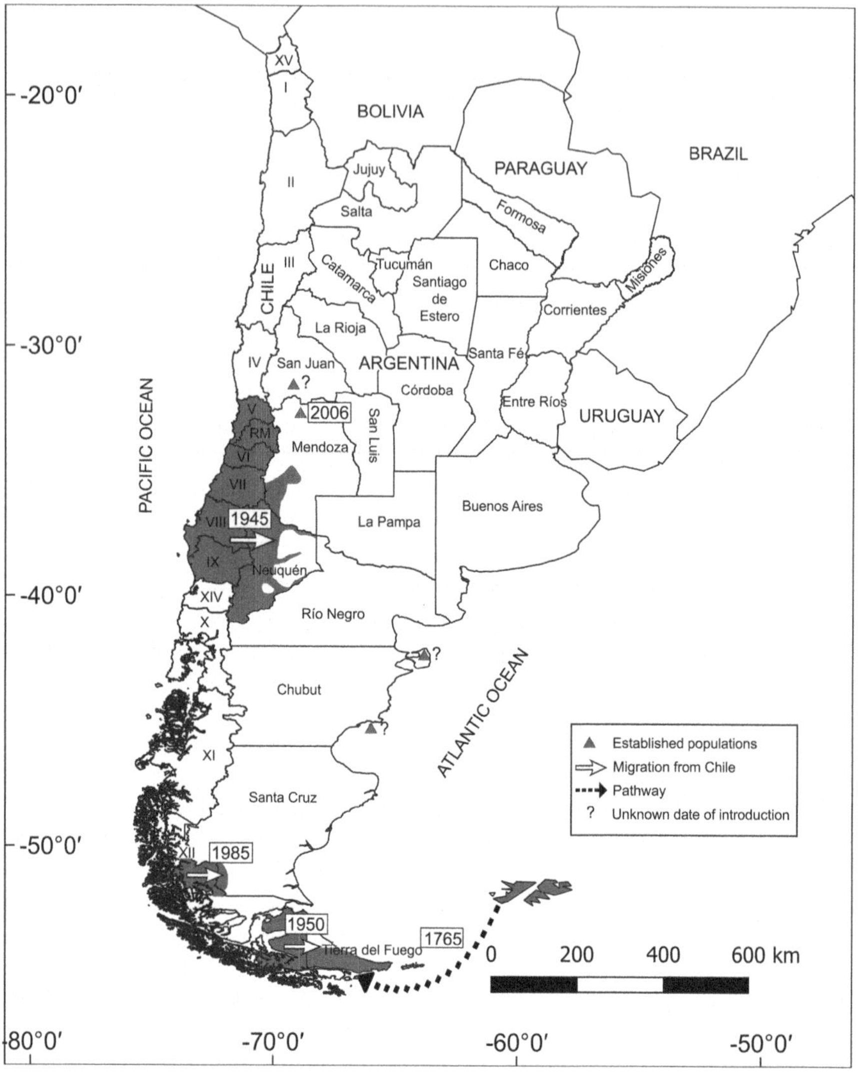

Fig. 9.1 Potential distribution of European rabbit in Argentina

1987). These individuals came from the Malvinas (or Falkland) islands, where they were introduced by French colonists around 1765 (Fig. 9.1). In addition, rabbits were introduced in Observatorio Island, near Ushuaia, in 1902 (Jaksic et al. 2002). A fourth introduction event was detailed by Amaya and Bonino (1981), Jaksic and Yáñez (1983), and Jaksic et al. (2002). In 1936, two pairs of rabbits were released near Porvenir city in Chile. They rapidly spread all over the northern part of Tierra del Fuego Island, arriving in Argentina's mainland and expanding east as far as San Sebastián Bay and beyond Río Grande city to the south (Fig. 9.1). In 1950,

Argentinian navy and private farmers released rabbits near Ushuaia city. Between 1950 and 1953, rabbit populations reached an abundance of 30 million individuals (30 individuals/ha). This was considered the worst invasion. Sheep were negatively affected by the impact of rabbits due to their building of burrows and to their grazing habits that leave the ground bare of grass. Therefore, sheep ranchers attempted to control these populations and started hunting and trapping them. Then, they used cyanide gas (that did not work). Later, in 1951, they introduced the gray fox (*Lycalopex griseus*) from the neighboring continental area, and finally, in 1954, they introduced myxoma virus from Brazil. All these actions reduced rabbit populations to a very low level. Another introduction event occurred in Tova and Tovita islands, Chubut province (Udrizar-Sauthier et al. 2017). However, the date of introduction, number of individuals, or status of populations is unknown (Fig. 9.1).

Populations of European rabbits established in five regions of continental Argentina: (1) south of San Juan, (2) north of Mendoza, (3) south of Mendoza and north of Neuquén, (4) east of Chubut, and (5) southwest of Santa Cruz province (Fig. 9.1).

(1) In the south of San Juan province, rabbits were detected in 2005 in El Leoncito National Park (Andino et al. 2008, in Laspina et al. 2013), but there is no information about the status of this species in the area.
(2) In 2006, the staff in the Villavicencio Nature Reserve intentionally released four rabbit couples which belonged to a breeding facility in the same Reserve (Cuevas et al. 2011). Rabbits spread along three corridors: Hornillos Ravine (northwest direction) which extends for 6 km, Darwin Ravine (west direction) which is 3.5 km long, and the third path towards Monte desert foothills (southeast direction) which stretches for 4 km. But after 4 years, rabbits were present only in the first two corridors. In this area, spread of rabbits would be related to movement along riverbed transects and across valleys, probably due to the influence of factors such as vegetation, soil, and moist places (Cuevas et al. 2011).
(3) Within the general area where the provincial borders of Neuquén and Mendoza meet, the first established rabbit population was seen between 1945 and 1950 in the headwaters of the Neuquén River, near Andacollo locality in Neuquén province (36° 80′ W) (Howard and Amaya 1975). These authors suggest that there is circumstantial evidence that rabbit populations in central Chile (brought from Spain) spread to Argentina crossing the Andes Cordillera through passes lower than 1800 m elevation. This dispersal probably occurred during summer, when environmental and habitat conditions are more suitable due to the presence of grass for food and shrubs for shelter (Jaksic et al. 2002). In 1969, Howard and Amaya (1975) recorded that rabbits crossed the Colorado River to the north (arriving in Mendoza province) and the Neuquén and Agrio rivers to the south. Thus, rabbits occupied 31,000 km^2 (with a dispersal rate of 16 km/year). The same authors also recorded that, in 1972, the invaded area increased by 3000 km^2 to the north, south, and east of the species' former range, reducing the rate of spread to 8 km/year. During 1975 and 1978, rabbits spread slowly to the east

and faster southwards reaching the Aluminé River, Neuquén province. Until 1978, the invaded area was 45,000 km^2 (83–114 individuals/ha; Bonino and Amaya 1984). Afterwards, Bonino and Gader (1987) found that, between 1982 and 1986, rabbits had advanced about 40 km eastwards and southeastwards. This is a dispersal rate of about 10 km/year. Towards the north, in Mendoza, rabbits arrived in the Grande and Malargüe rivers. By 1986, the area of expansion of rabbits covered 50,000 km^2 in Mendoza and Neuquén provinces. According to Bonino and Soriguer (2004), rabbits in Mendoza province occupied 11,000 km^2 in 2003, 77% higher than in 1986. A new update of the distribution of rabbits in this area was performed by Bonino and Soriguer (2009). They found that between 2006 and 2008, dispersal of this species occurred mainly along the banks of the Colorado, Neuquén, and Picún Luefú rivers, generally from east to west. In each case, the dispersal rate was 5 km/year. Furthermore, to the south of their range, rabbits spread 175 km farther than the distribution described in 1986 (dispersal rate: 6–9 km/year). The same authors also noticed the importance of rivers in the spread of rabbits in this region, especially in unfavorable habitats such as semiarid environments. Afterwards, Galende (2014) recorded the presence of rabbits in Nahuel Huapi National Park, 15 km farther south than the last record. Finally, between 2014 and 2015, Guichón et al. (2016) updated the distribution of rabbits in Neuquén province. In contrast to findings by Bonino and Soriguer (2009), they found that the rabbit expanded its range towards east and southeast, including in it the Nahuel Huapi and Lanín National Parks, and the Limay River valley; whereas rabbits moved to the central east region through the Neuquén River. Unlike Bonino and Soriguer (2009), Guichón et al. (2016) indicated absence of rabbits in the northwest of Neuquén province. In 2015, the total invaded area was 58,928 km^2, which means a 32% increase since 2008. Thereby, in this area, we can notice that rabbits are still spreading out using rivers as the main dispersal corridors.

(4) Nabte et al. (2009) recorded a new introduction of rabbits in Estancia La Irma, Chubut province. This ranch used rabbits as food supply, and several individuals were released in 2001. The authors confirmed this species' expansion to neighboring ranches.

(5) In 1985, rabbits were detected in Río Turbio city, in the southwest of Santa Cruz province, probably coming from Puerto Natales, Chile, where several animals were released in 1970 (Bonino and Soriguer 2009). Apparently, the population of Río Turbio is still restricted to that area.

9.4 Main Contributions to Ecology and Impacts in Argentina

In addition to the information described in the previous section about the distribution and expansion of rabbits in Argentina, from the 1980s onwards, there has been an increase in research on rabbits aiming to provide information about its diet,

Table 9.2 Main literature published on European rabbit in Argentina: invaded range, ecological traits, impacts, and management

Invaded range	Ecological traits	Impacts	Management	References
Northern Patagonia	Distribution range expansion			Guichón et al. (2016), Galende (2014), Cuevas et al. (2011), Bonino and Soriguer (2009), Bonino and Soriguer (2004), Navas (1987), Bonino and Gader (1987), Bonino and Amaya (1984), Howard and Amaya (1975)
	Body parameters			Donadio et al. (2005), Bonino and Donadio (2010)
	Genetics			Bonino and Soriguer (2008)
	Diet			Bonino and Borrelli (2006)
		competition for resources		Bonino (2006), Galende (2014) Bobadilla et al. (2020)
		Exotic prey		Barbar et al. (2018), Barbar et al. (2016), Novaro et al. (2004), Donázar et al. (1997), Hiraldo et al. (1995)
		Plant community		Veblen et al. (2004)
Southern Patagonia	Distribution range expansion			Nabte et al. (2009), Udrizar Sauthier et al. (2017)
	Body parameters			Amaya and Bonino (1981)
	Genetics			Bonino and Soriguer (2008)
	Diet	Cultural		Amaya and Bonino (1981), Valenzuela et al. (2013)
			Disease (myxomatosis)	Aparicio et al. (2006), Aparicio et al. (2004)
San Juan province	Diet			Laspina et al. (2013)

reproduction, morphology, diseases (myxomatosis) and competition with other herbivores. A synthesis of research on the European rabbit in Argentina is presented in Table 9.2.

9.4.1 Ecological Traits

In Argentina, the only published study addressing some aspects of body parameters and sexual dimorphism of the European rabbit in Patagonia was conducted in the Andean region of Neuquén province (Bonino and Donadío 2010). The study showed a clear sexual dimorphism between adult females and adult males, with females being significantly heavier and larger than males. Additionally, rabbits from Argentina were significantly heavier than rabbits inhabiting both their original (Europe) and invaded ranges (Australia and Chile). The authors suggest that low predation rates and/or the differential expression of genetic traits could explain the observed pattern. These results are consistent with the large body size and highly variable coat coloration observed in Argentina's rabbits (Bonino and Soriguer 2008). In this sense, a study of mitochondrial DNA reveals that the foundational stock of the populations in Patagonia comes from the domestic rabbit *O. cuniculus* (Bonino and Soriguer 2008). However, deciding what factor explains the observed pattern will require additional studies. On the other hand, Donadio et al. (2005) presented a simple method to estimate body mass and relative age of European rabbits that will yield new insights on prey selection patterns and on the feeding ecology of native predators.

Various studies have addressed the botanical composition and seasonal variations in the diet of European rabbits in Argentina. In the central Andean region of Neuquén province, grasses represented the main basis of the diet throughout the year (46%), where *Poa pratensis* and *Festuca pallescens* were the most consumed species (Bonino and Borelli 2006). Graminoids were the second most important group (28%), with *Juncus balticus* and *Carex gayana* as major nutritional items. Forbs had moderate participation in the spring and summer diets, whereas shrubs and tree species were relatively important during the winter (Bonino and Borelli 2006). According to this study, the rabbit proved to be mainly a grazer with wet meadows or "mallines" being its main feeding areas, although its use of space varies according to the seasons of the year. Amaya and Bonino (1981) found similar results for Tierra del Fuego province. These authors recorded grasses as the most important food in the diet, followed by graminoid plants (*Cyperaceae* and *Juncaceae*). The genus *Poa* spp. was the most important item within the first group, and *Carex* spp. within the second (Amaya and Bonino 1981). In San Juan province, an arid environment, grasses and shrubs were the most consumed food items in both wet and dry seasons. In this environment, the European rabbit did not consume the most abundant plant categories and species (Laspina et al. 2013). In conclusion, in Argentina, rabbits feed mainly on grasses, but when these are scarce, they consume woody vegetation, behaving like opportunist grazers by adjusting their diet to the available food supply.

There are no specific studies published in Argentina addressing habitat use by rabbits. The available data show that, in the western part of the distribution range in Neuquén province, the European rabbit occupies sub-Antarctic forests and grasslands. However, in the eastern sector, in a semi-desert environment with scrub formations, the rabbit presents the most irregular distributional pattern and is found in places alongside rivers and small streams (Bonino and Soriguer 2009). This was also observed for an arid environment in Mendoza and San Juan provinces, where distribution of the rabbit is restricted to wetlands or places with streams or moister sites (Laspina et al. 2013; Cuevas et al. 2011). Similar results were found in the south of Mendoza where the rabbit positively selects wetlands or riparian habitats

(Bobadilla et al. unpublished data). An important point is to consider watercourses as spreading routes for this invasive species, especially in arid habitats (Bonino and Soriguer 2004).

9.4.2 Impacts

Just like on the global scale, diet studies in the northwest of Patagonia showed food competition with domestic livestock, increased grazing pressure, and, accordingly, a decrease in the carrying capacity of agricultural land (Bonino 2006; Williams et al. 1995; Travers et al. 2017). Studies conducted in native forests of *Nothofagus pumilio* in Argentina's Patagonia showed that European rabbit's browsing can prevent regeneration of these forests (Veblen et al. 2004, and references therein). On the other hand, there is trophic overlap with a native herbivore, the mountain viscacha (*Lagidium viscacia*), and the rabbit represents a threat to its colonies in situations of food scarcity (Galende 2014). Rabbits are prey for both native and exotic predators in Argentina and could indirectly contribute to the decline of native fauna by becoming a subsidy resource for predators when native prey is scarce (Valenzuela et al. 2013). For example, in the arid Patagonian landscapes, the European hare (*Lepus europaeus*) and rabbit account for 55% of the prey biomass of the Great Horned owl (*Bubo magellanicus*) and for a 58% frequency in the diet of the Gray Buzzard-eagle (*Geranoaetus melanoleucus*) (Hiraldo et al. 1995; Donázar et al. 1997). This finding was consistent with current studies confirming that these exotic lagomorphs account for a high percentage of the biomass consumed by native carnivores in South America (Novaro et al. 2004; Barbar et al. 2016). Besides, the European rabbit may play a role in facilitating establishment of exotic predators (Barbar and Lambertucci 2018). Finally, rabbits have cultural impacts given their preference for digging their burrows on archeological sites along the coastline of the Beagle Channel (Valenzuela et al. 2013, and references therein).

9.4.3 Management

Myxomatosis and rabbit hemorrhagic disease are the major viral diseases in the Iberian Peninsula and significantly affect natural populations of European rabbits (Pacios-Palma et al. 2016). Myxomatosis is a lethal disease for the European rabbit, and even though it is vectored by fleas (*Spilopsyllus cuniculi*) and mosquitoes (*Anopheles annulipes*, *Culex annulirostris*), it can also be transmitted between rabbits by respiratory tract (Fenner and Ross 1994). Given this fact, this disease can lead to decline and extinction of wild populations. This virus has been used as biological control of exotic rabbit populations in Australia and Europe (Kerr 2012). By 1954, the myxoma virus, brought from Brazil, was introduced in Chile and succeeded in bringing rabbits to low population levels on Tierra del Fuego Island

(Jaksic and Yáñez 1983). In Argentina, its use is not allowed by the National Food Safety and Quality Service (Senasa) (Bonino and Amaya 1984). However, there are studies on the dynamics of the myxoma-*Oryctolagus* system which provide guidelines determining potential strategies for control of European rabbits in Argentina (Aparicio et al. 2004; Aparicio et al. 2006). These studies showed that the usual strategy, consisting in introducing highly virulent strains into rabbit populations, might not be the optimal, whereas a control strategy based on introduction of intermediate virulence strains can be much more effective.

9.5 What happens with European Rabbits in Argentina?

9.5.1 Anomalous Situation in Argentina's Northern Patagonia

European rabbits have successfully established in the wild in central Chile. From there, they slowly expanded their distribution across the Andes into Neuquén and Mendoza provinces (see above). Thus, the rate of spread for Neuquén and Mendoza varies between 2 and 10 km/year, depending on the environmental conditions of each valley. The lowest rate (2 km/year) was recorded in the western area with mountainous topography and forest vegetation, while the highest rate (10 km/year) was recorded in the eastern part with flatter topography and along watercourses (Bonino and Gader 1987; Bonino and Soriguer 2004). The dispersal rate in Australia varies between 15 km/year in denser woodlands of the eastern and southern regions and 300 km/year along the drainage channels in the Simpson Desert (Myers et al. 1994). Henzell et al. (2008) consider that there exists an anomalous situation for European rabbits in South America, particularly in Argentina's Northern Patagonia, where there is a low rate of dispersal compared to Australia. According to these authors, this anomalous situation may have arisen as a result of several factors: (1) Possible competition between rabbits and South America's diverse caviomorph rodent fauna (including species similar to rabbits in size, dietary preferences, and burrowing habits) and presence of native predators adapted to catching those caviomorph rodents; (2) potential competition with European hares (*Lepus europaeus*) because both species might occupy most of the other's habitat in its absence; (3) presence in South America of an undiscovered Biological Control Agent in non-lagomorphs that can cross the species barrier into European rabbits (e.g., a coccidian species was recently described in domestic rabbits in Argentina); (4) high probability that the origin of wild rabbits in South America was domestic (feral populations of domestic rabbits did not irrupt dramatically); and (5) high prevalence of C4 grass species, disadvantageous to an herbivorous small mammalian *r*-strategist like the rabbit (C3 grasses predominate in Mediterranean Europe and Australia) (Henzell et al. 2008). However, little is known about support for any of these possible explanations. Updated reports on expansion of the invaded area in Neuquén province showed that European rabbits are the invasive mammals presenting the largest area of occupation at this site, showing an active process of geographical expansion (see Fig. 9.1 and related text) (Guichón et al. 2016).

9.5.2 *Successful Establishment in the Primary Source*

When propagules are transported to a novel range, there could be a match between their native habitat and at least one habitat in the area of introduction – habitat compatibility – that enables their survival at the initial stages of invasion (Steinmaus 2011). The abundance of an introduced organism or its geographical extent in the invaded range may be influenced by a combination of factors in the initial establishment phases (e.g., propagule pressure, minimum residence time, species ecological requirements) (Thuiller et al. 2006). In this way, Sax (2001) showed that the invaded range of an introduced species is determined by the same kind of ecological forces as those limiting the native range of such species. Particularly, alien mammals display good climate matching, that is to say, occupy similar ecoregions to their native ranges (Novillo and Ojeda 2008). A good example is the establishment of a European rabbit population in central Chile, where climate matches that of its native range (Mediterranean-type climate) (Fig. 9.2). This population expanded its geographic distribution both to the north and south of central Chile towards new habitat types. Their northernmost distribution boundary is Quebrada Honda Bay (28° S latitude, tropical ecoregion/subtropical steppe), and their southernmost one is Paillaco town (40° S latitude, subtropical ecoregion) (Jaksic et al. 2002). When the rabbit population expanded its range towards Argentina (see Fig. 9.1), it initially established two

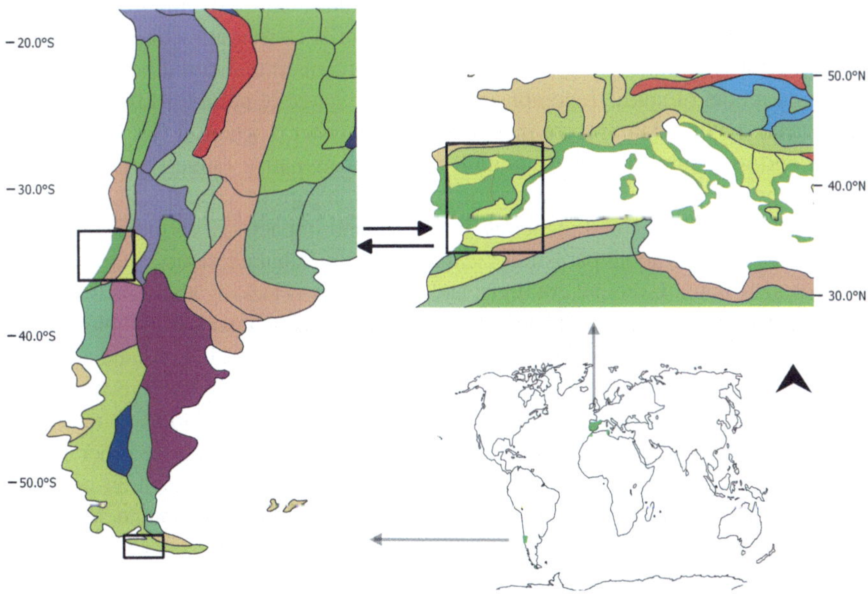

Fig. 9.2 Representation of native (Iberian Peninsula, south of France and north of Africa) and invaded (central Chile) ecoregions showing their good climate matching (Mediterranean-type climate)

different environments: one to the west where rainy Mediterranean climate prevails and another one to the east with semiarid Mediterranean characteristics (Bonino and Amaya 1984). Therefore, the main invaded range in Argentina also shows a climate regime similar to that of the native range (Fig. 9.2). A separate introduction of European rabbits occurred on Tierra del Fuego Island, despite this region lying at a higher latitude (54° S), and presents a cold, Steppe-like climate (Fig. 9.2). According to Flux (1994), the limiting factors appear to be snow cover depth at high latitudes and access to either water or green vegetation at low latitudes. As in its native range, in central Chile and Tierra del Fuego Island, growing seasons are cool and C3 plants predominate. In Australia, the success of European rabbits is believed to be in part the result of replacement of low-quality C4 plants by high-quality annual C3 plants (Henzell et al. 2008). These characteristics are repeated throughout Argentina's Patagonia, and, in this sense, the whole area is suitable for invasion by this species (Bonino and Amaya 1984).

9.6 Prospects for Research

The European rabbit probably is one of the most studied mammals, mainly in the Iberian Peninsula, England, Australia, and New Zealand. This is because, on the one hand, it is a species at conservation risk within its native range and, on the other hand, it is one of the most harmful invasive mammals. Currently, the flow of information is reciprocal between research for the rabbit's conservation and for its control. Thus, it is an interesting and very suitable organism that could be considered both as a model of study to demonstrate the wide range of complex effects that an introduced mammalian species may exert on ecosystems where it has been introduced and as a subject of conservation efforts in its native range (Lees and Bell 2008). However, currently, there are still information gaps in Argentina about its use of habitat, population trend, genetics, behavior, and impacts. Even more, there are no published works about parasitological and zoonotic characteristics, management strategies, or potential social conflicts, and knowledge is lacking about its introduction history and expansion (Bobadilla et al., unpublished manuscript) (Table 9.2). It is important to conduct more fundamental studies of its natural history and autecology, essential to answer complex ecological questions, anticipate their responses to invaded environments, and generate efficient protocols for its monitoring and management. Although there are some studies assessing niche overlap with other exotic herbivores, it is necessary to quantify more ecological impacts such as possible changes in abundance of native species (competition for resources, enhancing predator populations) and modification of vegetation and soil structure, which reduce pastoral productivity and eventually increase landscape degradation. Another of the associated problems, unknown in Argentina, is the possible risk of zoonosis, because rabbits are hosts of various parasites and can also act as natural reservoirs of infectious agents of some diseases (Shaughnessy et al. 2013; Ryll et al. 2018). In conclusion, future studies should be aimed to better understand the invasiveness of the

European rabbit in Argentina and should determine what key factors could limit or favor its expansion. Furthermore, it is important to quantify and categorize its impacts to improve management strategies and risk assessments.

References

Amaya J, Bonino NT (1981) El conejo silvestre europeo (*Oryctolagus cuniculus*) en Tierra del Fuego. Revista de Información sobre Investigación y Desarrollo Agropecuario 388:14–31

Aparicio JP, Solari HG, Bonino NA (2004) Competition and coexistence in host-parasite systems: the myxomatosis case. Popul Ecol 46:71–85

Aparicio JP, Solari HG, Bonino NA (2006) Perspectivas teóricas sobre la dinámica de la mixomatosis con aplicaciones en control biológico. Ecol Austral 16:15–28

Barbar F, Hiraldo F, Lambertucci SA (2016) Medium-sized exotic prey create novel food webs: the case of predators and scavengers consuming lagomorphs. PeerJ 4:e2273

Barbar F, Lambertucci SA (2018) The roles of leporid species that have been translocated: a review of their ecosystem effects as native and exotic species. Mammal Rev 48:245–260

Bobadilla SY, Marchetta A, Dacar MA, Ojeda RA, Cuevas MF (2020). Food habits of European rabbit and its role as seed dispersal of two Mosqueta roses: Facilitation among non-native species in a semiarid protected area of Argentina? Biological Invasions 22:1565–1571

Bonino NA, Amaya J (1984) Distribución geográfica, perjuicios y control del conejo silvestre europeo *Oryctolagus cuniculus* (L.) en la República Argentina. Revista de Información sobre Investigación y Desarrollo Agropecuario 429:25–50

Bonino NA, Gader R (1987) Expansión del conejo silvestre europeo (*Oryctolagus cuniculus* L.) en la República Argentina y perspectivas futuras. Anales Museo Historia Natural de Valparaíso 18:157–162

Bonino NA, Soriguer RC (2004) Distribución actual y dispersión del conejo europeo (*Oryctolagus cuniculus*) en Mendoza (Argentina). Mastozoología Neotropical 11:237–241

Bonino NA, Borrelli L (2006) Variación estacional de la dieta del conejo silvestre europeo (*Oryctolagus cuniculus*) en la región andina de Neuquén, Argentina. Ecol Austral 16:7–13

Bonino NA (2006) Interacción trófica entre el conejo silvestre europeo y el ganado doméstico en el noroeste de la Patagonia Argentina. Ecol Austral 16:135–142

Bonino NA, Soriguer R (2008) Genetic lineages of feral populations of the *Oryctolagus cuniculus* (Leporidae, Lagomorpha) in Argentina. Mammalia 72:355–357

Bonino NA, Soriguer R (2009) The invasion of Argentina by the European wild rabbit *Oryctolagus cuniculus*. Mammal Rev 39:159–166

Bonino NA, Donadio E (2010) Parámetros corporales y dimorfismo sexual en el conejo silvestre europeo (*Oryctolagus cuniculus*) introducido en Argentina. Mastozoología Neotropical 17:123–127

Bridges EL (1949) Uttermost part of the Earth. E. P. Dutton and Co., New York

Courchamp F, Langlais M, Sugihara G (2000) Rabbits killing birds: modelling the hyperpredation process. J Anim Ecol 69:154–164

Courchamp F, Chapuis JL, Pascal M (2003) Mammal invaders on islands: impact, control and control impact. Biol Rev 78:347–383

Crooks JA (2002) Characterizing ecosystem-level consequences of biological invasions: the role of ecosystem engineers. Oikos 97:153–166

Cuevas MF, Chillo V, Marchetta A, Ojeda RA (2011) Mammalia, Lagomorpha, Leporidae, *Oryctolagus cuniculus* Linnaeus, 1758: New record and its potential dispersal corridors for northern Mendoza, Argentina. Check List 7:565–566

Davey C, Sinclair ARE, Pech RP, Arthur AD, Krebs CJ, Newsome AE, Hik D, Molsher R, Allcock K (2006) Do exotic vertebrates structure the biota of Australia? An experimental test in New South Wales. Ecosystems 9:992–1008

Delibes-Mateos M, Delibes M, Ferreras P, Villafuerte R (2008) Key role of European rabbits in the conservation of the Western Mediterranean basin hotspot. Conserv Biol 22:1106–1117

Delibes R, Delibes-Mateos M (2015) Linking historical ecology and invasion biology: some lessons from European rabbit introductions into the new world before the nineteenth century. Biol Invasions 17:2505–2515

Donadio E, Pauli JN, Bonino NA (2005) A method to estimate body mass and relative age of exotic lagomorphs in the southern Neotropics. Acta Theriol 50:81–89

Donázar JA, Travaini A, Ceballos O, Delibes M, Hiraldo F (1997) Food habits of the great horned owl in northwestern Argentine Patagonia: the role of introduced lagomorphs. J Raptor Res 31:364–369

Eldridge DJ, Myers CA (2001) The impact of warrens of the European rabbit (*Oryctolagus cuniculus* L.) on soil and ecological processes in a semi-arid Australian woodland. J Arid Environ 47:325–337

Eldridge DJ, Koen TB (2008) Formation of nutrient-poor soil patches in a semi-arid woodland by the European rabbit (*Oryctolagus cuniculus* L.). Austral Ecol 33:88–98

Fenner F, Ross J (1994) Myxomatosis. In: Thompson HV, King CM (eds) The European rabbits, the history and biology of a successful colonizer. Oxford University Press, Oxford, pp 205–240

Flux JEC, Fullagar PJ (1992) World distribution of the Rabbit *Oryctolagus cuniculus* on islands. Mammal Rev 22:151–205

Flux JEC (1994) World distribution. In: Thompson HV, King CM (eds) The European rabbit: the history of a successful colonizer. Oxford University Press, Oxford, pp 8–21

Galende GI (2014) *Oryctolagus cuniculus* Linnaeus, 1758 (Mammalia: Lagomorpha: Leporidae): new record in the Nahuel Huapi National Park, Patagonia, Argentina. Check List 10:1179–1183

Gálvez-Bravo L, Belliure J, Rebollo S (2008) European rabbits as ecosystem engineers: warrens increase lizard density and diversity. Biodivers Conserv 18:869–885

Guichón ML, Monteverde M, Piudo L, Sanguinetti J, Di Martino S (2016) Mamíferos introducidos en la provincia de Neuquén: estado actual y prioridades de manejo. Mastozoología Neotropical 23:255–265

Henzell RP, Cooke BD, Mutze GJ (2008) The future biological control of pest populations of European rabbits, *Oryctolagus cuniculus*. Wildl Res 35:633–650

Hiraldo F, Donázar JA, Ceballos O, Travaini A, Bustamante J, Funes M (1995) Breeding biology of a grey eagle-buzzard population in Patagonia. The Wilson Bulletin 107:675–685

Holmgren M (2002) Exotic herbivores as drivers of plant invasion and switch to ecosystem alternative states. Biol Invasions 4:25–33

Howard WE, Amaya JN (1975) European rabbits invades western Argentina. J Wildl Manag 39:757–761

Jaksic FM, Yáñez J (1983) Rabbit and fox introductions in Tierra del Fuego: history and assessment of the attempts at biological control of the rabbit infestation. Biol Conserv 26:367–374

Jaksic FM (1998) Vertebrate invaders and their ecological impacts in Chile. Biodivers Conserv 7:1427–1445

Jaksic FM, Iriarte JA, Jiménez JE, Martínez DR (2002) Invaders without frontiers: cross-border invasions of exotic mammals. Biol Invasions 4:157–173

Kerr PJ (2012) Myxomatosis in Australia and Europe: a model for emerging infectious diseases. Antiviral Res 93:387–415

Mack RN, Simberloff D, Lonsdale MW, Evans H, Clout M, Bazzaz FA (2000) Biotic invasions: causes, epidemiology, global consequences, and control. Ecol Appl 10:689–710

Myers K, Parer I, Wood D, Cooke BD (1994) The rabbit in Australia. In: Thompson HV, King CM (eds) The European rabbit: the history of a successful colonizer. Oxford University Press, Oxford, pp 108–157

Nabte MJ, Saba SL, Monjeau A (2009) Mamíferos terrestres de la Península Valdés: lista sistemática comentada. Mastozoología Neotropical 16:109–120

Navas JA (1987) Los vertebrados exóticos introducidos en la Argentina. Revista del Museo Argentino de Ciencias Naturales, Serie Zoología XIV, pp 7–38

Novaro AJ, Funes MC, Jiménez JE (2004) Patagonian foxes selection for introduced prey and conservation of culpeo and chilla foxes in Patagonia. In: Macdonald DW, Sillero-Zubiri C (eds) The biology and conservation of wild canids. Oxford University Press Canada, Don Mills, pp 239–250

Novillo A, Ojeda RA (2008) The exotic mammals of Argentina. Biol Invasions 10:1333–1334

Laspina CA, Ortiz SG, Reus ML, De Los Ríos C (2013) Dieta de *Oryctolagus cuniculus* (conejo silvestre europeo) y disponibilidad de recursos tróficos en el Parque Nacional El Leoncito (San Juan, Argentina). Multequina (Argentina) 22:37–44

Lees AC, Bell DJ (2008) A conservation paradox for the 21st century: The European wild rabbit *Oryctolagus cuniculus*, an invasive alien and an endangered native species. Mammal Rev 38:304–320

Lowe S, Browne M, Boudjelas S, De Poorter M (2000) 100 of the world's worst invasive alien species: A selection from the Global Invasive Species Database. Published by The Invasive Species Specialist Group (ISSG) a specialist group of the Species Survival Commission (SSC) of the World Conservation Union (IUCN), Auckland

Pacios-Palma I, Santoro S, Bertó-Moran A, Moreno S, Rouco C (2016) Effects of myxoma virus and rabbit hemorrhagic disease virus on the physiological condition of wild European rabbits: Is blood biochemistry a useful monitoring tool? Res Vet Sci 109:129–134

Ryll R, Eiden M, Heuser E, Weinhardt M, Ziege M, Höper D, Groschup MH, Heckel G, Johne R, Ulrich RG, R. G. (2018) Hepatitis E virus in feral rabbits along a rural-urban transect in Central Germany. Infect Genet Evol 61:155–159

Sax DF (2001) Latitudinal gradients and geographic ranges of exotic species: implications for biogeography. J Biogeogr 28:139–150

Shaughnessy LJ, Smith LA, Evans J, Anderson D, Caldow G, Marion G, Low JC, Hutchings MR (2013) High prevalence of paratuberculosis in rabbits is associated with difficulties in controlling the disease in cattle. Vet J 198:267–270

Steinmaus S (2011) Habitat compatibility. In: Simberloff D, Rejmánek M (eds) Encyclopedia of biological invasions. University of California Press, Berkeley/Los Angeles, pp 305–309

Thompson HV, King CM (1994) The rabbit in Britain. In: The European rabbit: the history of a successful colonizer. Oxford University Press, Oxford, pp 64–107

Thuiller W, Richardson DM, Rouget M, Proches S, Wilson JR (2006) Interactions between environment, species traits, and human uses describe patterns of plant invasions. Ecology 87:1755–1769

Udrizar Sauthier DE, Pazos GE, Cheli GH, Coronato F (2017) Mamíferos terrestres en islas del Atlántico Sudoccidental, Patagonia, Argentina. Mastozoología Neotropical 24:251–256

Valenzuela AE, Anderson CB, Fasola L, Cabello JL (2013) Linking invasive exotic vertebrates and their ecosystem impacts in Tierra del Fuego to test theory and determine action. Acta Oecol 54:110–118

Vázquez DP (2002) Multiple effects of introduced mammalian herbivores in a temperate forest. Biol Invasions 4:175–191

Veblen TT, Kitzberger T, Villalba R (2004) Nuevos paradigmas en ecología y su influencia sobre el conocimiento de la dinámica de los bosques del sur de Argentina y Chile. In: Arturi MF, Frangi JL, Goya JF (eds) Ecología y Manejo de los Bosques de Argentina. Editorial de la Universidad Nacional de La Plata, La Plata, Argentina, p 19

Williams CK, Parer I, Coman BJ, Burley J, Braysher ML (1995) Managing vertebrate pests: rabbits. Australian Government Publishing Service, Canberra

Chapter 10
Wild Boar Invasion in Argentina and Chile: Ecology, Impacts, and Distribution

M. Fernanda Cuevas, Sebastián A. Ballari, Ricardo A. Ojeda, and Oscar Skewes

10.1 Ungulates as Invasive Species

About 2.6% of extant terrestrial mammals are "successful invaders." These include the artiodactyls (pigs, camels, deer, cattle, sheep, goats, and antelope), carnivores (canids, bears, mustelids, and cats), lagomorphs (rabbits and hares), and perissodactyls (equines). In particular, the large ungulates (artiodactyls and perissodactyls) represent the highest proportion of invasive mammals, the family Cervidae being the most successful because of their aesthetic and economic appeal as big game animals (Clout and Russell 2008).

The impact of introduced ungulates involves direct herbivory affecting not only plant composition and regeneration but also a decrease in above-ground biomass production causing a change in disturbance regimes in fire-prone ecosystems and changes in mineralization rates of soils (Mack and D'Antonio 1998; Frank and Groffman 1998; Blair et al. 2014). Furthermore, they generate soil disruption by the creation of wallows, trails, and hoof print (Friend and Taylor 1984). On the other hand, indirect effects implicate facilitation of other introduced and invasive species such as dispersal by endozoochory, resource competition, and alteration of native

M. F. Cuevas (✉) · R. A. Ojeda
Grupo de Investigaciones de la Biodiversidad (GiB), IADIZA, CCT Mendoza CONICET, Mendoza, Argentina
e-mail: fcuevas@mendoza-conicet.gob.ar; rojeda@mendoza-conicet.gob.ar

S. A. Ballari
Parque Nacional Nahuel Huapi (CENAC-APN). CONICET, San Carlos de Bariloche, Río Negro, Argentina

O. Skewes
Facultad de Ciencias Veterinarias, Universidad de Concepción, Campus Chillán, Chillán, Chile
e-mail: oskewes@udec.cl

F. M. Jaksic, S. A. Castro, *Biological Invasions in the South American Anthropocene*, https://doi.org/10.1007/978-3-030-56379-0_10

trophic webs (Roemer et al. 2002; Bergstrom et al. 2009; Forsyth et al. 2010; D'hondt et al. 2012; Nuñez et al. 2013; Blossey and Gorchov 2017).

Twenty-five out of 37 introduced non-domestic mammals in South America were successful in establishing wild populations, and most of them occur in temperate ecosystems of the South American southern cone (Long 2003; Novillo and Ojeda 2008; Ballari et al. 2016). The ungulates represent the most important group of introduced invasive mammals in Argentina (Novillo and Ojeda 2008; Guichón et al. 2016; Relva and Sanguinetti 2016). Among these, the red deer (*Cervus elaphus*) and wild boar (*Sus scrofa*) are among the most successful invaders regarding their population abundances and expansion of distributional ranges into different biomes (Flueck et al. 2003; Ballari et al. 2015a; Skewes and Jaksic 2015).

Sus scrofa is native from Eurasia and north of Africa (Oliver et al. 1993), and since European colonization, it has been established in each continent except in Antarctica (Long 2003). It has one of the most extensive distributions of all mammals (Oliver et al. 1993). Its presence in localities far from its original distribution has been due, mainly, by human activities. The wild boar is included in the list of the 100 most dangerous invasive species because of their impact on biodiversity, human activities, and in important issues associated with the process of biological invasions (Lowe et al. 2000). As an example, in the United States, invasive feral pigs occur in 35 states, and their economic impact represents billions of US dollars in annual losses (Pimentel et al. 2005; Corn and Jordan 2017; Beasley et al. 2018).

Wild boars have been reported as major agents of perturbation on livestock, agriculture, native plants and animals, and soil and as a vector of zoonotic viruses (i.e., hepatitis E virus, trichinellosis, and swine influenza virus, among others) (Bratton 1975; Massei and Genov 2004; Miller et al. 2017). Among the attributes for a good invader, we highlight its large size, a generalist and opportunistic omnivorous diet (Ballari and Barrios-García 2014), a breeding age under 1 year, more than one litter per year of 4–10 piglets (Novillo and Ojeda 2008), and behavioral plasticity (Podgórski et al. 2013). These intrinsic invasive traits combined with extrinsic factors of the invaded environment (i.e., reduced predator populations, absence of ecological equivalent species, and increased land irrigation and agricultural production, among others) have resulted into rapid population growth, colonization, and establishment in new habitats (Massei and Genov 2004; Barrios-García and Ballari 2012; Bevins et al. 2014). The aim of this chapter is to review the general ecology, distribution, habitat characteristics, and environmental impact of *Sus scrofa* (feral pigs and wild boar) in Argentina and Chile.

10.2 Feral Populations of Domestic Pigs in Argentina and Chile

Carpinetti et al. (2016) documented that domestic pigs arrived to Argentina in 1536, brought by the Spanish conquistadors. By the end of the sixteenth century, the number of animals was very large, so it was agreed that neighbors complied with the ordinance of having the pigs locked in a corral. The population kept raising free and eventually became feral pigs. In the mid-eighteenth century, feral pigs had spread southwards and reached the present town of Balcarce, Buenos Aires province.

In Chile, domestic pigs (*Sus scrofa domestica*) became feral in the southern region of Chilean Tierra del Fuego Island in the 1900s (Aravena et al. 2015). Jaksic and Castro (2010) citing Gay (1847) pointed out that domestic pigs were registered in Llanquihue between 1830 and 1841, 60 years before what has been considered. Currently, they inhabit part of Karukinka Natural Park (53° 42' S–69° 18' W), an ecotonal area of global conservation concern. The analysis of genetic variation in the mitochondrial DNA control region revealed two highly differentiated populations, one in the western and the other in the eastern area of the park, each harboring a different haplotype, suggesting no connectivity between populations. Comparison of these haplotypes with reference sequences from other countries and commercial breeds indicated that feral pigs from Chilean Tierra del Fuego are of European origin, very likely from two separate introduction events. The haplotype found in the western feral population was also identified in domestic pigs from a farm. This raises concerns regarding the possible connectivity between stocks from local farms and the wild population (Aravena et al. 2015).

10.3 Invasion History of Wild Boar in Argentina

Data of the introductions of wild boar in Argentina are scarce. The first records date from 1904 to 1906, when the owner of San Huberto ranch (36° 54' S–64° 15' W), La Pampa province, brought several individuals for hunting purposes and, subsequently, some of them managed to escape (Daciuk 1978; Navas 1987). The origin of these animals probably came from a Siberian stock characterized by large size and with rustic features (Daciuk 1978). From that location, several individuals were taken to Collun-có ranch (39° 58' S–71° 10' W), Neuquén province, between 1917 and 1922. It is not well known, but either the individuals escaped or were released and then dispersed towards Lanin and Nahuel Huapi National Parks (40° 09' S–71° 21' W and 41° 00' S–71° 30' W, respectively), Río Negro province. Other individuals were introduced between 1924 and 1926 by the owner of Huemul ranch (40° 58' S–71° 27' W), Río Negro province, with individuals brought from Uruguay (Daciuk 1978). In 1931, a pair of adults and their piglets escaped and then spread to Nahuel Huapi National Park, Río Negro province, and also to Los Alerces National Park (42° 48' S–71° 53' W), Chubut province (300 km away from their escape point) (Jaksic et al. 2002). A recent study about genetic variability focusing in the origin of local populations of wild boar in Argentina noticed that individuals from La Pampa and Neuquén descend from boars from France and Spain and animals from El Palmar National Park (Entre Ríos province) belong to Asian lineages and, furthermore, they appear connected with boars of Uruguay (Sagua et al. 2018). The authors also found that there was no evidence that animals from Neuquén and El Palmar National Park were translocated from La Pampa. Moreover, they have evidence that boars from Neuquén could be originated in the Caucasus region.

After those introductions, during the first part of twentieth century, wild boar reintroductions occurred several times in different parts of the country, but they were poorly documented. Furthermore, in the last 100 years, natural expansion, repeated unintentional releases from ranches, continuous installation of hunting resorts, deliberate inoculations, and translocation of individuals have converted the wild boar in the largest exotic omnivore mammal in many temperate and subtropical biomes in Argentina (Daciuk 1978; Bonino 1995; Ballari et al. 2015a; Guichón et al. 2016; Sanguinetti and Pastore 2016). At present, *Sus scrofa* (feral pigs and wild boar) is found in 20 of the 23 provinces of the country (Fig. 10.1).

A research platform to study the current distribution of wild boar in Argentina through its network of protected areas was studied by Ballari et al. (2015a). They found that this species is present in at least 10 of the 16 ecoregions of the country (Arid Chaco, Espinal, high Andean, Iberá Marshes, Monte of plains and plateaus, Monte of hills and valleys, Patagonian forests, Pampa, Patagonian steppe, Parana Flooded Savanna), but probably they present in other ecoregions. Moreover, these authors highlight the presence of *Sus scrofa* in ecoregions not previously mentioned such as high Andean, Parana Flooded Savanna, and Iberá Marshes, which indicates that wild boars are still expanding their geographic range in Argentina.

Natural expansion of wild boar's range was reported in Patagonian forest, where Pescador et al. (2009) found that between 1985 and 2005, boar increased its range with a spread rate of 3500 ha per year. In Pampa ecoregion, Merino and Carpinetti (2003) and Pérez Carusi et al. (2009) assessed feral pig populations using aerial counts in Bahia Samborombón (35° 26' S–57° 47' W), Buenos Aires province. Feral pigs were reported in that area for the first time in 1980 (Ballari et al. 2015a). Between 1995 and 1998, Merino and Carpinetti (2003) found that pig abundances showed an accelerated increase, from about 700 individuals at the beginning of the study to more than 2000 at the end, including several temporary peaks reaching over 4000 individuals. Pérez Carusi et al. (2009) observed a 400% increase of pig population during the period of study (from 2002 to 2005). However, it is important to mention that at the end of Merino and Carpinetti's (2003) study, pig population was over 2600 individuals, while in Pérez Carusi et al. (2009) study, it was 2690 individuals. Still, comparing pig density in both studies, there was a substantial increase from 1.59 ind/km^2 to 7.78 ind/km^2.

10.4 Origin and Expansion in Chile

The introduction of wild boars in Chile is documented in Skewes and Jaksic (2015), where they pointed out the following origins and years of establishment: (1) direct import from Germany (1948), (2) natural immigration from Argentina (1956–1970), (3) release of boars into the environment (2002–2009), and (4) escape from boar farms (2008–2009).

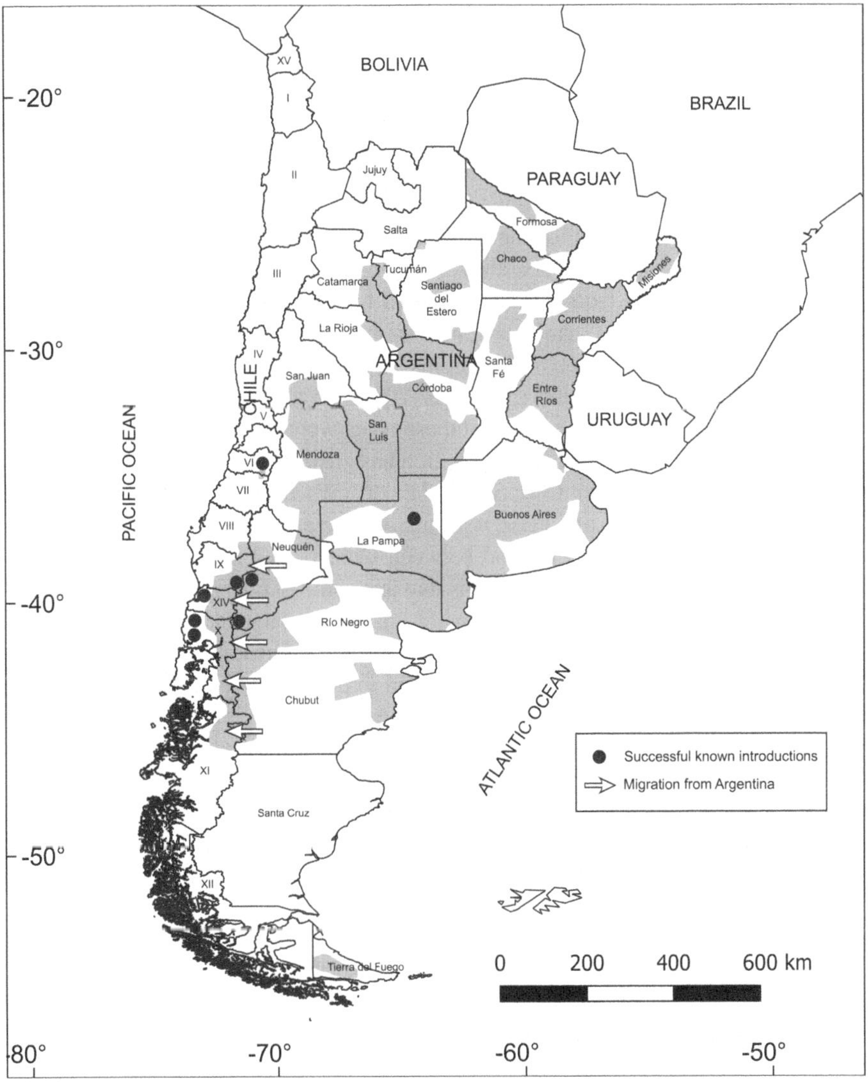

Fig. 10.1 Potential distribution of *Sus scrofa* in Argentina (adapted from La Sala et al. 2018) and Chile (adapted from Skewes and Jaksic 2015)

Wild boars imported from Germany in 1938 were settled in an enclosure in the Hacienda Allipén (38° 59' S–79°29' W), near the city of Villarrica (Skewes and Jaksic 2015). The boars escaped and caused damage in neighboring fields, which is why they were eliminated except for 25 boars that were transported by truck and released in 1948 in the area of Puesco in Villarrica National Park (39° 29' S–71° 43' W) (Gómez-Luna 1984). This release was at the root of the first sightings of boars on Chilean wild territory between 1950 and 1952 (Saavedra and Matus 2000).

In subsequent years, wild boars crossed the Chilean-Argentine border unimpeded by the Andes Range, probably using passes lower than 1000 m elevation (Fig. 10.1) (Jaksic et al. 2002). Wild boars were sighted from 1956 to 1958 in the lake port of Pirehueico (40° 01’ S–71° 43’ W) and in Palena (43° 37’ S–71° 48’ W), 500 km apart from each other. Years later (1967–1970), wild boars were observed for the first time in the National Reserve of Alto Biobío (38° 36’ S–70° 58’ W) (La Araucanía region) and in Coyhaique (45° 34’ S–72° 03’ W) (Aysén region) (Skewes and Jaksic 2015). From phylogenetic analyses of present wild boar inhabiting in Argentina (Sagua et al. 2018; see above), it can be deduced that much of the boar which migrate to Chile from Argentina would have the same origin. It remains to confirm genetic traces of German origin in animals released in Puesco, Chile.

In the administrative region of O’Higgins, wild boars were sighted between 2002 and 2004 in Sierras de Bellavista (34° 46′ S–70° 46′ W), 400 km from their recorded occupancy to the south. It is uncertain if these boars were originated from intentional release or escape from farms. This population has grown since and presently reaches the basin of the Teno river (34° 56’ S–71° 06’ W) in El Maule region. In the administrative region of Los Lagos, wild boars were released into the wild in 2008 and 2009 in the area of Río Frío and between Crucero and Riachuelo (Purranque) (40° 51’ S–73° 13’ W), areas that to this day are home to boars.

Around 2008–2009, wild boars escaped from a clandestine farm located upstream from the Valdivia river (39° 52’ S–73° 20’ W), in Los Ríos administrative region. Only part of this escape was hunted down, and to the present date, there are reports of sightings of boars with offspring (Skewes and Jaksic 2015).

Currently, wild boars are found in the wild in the central and southern regions of Chile, encompassing from north to south the administrative regions of O’Higgins, El Maule, Biobío, La Araucanía, Los Ríos, Los Lagos, and Aysén (Fig. 10.1). In the northern part of the distribution, the whole invaded area covers 500 km^2 on the piedmont of the Andes, from the Tinguiririca river (34°40′ S) in O’Higgins region, southwards to the Teno river (34°58’S) in El Maule region. The western border inhabited by these animals is marked by forests, and the eastern limit by mountains next to the depression of the central valley (Skewes and Jaksic 2015). In the southern part of the range, wild boars have expanded to the west in the regions of La Araucanía, Los Ríos, and Los Lagos and reached into the intermediate depression. In this general region, the species would have advanced 1.5–2.3 km/year in the last 50–60 years (Skewes and Jaksic 2015). More southwards, wild boars inhabit both the Andean and Coastal mountain ranges (between 39° 29’ S–72° 47’ W and 39° 39’ S–72° 53′ W) since 2007. In Los Lagos region, apart from having the oldest population in the Andes, there are newer populations of anthropogenic origin to the west of meridian 73°. In the southern part of this region, wild boars arrived early on 1988 at headwaters of Vodudahue and Reñihue rivers and later colonized part of Pumalín Private Park (42° 35’ S–72° 29’ W) (Gastó et al. 2000). In Aysén, the southernmost boundary of distribution of the species is the Simpson river basin. This area extends northwards through the mountain range, including part of the Queulat National Park (always east of meridian 72°W). Comparing the distribution reported by Skewes (1990), the area populated by wild boars in Chile has increased by 51.6% in the last 20 years and presently spans a surface of 2760 km^2 (Skewes and Jaksic 2015).

Intense hunting of wild boars in mountainous locations can be one of the causes of their relatively recent expansion from the Andean ranges to the intermediate depression in the regions of La Araucanía, Los Ríos, and Los Lagos. The number of licenses issued in Chile for the hunting of larger species – mainly wild boars and red deer (*Cervus elaphus*) – has practically doubled from 124 in 2008 to 240 in 2010 (Faúndez 2011). However, changes in land use in these regions must also be considered, since the landscape is characterized by a heterogeneous mosaic of a combination of agricultural areas, meadows, forests, and forestry plantations (Lara et al. 2012), which is a good habitat for the boar.

Colonization of the Coastal mountain range by wild boars is imminent in La Araucanía and Los Ríos regions, as there are already specimens inhabiting the intermediate depression. The extensive forestry plantations of *Pinus radiata* can favor their expansion. In Spain, changes in the landscape, such as the increase in forest surface and rural depopulation, favored the expansion of the wild boar (Sáez–Royuela and Tellería 1986; Rosell et al. 1998). Something similar happens in Chile, where there have been extensive reforestation and depopulation during the last decades in the Biobío and La Araucanía regions (Aguayo et al. 2009). Thus, given that the wild boar is a successful invader, it is highly likely that it will advance northwards along the Coastal mountain range (Skewes and Jaksic 2015).

10.5 Ecological Strategies in Invaded Habitats

Behavioral flexibility of a species is an adaptive response to environmental changes (Wright et al. 2010). Such behavioral traits are more labile than others and thus better able to respond quickly to new or changing environmental conditions (Snell-Rood 2013). In this sense, such traits can promote the colonization of new environments by an introduced species, giving it the ability to change or expand its ecological niche (Liebl and Martin 2014), through the exploitation of new foods, shelters, or habitats (Price et al. 2008). Therefore, studying these invasive species' traits could be useful to understand and explain invasiveness.

As we mentioned above, wild boars occupy an extensive range in Argentina that includes several ecoregions involving a wide range of climatic variability. This variability goes from humid subtropical climate in the northeast (Iberá Marshes) to oceanic climate in southwest (Patagonian forest) of the country (Table 10.1).

Ecological aspects of wild boar have been studied in four ecoregions: Pampa, Espinal, Patagonian forest, and Monte of plains and plateau (Cuevas et al. 2016). In Pampa ecoregions studies were described above. In this section we will discuss wild boar studies for Espinal and Patagonian forest ecoregions. Nevertheless, Monte ecoregion will be considered in a separate section below due to the challenge that survival implies for many species in such an arid zone (Cuevas et al. 2013a).

Table 10.1 Environmental characteristics of ecoregions of Argentina and Chile where *Sus scrofa* inhabits

Ecoregion	Climate	Annual precipitation	Vegetation
Argentina			
Pampa	Temperate to warm	600–1100 mm. Rainfall throughout the whole year that decreases from north to south and from east to west	Grasslands with a lower presence of halophyte steppes, marginal forests, and several types of hydrophilic shrubs
Patagonian forest	Temperate and wet	750–2000 mm. Decreasing from west to east	Pure or mixed forests of conifers of family Cupressaceae and evergreen species of family Nothofagaceae. These forests are mixed to a lesser extent with grasslands and peatlands
Espinal	Warm and wet in the northern part to temperate and dry in the central and southern areas	560–1200 mm. decreasing from north to south	Deciduous dry forests, palm groves, grasslands, savannahs, and shrub steppes
Monte of plains and plateaus	Subtropical to warm temperate desert and semidesert	80–250 mm.	Shrubland steppes of xerophytes, psammophytes, or halophytes, as well as marginal *Prosopis* woodlands
Chile			
Valdivian rainforest	Temperate from warm with rainy winters in the central zone to rainy and cold with no dry season in the south	777 to 2000–3000 mm from north to south	Four types of forests: (1) Deciduous forests of Mediterranean-type schlerophyllous forests and wet temperate forests farther south, with species of *Nothofagus* as dominant. (2) Valdivian broad-leaved forests dominated by a variety of tree species, including *Laureliopsis philippiana*, *Aextoxicon punctatum*, *Eucryphia cordifolia*, *Caldcluvia paniculata*, and *Weinmannia trichosperma*. (3) Northern Patagonian forests with a predominance of evergreen species such as *Nothofagus dombeyi*, *Podocarpus nubigena*, and *Drimys winteri*. (4) Patagonian Andean forests include *Araucaria araucana* and high Andean scrublands with deciduous *Nothofagus*, which are widely distributed in latitude

The southern part of the Espinal ecoregion was home to the first introduction of wild boar in Argentina (San Huberto ranch, La Pampa province). Despite having the oldest boar population in the country, there is no scientific information evaluating interactions neither with the invaded habitat and native species nor of the impact upon the invaded ecosystem. Instead, studies on the ecology and impacts of wild boar in this region have been conducted in El Palmar National Park (EPNP), Entre Ríos province (31° 50' S–58° 17' W) (Goveto 1999; Ballari 2013; Ballari et al. 2015b). The landscape is characterized by a heterogeneous mosaic of vegetation patches that includes gallery forests, shrublands, grasslands, and savannahs, with Yatay palms (*Butia yatay*) in highlands (Movia and Menvielle 1994). The climate is warm (annual mean temperature 28.9 °C) and wet throughout the year with no dry season (annual mean precipitation 1300 mm; Papadakis 1974). Wild boars have been reported in this protected area since 1950 (Ballari et al. 2015a). Based on stomach contents, Ballari et al. (2015b) found that 81.2% of the boar diet at EPNP was plant material and almost 18.8% was animal matter. They observed that during the fruiting of Yatay palm (an endemic and protected species), wild boars eat those fruits reaching to approximately 50% of their diet in summer. But during winter/autumn and spring, when those fruits are not available on the ground, boar fed mainly on corn (a supplemental feeding used as a bait in EPNP for controlling the species and to enhance hunting), where this item comprised between 40% and 50% of the diet. Only during spring, the bulk of the diet was corn (42%) and animal matter (27%). So, it appears that during the masting period of Yatay palm, wild boars prefer to eat it over the supplemental corn. The relatively high dietary content of animal matter in EPNP could be related with the ingestion of corn and fruits of *Butia* spp., high in carbohydrates but low in protein (Schley and Roper 2003; Hoffmann et al. 2014). So their gain of caloric requirement from corn may cause wild boar to compensate for lack of protein by eating more animal matter (Schley and Roper 2003). Habitat use of wild boars through signs revealed that there was a high density in Yatay palm forest during the masting period (February and March) (Goveto 1999). Later, Ballari (2013) also found that wild boar prefers habitats with a dominant tree canopy, e.g., Yatay palm forest and forest of exotic xerophytes (*Melia azedarach*, *Pyracantha atalantoides*, *Gleditsia triacanthos*, *Ligustrum lucidum*, *Ligustrum sinense*). Agricultural lands that surround the park were not preferred by wild boars, possibly due to abundant food resources present in the park, including the supplemental feeding (corn) for hunting practices (Ballari et al. 2015b). On the other hand, Caruso et al. (2018) reported wild boar's habitat use and activity patterns at landscape level in a mosaic of natural xerophytic deciduous woodlands, grasslands, and shrublands, mixed with agriculture and ranching activities. They found that wild boars were crepuscular and nocturnal with a peak at midnight, maybe related to the behavior of avoiding high temperatures during daytime. Habitat use was negatively related to open areas such as grasslands and positively associated with shrublands and shrublands mixed with grasslands. They also found that temperature and availability of water were important factors affecting wild boar distribution and its abundance. The authors suggest that pampas fox (*Lycalopex gymnocercus*) and wild boar not only use similar habitat but also have a highly overlap of activity patterns during crepuscular hours. The opposite occurs with puma (*Puma concolor*), whose activity patterns with wild boar have very low overlap.

In Patagonian forest ecoregion, wild boars have been introduced since 1917 (Daciuk 1978). The bulk of the studies of boar have focused on habitat use, distribution range, and impact on soil properties and vegetation. These last two topics, we will discuss below (Pescador et al. 2009; Sanguinetti and Kitzberger 2010; Schiaffini and Vila 2012; Barrios-García 2012; Barrios-García and Simberloff 2013; Barrios-García et al. 2014; Gantchoff and Belant 2015).

In terms of habitat use, several studies found that wild boars used more frequently *Nothofagus dombeyi* and *N. antarctica* forests compared with *N. pumilio* forests and grasslands (Schiaffini and Vila 2012; Gantchoff and Belant 2015). Furthermore, Schiaffini and Vila (2012) found that between 600 and 700 m elevation had the highest abundance of wild boar signs and that at elevation 1200 m there was no evidence of its presence. The authors concluded that the increased presence of boar in intermediate elevations was associated with the dense understory vegetation of *Nothofagus* forests, which provides warmth and moisture conditions that boars need to meet thermal requirements. Also, the high canopy (40 m) provides shelter from hot summer temperatures (~30 °C). Furthermore, the dense understory of bamboo (*Chusquea culeou*) affords protection from frost during cold days, allowing boars to find food under it. On the other hand, a study conducted in tourist sites in Nahuel Huapi National Park revealed that anthropogenic factors influence wild boar occurrence (Gantchoff and Belant 2015). They found that longer distance to human settlements and closer distances to roads were the most important variables that explain the occurrence of wild boar in the park. This could indicate that boars use roads as corridors to move across forests while avoiding humans and their hunting pressure. In mixed forests of *Araucaria*, wild boars preferred for feeding mixed *Araucaria-N. pumilio* over *Araucaria-N. antarctica* forests (Sanguinetti and Kitzberger 2010). They also found that wild boar consumed between 10% and 30% of available seeds. Predation was greater in places with low plant cover and close to seeding trees. The experimental exclusion of wild boars demonstrated that the number of surviving seeds increased, resulting in higher seedling establishment during non-masting years (boars ate proportionally more seeds during such periods than during masting ones). On the other hand, Lantschner et al. (2013) found that wild boar preferred pine plantations (*Pinus* spp.) over native vegetation (transition between *Austrocedrus chilensis* forest and arid steppe of bunchgrasses, such as *Poa*, *Festuca*, and *Stipa* spp., and sparse patches of woodlands). The authors proposed that pine plantation (closed habitat) offers better shelter against climate and predators than native vegetation (open habitat).

In Chile, the environmental characteristics of the areas inhabited by the wild boar are those of the ecoregion of Valdivia (see Table 10.1) (Hajek and Di Castri 1975; Gajardo 1994). Information about wild boar ecology is scarce. There are two studies that evaluated diet of wild boar in Valdivian rainforest in Los Ríos and Los Lagos regions. In Mocho-Choshuenco volcano (39° 54' S–72° 02' W) and Vicente Pérez Rosales National Park (41° 03' S–71° 54' W), Skewes et al. (2007) found that wild boar eats mainly plant material (between 80% and 100%) and 16% of animal matter. Grasses, fungi, and *Gunnera tinctoria* (food item with high energy value) were predominant foods in the diet during spring, autumn, and winter. On the other hand,

Hernandez et al. (Hernández et al. 2017) in Puyehue National Park (40° 38' S–72° 05' W) found that wild boars eat manly plants (~60%), followed by invertebrates (~28%), mammals (~8%), and fungi (~4%). They also found high content of seed and fruit of *Myrceugenia planipes,* which is a native species, and the presence of *Chiasognathus grantii*, an endemic species of Chile and Argentina. The authors proposed a positive potential impact of wild boar on plant's seed dispersal and a negative potential impact on insect's populations.

Bonacic et al. (2010) modeled the geographical potential distribution of wild boar in the south central zone of Chile and found that areas climatically suitable for boar extend from 37° 16′ 49″ S to 48° 48′ 28″ S and from 70° 55′ 38″ W to 73° 13′ 26″ W. Seasonal precipitation was the most important variable for the model. This result represents an area larger than the one presently inhabited by the wild boar according to Skewes and Jaksic (2015), so an increase in its distribution is expected from south to north along the Coastal mountain range.

10.6 Wild Boar's Expansion on Arid Lands of Argentina

Arid lands are characterized by high temperatures, water deficit, and low plant productivity, generating a great challenge to the survival of plants and animals in these environments (Cloudsley-Thompson 1975; Brown et al. 1979; Polis 1995). In Argentina, 57% of the territory consists of arid lands (Verbist et al. 2010). As a result of human activities (agriculture, grazing, logging, etc.), desertification, and salinization, these arid lands are undergoing rapid habitat conversion (Ojeda and Mares 1982). To these challenges, environmental changes driven by climate change as well as changes caused by invasive species are added (Boulanger et al. 2007; Cuevas et al. 2012).

In Monte of plains and plateau, wild boar has been studied from a great variety of aspects, including habitat use, diet, climatic influence, and impacts on vegetation composition and on physical, chemical, and microbiological soil properties (Campos and Ojeda 1997; Cuevas et al. 2010; Cuevas 2012; Cuevas et al. 2012, 2013a, b). Although wild boar is not physiologically adapted to arid environments (Baber and Coblentz 1986), they have successfully colonized them worldwide, such as the deserts of United States, Australia, and Argentina (Barrett 1978; Saunders and Giles 1995; Cuevas et al. 2010). For that reason, studies that help us understand the ecological strategies that boars use in environments quite different from their native range may yield insights about the traits and factors that constraint or facilitate the expansion of invasive species.

Wild boar studies have been conducted in the Man and Biosphere (MaB) Reserve of Ñacuñán (34° 02′ S–67° 58′ W), Mendoza province. The landscape is characterized by a heterogeneous mosaic of vegetation patches of *Prosopis* woodland or algarrobal (*Prosopis flexuosa*, Fabaceae), *Larrea* shrubland or jarillal (*Larrea cuneifolia*, Zygophyllaceae), and sand dunes. The climate is semiarid and strongly seasonal, with hot, humid summers and cold, dry winters. Mean annual precipitation and temperature are 326 mm and 15.6 °C, respectively, with a maximum annual mean of 23.8 °C and a minimum annual mean of 7.6 °C (Estrella et al. 2001; Labraga and Villalba 2009). Wild boar was first sighted in this area in the 1980s (Cuevas et al. 2010).

Cuevas et al. (2013a) studied habitat use by wild boar through signs such as tracks (used as moving from one place to another) and rooting (used as place for eating). The authors found that for displacement boars used the different habitats in proportion to their availability. Regarding rooting activity, they found that wild boar positively selected *Larrea* shrubland and avoided *Prosopis* woodland. At microhabitat level, herb cover was the most important variable affecting boar presence, showing a positive association between these plants and the abundance of signs.

In another study, evaluating wild boar diet through feces, Cuevas et al. (2013b) found that 96% consisted on plant and 4% was animal matter. Herbs were the most frequently consumed food item (~50%) followed by woody species. Aerial parts were consumed more frequently during the dry season, whereas fruits and animal tissue were more frequent during the wet season. In terms of trophic selection, herbs were the only food item selected by wild boars, while trees like *Prosopis flexuosa* were consumed as available only during the wet season, which is the season of fruiting. They also found a broader trophic niche with higher plant diversity during the wet season. Therefore, Cuevas et al. (2013a) found that *Larrea* shrubland (habitat) and herbs species (diet) were the only two ecological aspects selected by wild boar in Monte. Furthermore, *Larrea* shrubland is associated with high herb cover, so the authors concluded that boars used the habitat as a function of food availability.

In arid conditions the majority of plants have high fiber content and low nutritional value (Noy-Meir 1973). Because of this, added to that around the study area there were no croplands, the ingestion of energy-rich food is crucial for boar survival. In their study, Cuevas et al. (2013a) found that the most consumed food items (fruits of *Prosopis flexuosa*, leaves of *Sphaeralcea miniata,* and bulbs of *Pitraea cuneato-ovata*) had high forage quality and high carbohydrate contents, which means immediate energy for the organism. High carbohydrate input is considered important in the diet of an individual as it is an essential component in keeping the body in good physical condition and also for the accumulation of reserves to be used during more critical periods (food scarcity) and/or periods of highest energy demand (reproduction) (Abaigar 1993). This foraging strategy enables wild boar to maximize energy budget through food selection. Besides being a generalist species (Rosell et al. 2001), in the semiarid environment of the Monte, the wild boar appears as a species that selects both space (habitat use) and food (herbs) (Cuevas et al. 2013a, b).

Regarding climatic influences on wild boar activity at local scales, Cuevas et al. (2013a) found a positive association between the number of days with low temperature and the number of wild boar signs recorded in the Reserve. This means that boars reduced their seasonal activity and/or daily movements in periods or seasons of high temperature. Due to the wild boar's lack of sweat glands or other cooling physiological mechanisms for maintaining hydric and thermal balance, temperature could be a limiting factor for boar activity, especially in arid lands. They require free water, shade, a diet rich in water, and/or a behavioral response to increased environmental temperatures (Rosell et al. 2001; Dexter 2003). In the Monte, Cuevas et al. (2013a) found that wild boars showed a behavioral response related with daily movement patterns to increased environmental temperature, but they did not find a strong association with free water.

To sum up, ecological strategies of wild boar in arid lands of Argentina–where water resource is scarce and exposure to the sun is high–implicate that shade is essential for surviving. Therefore, boars must minimize the exposure to high temperatures and maximize the food intake of high-quality forage to maximize their energy input.

10.7 Impacts of Wild Boar in Argentina and Chile

The impacts of the wild boar are very diverse and have repercussions at different levels of the ecosystem, being able to affect from soil nutrient processes to affect populations of large terrestrial mammals (Barrios-García and Ballari 2012). Wild boar impact affects physical (e.g., Cushman et al. 2004), chemical (e.g., Moody and Jones 2000), and biological soil properties (e.g., Siemann et al. 2009); communities of plant species (e.g., Heinken and Raudnitschka 2002); animal species (e.g., Van and Scott 2001); fungi species (e.g., Fournier-Chambrillon et al. 1995); aquatic biodiversity (e.g., Arrington et al. 1999); and water quality and chemistry (e.g., Doupé et al. 2010).

The wild boar is an omnivorous and opportunistic mammal whose diet is mostly vegetable that is determined by the seasonality and availability of food resources (Schley and Roper 2003; Ballari and Barrios-García 2014). Wild boar consumes an important diversity of plant species, many sensitive to disturbances or herbivory caused by it. In protected areas of Argentina, the boar affects native and emblematic tree species. For example, in El Palmar National Park, boar consumed palm tree seedlings, whose conservation problems were the main reason for the creation of the protected area, being the last great relict of this species in Argentina (Ballari 2013; Gürtler et al. 2017). In Monte ecoregion, Campos and Ojeda (1997) found that wild boar causes damages by chewing nearly 100% of ingested seeds of *Prosopis flexuosa*, a key species of the desert. Predation by boar of seed of *Araucaria araucana* reduced seedling establishment during non-masting years (Sanguinetti and Kitzberger 2010). In southern Chile, the *Gunnera tinctoria*, an important cultural species with traditional use, is heavily consumed by the boar (Skewes et al. 2007).

Wild boar also affects animal wildlife by predation, nest or refuge destruction, resource competition, and habitat destruction (Long 2003; Cruz et al. 2005; Wilcox and van Vuren 2009). Animal material is a minority part of the diet of wild boar; however, in Argentina and Chile, volumes of animal matter have been recorded much higher than in other regions where the species is native or has been introduced (Skewes et al. 2007; Ballari et al. 2015b). Although part of this animal matter may be taken as carrion (Ballari et al. 2015b), the wild boar is recognized as a predator of a great variety of invertebrates, mammals, and birds (Ballari and Barrios-García 2014). Some of the prey of wild boar are species with threatened status for their conservation, or they are endemic (Giménez-Anaya et al. 2008; Jolley et al. 2010). In Chile, for example, the consumption of endemic birds (*Scelorchilus rubecula* and *Pteroptochos tarnii*) and endemic insects (*Chiasognathus grantii*) of temperate forests has been recorded (Skewes et al. 2007). The impacts of the boar on biodiversity can often be indirect. For example, in Bahía Samborombón Wildlife Refuge (Argentina), Pérez Carusi et al. (2009) found a possible negative association between the presence of wild boar and pampas deer (*Ozotoceros bezoarticus*), in risk of extinction.

Around the world and in all the places where the wild boar was introduced, as well as in Chile and Argentina, the boar represents an important source of food for predators and scavengers (Novaro et al. 2000; Skewes et al. 2012). The presence and management of wild boar (and other introduced species) can have unexpected or unintended consequences (Ballari et al. 2016) in the native ecosystems of Chile and Argentina. The availability of exotic prey such as wild boars can cause changes in the abundance of exotic species, thus not only affecting native species but also strongly modifying the food web. This last can result in novel interactions that could potentially favor the ecological extinction of native species (Jaksic 1998; Novaro et al. 2000). Finally, the wild boar as a puma's prey may have implications not only ecological but also sanitary because it can give rise to a new wild cycle of trichinosis (Skewes et al. 2012).

10.8 Socioeconomic Impacts of Wild Boar in the Southern Cone of South America

The impacts of the boar are not exclusively ecological. This species is often associated with human productive activities (e.g., agriculture and livestock) which cause considerable economic damage in different regions of the world (Pimentel et al. 2005).

Another type of latent impact of *Sus scrofa* is the cross-species disease transmission between wildlife, domestic animals, and humans, due to wild boar is a reservoir of many viral and bacterial diseases and parasites which can be transmitted by direct contact with wild boar or their feces, or by consuming contaminated food or uncooked boar meat (Barrios-García and Ballari 2012). Wild boars are increasingly a potential veterinary and public health threat. For example, in Argentina, trichinellosis was found in boar's meat in Neuquén province (Tesón et al. 1997). Miller et al. (2017) identified 34 World Organization for Animal Health (OIE) listed swine pathogens that cause clinical disease in livestock, poultry, wildlife, and humans.

The Chilean central zone hosts more than 2.6 million domestic pigs, 89% of all pigs in Chile (INE 2007), and so the presence of some of these swine pathogens in areas close to the pig industry represents a potential epidemiologic danger (Kukushkin et al. 2008; Ruiz-Fons et al. 2008; Costard et al. 2013) that could have serious sanitary and economic consequences.

Recent data submitted by the OIE inform about wild boars positive to diagnosis of African swine fever (ASF) in Belgium, which places the European pig industry on high alert. ASF is an infectious viral disease of pigs, including wild boar, which can act as reservoir. No vaccine or treatment against ASF is available (OIE 2018). The situation is more complex than before since over the past 30 years, the population of wild boar in Europe has systematically increased in size and in its distribution range (Melis et al. 2006).

Socioeconomic studies of the impacts of wild boar have been growing in the last decades globally. However, and in particular in South America, the studies tending to evaluate the cost or economic losses in agriculture and livestock production are very scarce (e.g., in Brazil, Pedrosa et al. 2015). In a study carried out in southern Chile, 75% of survey respondents acknowledged the wild boar as a harmful agent for agriculture, especially on sown fields, meadows, and fruit trees (Bonacic et al. 2010). Thus, the expansion of the wild boar to agricultural areas – as is happening in La Araucanía region – will probably have a detrimental impact on this activity and also on the protected native coastal areas in the south of Chile.

Cerda et al. (2017) evaluated the impacts of some invasive exotic species in Chile. In the case of wild boar, it was possible to assess impacts on forestry and agricultural activity and a preliminarily approach of biodiversity's components affected by wild boar. These authors calculated a minimum annual loss of approximately USD $ 38,278,724. Given the difficulty to assess the monetary value of the impacts on biodiversity, Cerda et al. (2017) recommended treating these results with extreme caution. The values obtained should not represent in any way the main engine for decision-making and constitute a first image that clearly presents the information gaps.

10.9 Rooting as a Main Disturbance

The most common disturbance associated with the wild boar is rooting, where this species uses their snout to turn over soil in search for below-ground feeding resources, mounding it up in places and creating depressions in others (see Fig. 10.2; Arrington et al. 1999). Rooting causes several impacts, including the alteration of nutrient cycles or structure and composition of soil biodiversity (Massei and Genov 2004; Bueno et al. 2013; Burrascano et al. 2015).

Due to the feeding habits, one of the main attractants for wild boar rooting activity is the abundance of a valuable food item (Bueno and Jiménez 2014). Rooting has been recorded in different environments where the boar is native or has been introduced, including forests, grasslands, wetlands, and deserts, which demonstrates the

Fig. 10.2 Wild boar rooting in different ecoregions from Argentina: (**a**) Patagonian forest ecoregion, (**b**) Monte ecoregion and Chile, and (**c**) Valdivian rainforest

Table 10.2 Effects of wild boar rooting on soil properties in both Monte and Patagonian forest ecoregions

Soil properties			Monte desert		Patagonian forest	
			Rooted	Unrooted	Rooted	Unrooted
Physical	Hardness		–	+	–	+
	Moisture		+	–	No change	
	Temperature				No change	
	Texture	Silt	–	+		
		Clay	–	+		
		Sand	+	–		
Chemical	Total nitrogen		No change		No change	
	Mineral nitrogen		+	–	No change	
	Nitrate + nitrite		+	–		
	NH_4		No change			
	Organic carbon		No change			
	C/N ratio		+	–		
	Organic matter		No change			
	pH		No change		No change	
	Total carbon				No change	
	Extractable P				No change	
Microbiological	Soil respiration		–	+	No change	
	Ammonifiers		No change			
	Cellulolytics		No change			
	N fixers		No change			
	Nitrifiers		No change			

Table extracted and modified from Cuevas et al. (2016)

ecological plasticity of the boar to search for food in very diverse environments (Arrington et al. 1999; Bueno et al. 2009; Desbiez et al. 2009; Cuevas et al. 2012). This particular disturbance can be observed throughout the year, although there may be peaks of seasonal activity when above-ground resources are scarce (e.g., in winter and early spring; Scott 1973; Barrett 1978; Baron 1982).

In Argentina, rooting has been widely reported as one of the main disturbances in protected areas. However, because rooting is one of the most visible and characteristic effects of wild boar, the perception of their frequency and intensity over other negative impacts (e.g., predation, competition) could be overestimated by protected area managers (Ballari et al. 2015a). However, different studies have shown the negative impacts that rooting causes on ecosystems. In the Monte ecoregion of Argentina, Cuevas et al. (2010) showed that rooting alters the plant cover reducing plant richness and diversity. Likewise, rooted soils have less compaction, high soil moisture, high C/N ratio, and a high content of mineral nitrogen (Table 10.2) (Cuevas et al. 2012). The implications of the effect of wild boar rooting in desert soils are key for the conservation of these environments because, for example, the new characteristics of soils created by rooting could be responsible for the reduction of plant cover and less soil bulk density, which can accelerate soil degradation by wind erosion (Cuevas et al. 2013b). In a different scenario, in Patagonian forest (Argentina), wild boar rooting produces also a wide range of impacts affecting the structure and composition of plant communities as a result of boar disturbances. Barrios-García et al. (2014) found that the aerial parts of the plants are strongly reduced by the rooting and the structure of the plant community is modified decreasing both grass and herb cover relative to areas where wild boars were excluded. Also, decomposition rates and soil compaction declined in rooting areas but with no strong consequences on soil nutrient stocks and cycling (Table 10.2). Additionally, wild boar rooting promotes the establishment and growth of nonnative plants in Patagonian forest ecoregion, which shows the importance of the role of wild boar as a facilitator of other invasions (Barrios-García and Simberloff 2013). These results could have serious implications, not only in areas where the species was introduced (such as Chile and Argentina) but also in their native range, where rooting disturbance may facilitate community composition shifts (Barrios-García and Simberloff 2013).

In Chile, the specific impacts of rooting have not been studied in depth. However, diet studies (see Skewes et al. 2007; Hernández et al. 2017) have shown direct impacts through wild boars' feeding behavior since they affect plant species of socio-ecological importance (e.g., *Gunnera tinctoria*) and consume earthworms and insect larvae, thus being able to cause impacts on the ecosystems' conservation due to significant changes in native biodiversity.

10.10 Wild Boar as a Prey for Native Predators

The incidence of wild boar as puma's (*Puma concolor*) prey was studied in 1988 and 2004 in the foothills of the volcano Mocho-Choshuenco, Valdivian rainforest, southern Chile. The studies were made through the identification of prey items in puma's feces and field surveys of boar carcasses (Skewes et al. 2012). The puma preys mainly on juveniles, and the frequency of occurrence of wild boar remains in puma scats was 27.3% and 53.8% for years 1998 and 2004, respectively. The percent of occurrence in the same period was quite stable (17.1–15.9%), which would reflect the territorial stability exhibited by sounders (family groups) in Europe (Keuling et al. 2009). In this case, the irruption of larger prey, and with a high reproductive rate for high trophic level predators, may lead to the establishment of predators in areas where predators were not previously supported (Roemer et al. 2002), the so-called facilitation hypothesis (Rodriguez 2006; White 2008). In addition, the wild boar as prey could imply, in part, a better puma's body condition, smaller territories, and higher reproductive rates that could have consequences for both the conservation of the puma and the abundance of native prey. One effect that could have biological implications would be that predation by pumas on wild boar herds can cause that they be distributed homogeneously in the available habitat, precisely as found by Hornocker (1970) for cervids depredated by puma in the United States (Skewes et al. 2012).

10.11 Conclusions and Prospects for Further Research

In this chapter, we summarized *Sus scrofa* (feral pigs and European wild boar) information in Argentina and Chile. Invasive wild boar distribution in Argentina is approximately from 22° 46′ S to 54° 42′ S and in Chile is approximately from 37° 55′ S to 45° 45′ S. Not only the continuous installation of hunting lodges but also natural expansion has led wild boar to colonize a great part of Argentina, occupying different habitats from humid subtropical climate to semiarid conditions. Besides, the current situation of wild boar in Chile represents an expanding population that has not colonized the entire potential habitat available (Bonacic et al. 2010): its distribution seems to be limited to Valdivian ecoregion in Chilean Patagonia. This could be related to unsuitable environments probably associated to climatic barriers towards the regions of the north of the country, dominated by arid conditions (approximately 31–32° S and 200–300 mm annual precipitation).

As we could see, wild boar occupies a wide variety of anthropogenic and natural habitats. In spite of boar not being physiologically adapted to arid environments (Baber and Coblentz 1986), they have successfully colonized them worldwide (Barrett 1978; Saunders and Giles 1995; Cuevas et al. 2010). While in Argentine Patagonia wild boar prefers mainly *Nothofagus* spp. forests than others such as grasslands, arid steppe, shrublands, and agricultural lands (Sanguinetti and Kitzberger 2010; Ballari 2013; Schiaffini and Vila 2012; Gantchoff and Belant 2015), in Chilean Patagonia they only occupy Valdivian forest. In its native range, wild boars find food in the agricultural areas and find shelter, thermal comfort, and bushy spots to nest while farrowing in adjoining woods or shrubs (Dardaillon 1987; Rosell et al. 2001; Fruziski and Labudzki 2002; Lemel et al. 2003; Sodeikat and Pohlemyer 2007). Opposite to that, in El Palmar National Park, wild boar prefers habitat with high canopy cover rather than agricultural lands that surround the park, maybe due to abundant food resources present in the park, including the supplemental feeding (corn) for hunting practices related to its control (Ballari et al. 2015b). But when climate conditions are more arid such as in the Monte ecoregion, wild boar selects shrublands rather than woodlands (Cuevas et al. 2013a). In all cases, boar habitat use is strongly related to food availability (Barrett 1982; Welander 2000; Cuevas et al. 2013a). In the Patagonian forest ecoregion of Argentina, future studies evaluating wild boar feeding habits, including their seasonal variation, could help us to understand why this species prefers *Nothofagus* forests to other types of environment. Similar information on diet and habitat use of wild boar is also needed in the Pampean ecoregion, where niche interactions and possible competition between this species with pampas deer, puma, and pampas fox require evaluation.

On the other hand, temperature and the availability of free water are two important factors that affect wild boar distribution and abundance. As occurred in Monte ecoregion when daily movements of wild boars were reduced in periods or seasons of high temperature (Cuevas et al. 2013b), and in Espinal ecoregion during summer, wild boars were active during nighttime to avoid high daytime temperatures (Caruso et al. 2018).

Wild boar is an omnivorous species whose diet consists mainly of plant matter. In Argentina and Chile, it prefers food items rich in energy such as bulbs and fruits of *Pitraea cuneato-ovata, Prosopis flexuosa, Araucaria araucana, Butia yatay*, and *Gunnera tinctoria*, when they are available. High dietary intake of carbohydrates (e.g., in the fruits mentioned above) is expected to be compensated with high intake of protein (Schley and Roper 2003). So, wild boar compensates the lack of protein by eating more animal matter, especially during masting periods (Skewes et al. 2007; Cuevas et al. 2013b; Ballari et al. 2015b). This was observed in Espinal and Monte ecoregions and in Chile. This strategy was also observed in places where the species is native and where it is introduced (Barrett 1978; Abaigar 1993; Massei et al. 1996; Schley and Roper 2003).

Disturbances generated by wild boars vary among habitats and climates. In El Palmar National Park, boars may serve dual roles as possible seed dispersers of Yatay palms and as predators upon Yatay seedlings. This is because during non-masting periods, wild boars not only dig around the plant to feed on the fleshy parts, but also they seek in the soil the remnant seeds and destroy them by chewing, thus affecting their survival (Ballari 2013). Therefore, wild boar may in fact reduce the recruitment of Yatay palm in EPNP, and future studies about its role in the predation or dispersal of the palm are crucial to ensure the conservation of this species.

Regarding the impact of wild boar rooting behavior, we found evidence suggesting no substantial changes in soil properties in the Patagonian forest ecoregion, whereas in the Monte there were soil changes in physical, chemical, and microbiological properties leading to wind erosion of soil (Table 10.2). These differences in the impact of rooting may be tied to soil characteristics and resilience. While soils of Patagonian forest are derived from volcanic ashes with high capacity of stabilizing soil organic matter, buffering pH, and retaining P and water – which confers high resistance to nutrient loss (Diehl et al. 2003), soils in Monte have an inherent tendency to desertification attributable to an interaction between the system's own fragility due to aridity, erosive forces from water and wind, salinization processes, and anthropic actions such as livestock pressure, logging, and fire regime modifications (Villagra et al. 2009). Therefore, soils in the Patagonia forest may be better buffered to short-term disturbances in comparison with Monte Desert soils, where the presence of a new disturbance factor (wild boar rooting) could have consequences such as increasing desertification.

In Chile, more ecological studies of wild boar are needed. Disturbances caused by boars on biodiversity have not been studied at all. It is necessary to have well-designed field experiments focusing on the effect of rooting over the vegetation and composition of the soil and on native fauna. For example, the relatively high consumption by the wild boar of hypogeous larvae of *Chiasognathus grantii* – an endemic to the forests of Chile (Skewes et al. 2007) – must be considered. It also urges to know what is the real impact of wild boar on the invertebrate soil fauna. Furthermore, another important issue to take into account is the consumption by wild boar of hypogeous and epigeous fungi. These organisms are not only involved in the mobilization, uptake, and translocation of nutrients in forest soils, but also they are essential for nutrient uptake of green plants through mycorrhizal symbiosis (Steiner et al. 2002).

To improve the understanding of biological invasions, it is fundamental to consider interactions of exotic invaders with their new environment (Jeschke et al. 2012). So, to design a management plan for wild boar, it is necessary to know not only the impact that this species generates on the environment but also its ecological strategies in each particular area. Studies focused on movements, activity pattern, home range, and reproductive capacity under different climatic conditions are needed to understand why this species is so successful. Doing so will also allow us to better understand their potential future expansion to new areas. Since wild boar populations between Argentina and Chile are connected, a binational cooperation between these countries is needed.

We conclude that the impacts of wild boar in Argentina and Chile are mostly negative, demanding interactions among different players (scientists, government officials, managers of protected areas, landowners) to plan a strategy with concrete actions to control its populations in both countries, in this way, to avoid the expansion of *Sus scrofa* to new areas and to mitigate its damage on native ecosystem and the productive systems of the countries.

References

Abaigar T (1993) Régimen alimentario del jabalí (*Sus scrofa*) en el sureste Ibérico. Doñana Acta Vertebrata 20:35–48

Aguayo M, Pauchard A, Azócar G, Parra O (2009) Cambio del uso del suelo en el centro sur de Chile a fines del siglo XX. Entendiendo la dinámica espacial y temporal del paisaje Revista Chilena de Historia Natural 82:361–374

Aravena P, Skewes O, Gouin N (2015) Mitochondrial DNA diversity of feral pigs from Karukinka Natural Park, Tierra del Fuego Island, Chile. Genet Mol Res 14:4245–4257

Arrington DA, Toth LA, Koebel JW (1999) Effects of rooting by feral hogs *Sus scrofa* L. on the structure of a floodplain vegetation assemblage. Wetlands 19:535–544

Baber DW, Coblentz BE (1986) Density, home range, habitat use, and reproduction in feral pigs on Santa Catalina Island. J Mammal 67:512–525

Ballari SA (2013) El jabalí (*Sus scrofa*) en el Parque Nacional El Palmar, Entre Ríos: uso de hábitat, dieta, impactos y manejo. PhD thesis, University of Córdoba, Argentina

Ballari SA, Barrios-García MN (2014) A review of Wild boar *Sus scrofa* diet and factors affecting food selection in native and introduced ranges. Mammal Rev 44:124–134

Ballari SA, Cuevas MF, Cirignoli S, Valenzuela AEJ (2015a) Invasive Wild boar in Argentina: using protected areas as a research platform to determine distribution, impacts and management. Biol Invasions 17:1595–1602

Ballari SA, Cuevas MF, Ojeda RA, Navarro JL (2015b) Diet of Wild boar (*Sus scrofa*) in a protected area of Argentina: the importance of baiting. Mammal Res 60:81–87

Ballari SA, Kuebbing SE, Nuñez MA (2016) Potential problems of removing one invasive species at a time: a meta-analysis of the interactions between invasive vertebrates and unexpected effects of removal programs. PeerJ 4:e2029

Baron J (1982) Effects of feral hogs (*Sus scrofa*) on the vegetation of Horn Island, Mississippi. Am Midl Nat 107:202–205

Barrett RH (1978) The feral hogs on the Dye Creek Ranch, California. Hilgardia 46:281–346

Barrett RH (1982) Habitat preferences of feral hogs, deer, and cattle on a Sierra foothill range Journal of Range Management 35: 342–346

Barrios-García MN (2012) Multi-level impacts of introduced Wild boar on Patagonian ecosystems. PhD thesis, University of Tennessee, Knoxville, USA

Barrios-García MN, Ballari SA (2012) Impact of Wild boar (*Sus scrofa*) in its introduced and native range: a review. Biol Invasions 14:2283–2300

Barrios-García MN, Simberloff D (2013) Linking the pattern to the mechanism: How an introduced mammal facilitates plant invasions. Austral Ecol 38:884–890

Barrios-García MN, Classen AT, Simberloff D (2014) Disparate responses of above- and belowground properties to soil disturbance by an invasive mammal. Ecosphere 5:1–13

Beasley JC, Ditchkoff SS, Mayer JJ, Smith MD, Vercauteren KC (2018) Research priorities for managing invasive Wild pigs in North America. J Wildl Manag 82:674–681

Bergstrom DM, Lucieer A, Kiefer K, Wasley J, Belbin L, Pedersen TK, Chown SL (2009) Indirect effects of invasive species removal devastate World Heritage Island. J Appl Ecol 46:73–81

Bevins SN, Pedersen K, Lutman MW, Gidlewski T, Deliberto TJ (2014) Consequences associated with the recent range expansion of nonnative feral swine. Bioscience 64:291–299

Blair J, Nippert J, Briggs J (2014) Grassland ecology. In: Ecology and the environment. Springer, New York

Blossey B, Gorchov DL (2017) Introduction to the special issue: Ungulates and invasive species: quantifying impacts and understanding interactions. AoB Plants 9:1–6

Bonacic C, O Ohrens & F Hernández (2010) Informe Final, Estudio de distribución y estimación poblacional de las especies exóticas invasoras: jabalí y ciervo rojo en Chile. Servicio Agrícola y Ganadero SAG Dpto. Recursos Naturales Renovables, Santiago

Bonino NA (1995) Introduced mammals in Patagonia, southern Argentina: consequences, problems, and management considerations. In Proceedings of the First International Wildlife Management Congress. The Wildlife Society, Maryland, EE. UU. pp. 406–409

Boulanger JP, Martinez F, Segura EC (2007) Projection of future climate change conditions using IPCC simulations, neural networks and Bayesian statistics. Part 2: Precipitation mean state and seasonal cycle in South America. Climatic Dynamics 28:255–271

Bratton SP (1975) The effect of the European Wild boar, *Sus scrofa*, on gray beech forest in the Great Smoky Mountains. Ecology 56:1356–1366

Brown JH, Davidson D, Reichman OJ (1979) Granivory in desert ecosystems. Annu Rev Ecol Syst 10:210–227

Bueno CG, Jiménez JJ (2014) Livestock grazing activities and Wild boar rooting affect alpine earthworm communities in the Central Pyrenees (Spain). Appl Soil Ecol 83:71–78

Bueno CG, Alados CL, Gómez-García D, Barrio IC, García-Gonzalez R (2009) Understanding the main factors in the extent and distribution of Wild boar rooting on alpine grasslands. J Zool 279:195–202

Bueno CG, Azorín J, Gómez-García D, Alados CL, Badía D (2013) Occurrence and intensity of Wild boar disturbances, effects on the physical and chemical soil properties of alpine grasslands. Plant and Soil 373:243–256

Burrascano S, Copiz R, Del Vico E, Fagiani S, Giarrizzo E, Mei M, Blasi C (2015) Wild boar rooting intensity determines shifts in understory composition and functional traits. Community Ecol 16:244–253

Campos CM, Ojeda RA (1997) Dispersal and germination of *Prosopis flexuosa* (Fabaceae) seeds by desert mammals in Argentina. J Arid Environ 35:707–714

Carpinetti B, Di Guirolamo G, Delgado JV, Martínez RD (2016) El Cerdo Criollo Costero: Valioso recurso zoogenético local de la provincia de Buenos Aires Argentina. Archivos de Zootecnia 65:403–407

Caruso N, Valenzuela AE, Burdett CL, Vidal EML, Birochio D, Casanave EB (2018) Summer habitat use and activity patterns of Wild boar *Sus scrofa* in rangelands of central Argentina. PLoS One 13:1–15

Cerda C, Cruz G, Skewes O, Araos A, Tapia P, Baeriswyl F, Critician P (2017) Especies exóticas invasoras en Chile como un problema económico: valoración preliminar de impactos. Revista Chagual 15:12–22

Cloudsley-Thompson JL (1975) The desert as a habitat. In: Rodents in desert environments. Prakash I & PK Ghosh (Eds.) Dr. W. Junk b.v. Publishers The Hague. pp. 1–14

Clout MN, Russell JC (2008) The invasion ecology of mammals: a global perspective. Wildl Res 35:180–184
Corn JL, Jordan TR (2017) Development of the national feral swine map, 1982–2016. Wildl Soc Bull 41:758–763
Costard S, Jones BA, Martínez-López B, Mur L, De La Torre A (2013) Introduction of African Swine Fever into the European Union through illegal importation of pork and pork products. PLoS One 8:e61104
Cruz F, Donlan CJ, Campbell K, Carrion V (2005) Conservation action in the Galapagos: feral pig (*Sus scrofa*) eradication from Santiago Island. Biol Conserv 121:473–478
Cuevas MF (2012) Ecología del jabalí (*Sus scrofa*) en el Desierto del Monte central, Argentina. PhD thesis, University of Río Cuarto, Argentina
Cuevas MF, Novillo A, Campos C, Dacar MA, Ojeda RA (2010) Food habits and impact of rooting behaviour of the invasive Wild boar, *Sus scrofa*, in a protected area of the Monte Desert, Argentina. J Arid Environ 74:1582–1585
Cuevas MF, Mastrantonio L, Ojeda RA, Jaksic FM (2012) Effects of Wild boar disturbance on vegetation and soil properties in the Monte Desert, Argentina. Mammalian Biology-Zeitschrift für Säugetierkunde 77:299–306
Cuevas MF, Ojeda RA, Jaksic FM (2013a) Multi-scale patterns of habitat use by Wild boar in the Monte Desert of Argentina. Basic & Appl Ecol 14:320–328
Cuevas MF, Ojeda RA, Dacar MA, Jaksic FM (2013b) Seasonal variation in feeding habits and diet selection by Wild boars in a semi-arid environment of Argentina. Acta Theriol 58:63–72
Cuevas MF, Ojeda RA, Jaksic FM (2016) Ecological strategies and impact of Wild boar in phytogeographic provinces of Argentina with emphasis on aridlands. Mastozoología neotropical 23:239–254
Cushman JH, Tierney TA, Hinds JM (2004) Variable effects of feral pig disturbances on native and exotic plants in a California grassland. Ecol Appl 14:1746–1756
D'Hondt B, D'Hondt S, Bonte D, Brys R, Hoffmann M (2012) A data-driven simulation of endozoochory by ungulates illustrates directed dispersal. Ecol Model 230:114–122
Daciuk J (1978) Estado actual de las especies de mamíferos introducidos en la Subregión Araucana (Rep. Argentina) y grado de coacción ejercido en algunos ecosistemas surcordilleranos. Anales de Parques Nacionales 14:105–130
Dardaillon M (1987) Seasonal feeding habits of the Wild boar in a Mediterranean wetland: the Camargue (southern France). Acta Theriol 32:381–401
Desbiez ALJ, Aparecida Santos S, Keuroghlian A (2009) Predation of young palms (*Atalea phalterata*) by feral pigs in the Brazilian Pantanal. Suiform Soundings 9:35–40
Dexter N (2003) The influence of pasture distribution, and temperature on adult body weight of feral pigs in a semi-arid environment. Wildl Res 30:75–79
Diehl P, Mazzarino MJ, Funes F, Fontenla S, Gobbi M, Ferrari J (2003) Nutrient conservation strategies in native Andean-Patagonian forests. J Veg Sci 14:63–70
Doupé RG, Mitchell J, Knott MJ, Davis AM, Lymbery AJ (2010) Efficacy of exclusion fencing to protect ephemeral floodplain lagoon habitats from feral pigs (*Sus scrofa*). Wetl Ecol Manag 18:69–78
Estrella H, J Boshoven & M Tognelli (2001) Características del clima regional y de la Reserva de Ñacuñán. In: El desierto del Monte: La Reserva de Biosfera de Ñacuñán. Claver S & S Roig-Juñent (Eds). IADIZA, MAB, UNESCO. Ed. Triunfar. Argentina. pp. 25–33
Faúndez NS (2011) Revisión de los aspectos cinegéticos del jabalí europeo (*Sus scrofa* L.) en Chile. Memoria de título presentada a la Facultad de Ciencias Veterinarias de la Universidad de Concepción, Concepción, Chile
Flueck WT, Smith-Flueck JA, Naumann CM (2003) The current distribution of red deer (*Cervus elaphus*) in southern Latin America. Eur J Wildl Res 49:112–119
Forsyth DM, Wilmshurst JM, Allen RB, Coomes DA (2010) Impacts of introduced deer and extinct moa on New Zealand ecosystems. N Z J Ecol 34:48–65

Fournier-Chambrillon C, Maillard D, Fournier P (1995) Diet of Wild boar (*Sus scrofa* L.) inhabiting the Montpellier garrigue. J Mount Ecol (Ibex) 3:174–179

Frank DA, Groffman PM (1998) Ungulate vs. landscape control of soil C and N processes in grasslands of Yellowstone National Park. Ecology 79:2229–2241

Friend GR, Taylor JA (1984) Ground surface features attributable to feral buffalo, *Bubalus bubalis*. II. Their relationship to the abundance of small animal species. Aus Wildlife Res 11:311–323

Fruziski B, Labudzki L (2002) Management of Wild boar in Poland Bewirtschaftung von Wildschweinen in Polen. Z Jagdwiss 48:201–207

Gajardo R (1994) La vegetación natural de Chile. Clasificación y distribución. Editorial Universitaria, Santiago, Chile

Gantchoff MG, Belant JL (2015) Anthropogenic and environmental effects on invasive mammal distribution in northern Patagonia, Argentina. Mamm Biol 80:54–58

Gastó J, Retamal A, Guzmán D (2000) Proyecto Pumalín. Informe Técnico Santuario de la Naturaleza. Pontificia Universidad Católica de Chile, Facultad de Agronomía e Ingeniería Forestal. Programa de Ecología y Medio Ambiente, Santiago, Chile

Gay C (1847) Historia física y política de Chile. Zoología. Santiago, Museo de Historia Natural de Santiago, 126 pp

Giménez-Anaya A, Herrero J, Rosell C, Couto S, A García-Serrano A. (2008) Food habits of Wild boars (*Sus scrofa*) in a Mediterranean coastal wetland. Wetlands 28:197–203

Gómez-Luna S (1984) Plan de manejo del Parque Nacional Kilapillan. CONAF Santiago, Chile

Goveto L (1999) Manejo adaptativo de las poblaciones de jabalíes en las áreas protegidas. Administración de Parques Nacionales. Technical report

Guichón ML, Monteverde M, Piudo L, Sanguinetti J, Di Martino S (2016) Mamíferos introducidos en la provincia de Neuquén: estado actual y prioridades de manejo. Mastozoología Neotropical 23:255–265

Gürtler RE, Izquierdo VM, Gil G, Cavicchia M, Maranta A (2017) Coping with Wild boar in a conservation area: impacts of a 10-year management control program in north-eastern Argentina. Biol Invasions 19:11–24

Hajek ER, Di Castri F (1975) Bioclimatografía de Chile. Dirección de Investigación, Vice-Rectoría Académica, Universidad Católica de Chile, Santiago, Chile

Heinken T, Raudnitschka D (2002) Do Wild ungulates contribute to the dispersal of vascular plants in central European forests by epizoochory? A case study in NE Germany. Forstwissenschaftliches Centralblatt 121:179–194

Hernández C, Sade S, Rau J (2017) Dieta del jabalí (*Sus scrofa*), invasor biológico reciente del Parque Nacional Puyehue, sur de Chile. Mastozoología Neotropical 24:467–473

Hoffmann JF, Barbieri RL, Rombaldi CV, Chaves FC (2014) *Butia* spp. (Arecaceae): an overview. Sci Hortic 179:122–131

Hornocker MG (1970) An analysis of mountain lion predation upon mule deer and elk in the Idaho Primitive Area. Wildl Monogr 21:1–39

INE (2007) Instituto Nacional de Estadísticas, Censo Agropecuario y Forestal por comunas. URL: http://www.ine.cl/canales/chile_estadistico/censos_agropecuarios/censo_agropecuario_07_comunas.php

Jaksic FM (1998) Vertebrate invaders and their ecological impacts in Chile. Biodivers Conserv 7:1427–1445

Jaksic FM, Iriarte JA, Jiménez JE, Martínez DR (2002) Invaders without frontiers: cross-border invasions of exotic mammals. Biol Invasions 4:157–173

Jaksic FM, Castro SA (2010) Ecología y biodiversidad de vertebrados de Chile: Análisis comentado de la Zoología de Claude Gay. Revista Chilena de Historia Natural 83: 323–333

Jeschke JM, Gómez Aparicio L, Haider S, Heger T, Lortie CJ, Pyšek P, Strayer DL (2012) Support for major hypotheses in invasion biology is uneven and declining. Neobiota 14:1–20

Jolley DB, Ditchkoff SS, Sparklin BD, Hanson LB, Mitchell MS, Grand JB (2010) Estimate of herpetofauna depredation by a population of Wild pigs. J Mammal 91:519–524

Keuling O, Stier N, Roth M (2009) Commuting, shifting or remaining? Different spatial utilization patterns of Wild boar *Sus scrofa* L. in forest and field crops during summer. Mamm Biol 74:145–152

Kukushkin S, Kanshina A, Timina A, Baybikov T, Mikhalishin V (2008) Investigation of Wild boar (*Sus scrofa*) for porcine reproductive and respiratory syndrome in some territories of Russia. Europ J Wildlife Res 54:515–518

La Sala L, Ballari SA, Merino M (2018) Distribución e impacto de los jabalíes y cerdos cimarrones en Argentina. III Seminario Internacional sobre Manejo de Jabalí y cerdos cimarrones. XXXI Jornadas Argentinas de Mastozoología. La Rioja, Argentina

Labraga JC, Villalba R (2009) Climate in the Monte Desert: Past trends, present conditions and future projections. J Arid Environ 73:154–163

Lantschner MV, Rusch V, Hayes JP (2013) Do exotic pine plantations favour the spread of invasive herbivorous mammals in Patagonia? Austral Ecol 38:338–345

Lara A, Solari ME, Prieto MR, Peña MP (2012) Reconstrucción de la cobertura de la vegetación y uso del suelo hacia 1550 y sus cambios a 2007 en la ecorregión de los bosques valdivianos lluviosos de Chile (35°–43° 30′S). Bosque 33:13–23

Lemel J, Truvé J, Söderberg B (2003) Variation in ranging and activity behaviour of European Wild boar *Sus scrofa* in Sweden. Wildl Biol 9:29–36

Liebl AL, Martin LB (2014) Living on the edge: range edge birds consume novel foods sooner than established ones. Behav Ecol 25:1089–1096

Long JL (2003) Introduced mammals of the world: their history, distribution and influence. CABI Publishing, United Kingdom. CSIRO Publishing, Collingwood

Lowe S, Browne M, Boudjelas S, De Poorter M (2000) 100 of the World's worst invasive alien species: a selection from the Global Invasive Species Database. Published by The Invasive Species Specialist Group (ISSG) a specialist group of the Species Survival Commission (SSC) of the World Conservation Union (IUCN), Auckland

Mack MC, D'Antonio CM (1998) Impacts of biological invasions on disturbance regimes. Trends Ecol Evol 13:195–198

Massei G, Genov PV (2004) The environmental impact of Wild boar. Galemys 16:135–145

Massei G, Genov PV, Staines BW (1996) Diet, food availability and reproduction of Wild boar in a Mediterranean coastal area. Acta Theriol 41:307–320

Melis C, Szafrańska PA, Jędrzejewska B, Bartoń K (2006) Biogeographical variation in the population density of Wild boar (*Sus scrofa*) in western Eurasia. J Biogeogr 33:803–811

Merino ML, Carpinetti BN (2003) Feral pig *Sus scrofa* population estimates in Bahía Samborombón Conservation Area, Buenos Aires province, Argentina. Mastozoología Neotropical 10:269–275

Miller RS, Sweeney SJ, Slootmaker C, Grear DA, Di Salvo PA, Kiser D, Shwiff SA (2017) Cross-species transmission potential between Wild pigs, livestock, poultry, Wildlife, and humans: implications for disease risk management in North America. Sci Rep 7:7821

Moody A, Jones JA (2000) Soil response to canopy position and feral pig disturbance beneath *Quercus agrifolia* on Santa Cruz Island, California. Appl Soil Ecol 14:269–281

Movia C, Menvielle MF (1994) Mapa de vegetación del P.N. El Palmar. In: Plan de Manejo del P.N. El Palmar. Administración de Parques Nacionales. Argentina

Navas JA (1987) Los vertebrados exóticos introducidos en la Argentina. Revista del Museo Argentino de Ciencias Naturales, Serie Zoología XIV: 7–38

Novaro AJ, Funes MC, Walker RS (2000) Ecological extinction of native prey of a carnivore assemblage in Argentine Patagonia. Biol Conserv 92:25–33

Novillo A, Ojeda RA (2008) The exotic mammals of Argentina. Biol Invasions 10:1333–1344

Noy-Meir I (1973) Desert ecosystems: environment and producers. Annu Rev Ecol Syst 4:25–51

Nuñez MA, Hayward J, Horton TR, Amico GC, Dimarco RD, Barrios-García MN, Simberloff D (2013) Exotic mammals disperse exotic fungi that promote invasion by exotic trees. PLoS One 8:e66832

OIE (2018) https://www.oie.int/wahis_2/public/wahid.php/Reviewreport/Review?page_refer=MapFul EventReport&reportid=27948

Ojeda RA, Mares MA (1982) Conservation of South American mammals: Argentina as a paradigm. In: Mammalian Biology in South America. Pymatuning Symposia in Ecol 6:505–521
Oliver WLR, Brisbin IL Jr, Takahashi S (1993) The Eurasian wild pig (*Sus scrofa*). In: WLR O (ed) Pigs, peccaries and hippos. Status survey and conservation action plan. IUCN/SSC Pigs and Peccaries Specialist Group and IUCN/SSC Hippo Specialist Group, Gland, pp 112–121
Papadakis J (1974) Ecología y posibilidades agropecuarias de las provincias argentinas. Enciclopedia Argentina de Agricultura y Ganadería 2:1–86
Pedrosa F, Salerno R, Borges Padilha FV, Galetti M (2015) Current distribution of invasive feral pigs in Brazil: economic impacts and ecological uncertainty. Natureza & Conservação 13:84–87
Pérez Carusi LC, Beade MS, Miñarro F, Vila AR, Giménez-Dixon M, Bilenca DN (2009) Relaciones espaciales y numéricas entre venados de las pampas (*Ozotoceros bezoarticus celer*) y chanchos cimarrones (*Sus scrofa*) en el Refugio de Vida Silvestre Bahía Samborombón, Argentina. Ecol Austral 19:63–71
Pescador M, Sanguinetti J, Pastore H, Peris S (2009) Expansion of the introduced Wild boar (*Sus scrofa*) in the Andean region, Argentinean Patagonia. Galemys 21:121–132
Pimentel D, Zuniga R, Morrison D (2005) Update on the environmental and economic costs associated with alien-invasive species in the United States. Ecol Econ 52:273–288
Podgórski T, Baś G, Jędrzejewska B, Sönnichsen L, Śnieżko S, Jędrzejewski W, Okarma H (2013) Spatiotemporal behavioral plasticity of Wild boar (*Sus scrofa*) under contrasting conditions of human pressure: primeval forest and metropolitan area. J Mammal 94:109–119
Polis GA (1995) Desert communities: an overview of patterns and processes. In: Polis GA (ed) The ecology of desert communities. University of Arizona, Tucson, pp 1–26
Price TD, YEH PJ, Harr B (2008) Phenotypic plasticity and the evolution of a socially selected trait following colonization of a novel environment. Am Nat 172:49–62
Relva MA, Sanguinetti J (2016) Ecología, impacto y manejo del ciervo colorado (Cervus elaphus) en el noroeste de la Patagonia, Argentina. Mastozoología Neotropical 23: 221–238
Rodriguez L (2006) Can invasive species facilitate native species? Evidence of how, when, and why these impacts occur. Biol Invasions 8:927–939
Roemer GW, Donlan CJ, Courchamp F (2002) Golden eagles, feral pigs, and insular carnivores: How exotic species turn native predators into prey? Proc Nat Acad Sci USA 99:791–796
Rosell C, Carretero MA, Bassols E (1998) Seguimiento de la evolución demográfica del jabalí (*Sus scrofa*) y efectos del incremento de presión cinegética en el Parque Natural de la zona volcánica de la Garrotxa. Galemys 10:59–74
Rosell C, Fernández-Llario P, Herrero J (2001) El Jabalí (*Sus scrofa* Linnaeus, 1758). Galemys 13:1–25
Ruiz-Fons F, Vidal D, Vicente J, Acevedo P, Fernández-de-Mera IG, Montoro V, Gortázar C (2008) Epidemiological risk factors of Aujeszky's disease in Wild boars (*Sus scrofa*) and domestic pigs in Spain. Europ J Wildlife Res 54:549–555
Saavedra M, Matus A (2000) Impacto de poblaciones de jabalí (*Sus scrofa*) y ciervo rojo (*Cervus elaphus*) en los Parques Nacionales Villarrica y Conguillío IX región, Chile. Informe de la Corporación Nacional Forestal Ug. Patrimonio Silvestre IX Región, Unidad Planificación/Operación Sección Flora – Fauna
Sáez–Royuela C, Tellería JL (1986) The increased population of the Wild boar (*Sus scrofa* L.) in Europe. Mammal Rev 16:97–101
Sagua MI, Figueroa CE, Acosta DB, Fernández GP, Carpinetti BN, Birochio D, Merino ML (2018) Inferring the origin and genetic diversity of the introduced Wild boar (*Sus scrofa*) populations in Argentina: an approach from mitochondrial markers. Mammal Res 63:467–476
Sanguinetti J, Kitzberger T (2010) Factors controlling seed predation by rodents and non-native *Sus scrofa* in *Araucaria araucana* forests: potential effects on seedling establishment. Biol Invasions 12:689–706
Sanguinetti J, Pastore H (2016) Abundancia poblacional y manejo del jabalí (*Sus scrofa*): una revisión global para abordar su gestión en la Argentina. Mastozoología Neotropical 23:305–323
Saunders G, Giles J (1995) Ecological comparison of two Wild pig populations in semi-arid and sub-alpine Australia. Ibex JME 3:152–155

Schiaffini MI, Vila AR (2012) Habitat use of the Wild boar, *Sus scrofa* Linnaeus 1758, in Los Alerces National Park, Argentina. Stud Neotropical Fauna Environ 47:11–17

Schley L, Roper TJ (2003) Diet of Wild boar, *Sus scrofa*, in Western Europe, with particular reference to consumption of agricultural crops. Mammal Rev 33:43–56

Scott C (1973) Seasonal food habits of European Wild hogs (*Sus scrofa*) in the Great Smoky Mountains National Park. MSc thesis, University of Tennessee, USA

Siemann E, Carrillo JA, Gabler CA, Zipp R, Rogers WE (2009) Experimental test of the impacts of feral hogs on forest dynamics and processes in the southeastern US. For Ecol Manage 258:546–553

Skewes O (1990) Status des Wildschweines, *Sus scrofa* L., in Chile. Dokumentation einer Fremdansiedlung von europäischem Wild. Dissertation Doktorgrades der Forstl. Fakultät, Georg-August Universität zu Göttingen, Germany

Skewes O, Jaksic FM (2015) History of the introduction and present distribution of the European Wild boar (*Sus scrofa*) in Chile. Mastozoología Neotropical 22:113–124

Skewes O, Rodriguez R, Jaksic FM (2007) Ecología trófica del jabalí europeo (*Sus scrofa*) silvestre en Chile. Rev Chil Hist Nat 80:295 307

Skewes O, Moraga CA, Arriagada P, Rau JR (2012) El jabalí europeo (*Sus scrofa*): Un invasor biológico como presa reciente del puma (*Puma concolor*) en el sur de Chile. Rev Chil Hist Nat 85:227–232

Snell-Rood EC (2013) An overview of the evolutionary causes and consequences of behavioural plasticity. Anim Behav 85:1004–1011

Sodeikat G, Pohlemyer K (2007) Impact of drive hunts on daytime resting site areas of Wild boar family groups (*Sus scrofa* L.). Wildlife Biol Practice 1:28–38

Steiner M, Linkov I, Yoshida S (2002) The role of fungi in the transfer and cycling of radionuclides in forest ecosystems. J Environ Radioact 58:217–241

Tesón M, Regis A, Huici N, Novak F (1997) Triquinelosis en jabalíes (*Sus scrofa*) en Dpto. Lacar, Neuquén, República Argentina. Revista Veterinaria Argentina 14:187–190

Van Riper CIII, Scott J (2001) Limiting factors affecting Hawaiian native birds. Stud Avian Biol 22:221–233

Verbist K, Santibañez F, Gabriels D, Soto G (2010) Atlas de zonas áridas de América Latina y El Caribe. CAZALAC. Documentos Técnicos del PHI-LAC N° 25, UNESCO

Villagra PE, Defossé GE, del Valle HF, Tabeni S, Rostagno M, Cesca E, Abraham E (2009) Land use and disturbance effects on the dynamics of natural ecosystems of the Monte Desert: Implications for their management. J Arid Environ 73:202–211

Welander J (2000) Spatial and temporal dynamics of Wild boar (*Sus scrofa*) rooting in a mosaic landscape. J Zool 252:263–271

White TCR (2008) The role of food, weather and climate in limiting the abundance of animals. Biol Rev 83:227–248

Wilcox JT, van Vuren DH (2009) Wild pigs as predators in oak woodlands of California. J Mammal 90:114–118

Wright TF, Eberhard JR, Hobson EA, Avery ML, Russello MA (2010) Behavioral flexibility and species invasions: the adaptive flexibility hypothesis. Ethol Ecol Evol 22:393–404

Chapter 11
Reconceiving the Biological Invasion of North American Beavers (*Castor canadensis*) in Southern Patagonia as a Socio-ecological Problem: Implications and Opportunities for Research and Management

Christopher B. Anderson, J. Cristóbal Pizarro, Alejandro E. J. Valenzuela, Natalia Ader, Sebastián Ballari, José Luis Cabello Cabalín, Valeria Car, Mara Dicenta, Erik A. Nielsen, Catherine Roulier, and Peter Van Aert

11.1 Biological Invasions Even at the End of the World

In the past few decades, natural scientists and natural resource managers have come to recognize that environmental problems have causes and consequences that are both ecological and social (Carpenter et al. 2009). For example, as an academic discipline, ecology increasingly acknowledges that ecosystems are not only

C. B. Anderson (✉)
Centro Austral de Investigaciones Científicas (CADIC), Consejo Nacional de Investigaciones Científicas y Técnicas (CONICET), Ushuaia, Tierra del Fuego, Argentina

Instituto de Ciencias Polares, Ambiente y Recursos Naturales (ICPA), Universidad Nacional de Tierra del Fuego (UNTDF), Ushuaia, Tierra del Fuego, Argentina

J. C. Pizarro
Facultad de Ciencias Forestales, Universidad de Concepción, Concepción, Bío Bío, Chile

Millennium Nucleus Center for the Socioeconomic Impact of Environmental Policies (CESIEP), Santiago de Chile, Chile

A. E. J. Valenzuela
Instituto de Ciencias Polares, Ambiente y Recursos Naturales (ICPA), Universidad Nacional de Tierra del Fuego (UNTDF), Ushuaia, Tierra del Fuego, Argentina

Consejo Nacional de Investigaciones Científicas y Técnicas (CONICET), Ushuaia, Tierra del Fuego, Argentina

N. Ader · V. Car · P. Van Aert
Instituto de Sociedad, Cultura y Estado (ICSE), Universidad Nacional de Tierra del Fuego (UNTDF), Ushuaia, Tierra del Fuego, Argentina

F. M. Jaksic, S. A. Castro, *Biological Invasions in the South American Anthropocene*, https://doi.org/10.1007/978-3-030-56379-0_11

composed of biotic and abiotic elements but also that humans form an integral part of what are ostensibly socio-ecological systems (Pickett and Ostfeld 1995; Anderson et al. 2015). As a result, efforts are being made to expand the ways that issues, such as biological invasions, ecological restoration, or biodiversity conservation, are studied and managed to address them not only as ecological systems but also their social domain and human dimensions (Collins et al. 2011; Díaz et al. 2015; Pascual et al. 2017). Yet, significant work still remains to achieve this goal. For example, studies about invasive exotic species have a clear biological bias at the national level in Chile (Quiroz et al. 2009), at the regional scale in Patagonia (Anderson and Valenzuela 2014) and the Southern Cone (Ballari et al. 2016), across the Latin American continent (Pauchard et al. 2011) and even globally (Estévez et al. 2015; Vaz et al. 2017). Therefore, the relative dearth of interdisciplinary, applied, and social studies about biological invasions explains why they continue to increase and currently constitutes a barrier to addressing it as a socio-ecological problem.

Even places that in our collective social imaginary are considered remote, pristine, or wilderness areas, such as Patagonia, actually have a long history of alteration resulting from complex and changing human relationships with nature (Moss 2008). For example, southern Patagonia and the Tierra del Fuego Archipelago have been home to diverse human settlers with different views of nature that have impacted its biodiversity. One current environmental issue that has captured the attention of researchers and some authorities is biological invasions (Anderson et al. 2006a; Valenzuela et al. 2014). However, for invasive species to be constructed as a socio-ecological problem it must be reconceived under a new paradigm that goes beyond the biological emphasis and incorporates these multiple human relationships to facilitate the implementation of more effective policies and actions.

S. Ballari
Parque Nacional Nahuel Huapi (CENAC-APN), Consejo Nacional de Investigaciones Científicas y Técnicas (CONICET), San Carlos de Bariloche, Río Negro, Argentina

J. L. Cabello Cabalín
Patagonia Wildlife Ltda, Punta Arenas, Magallanes y Antártica Chilena, Chile

Centro de Estudios del Cuaternario,
Punta Arenas, Región de Magallanes y de la Antártica Chilena, Chile

M. Dicenta
Department of Science and Technology Studies, Rensselaer Polytechnic Institute, New York, USA

E. A. Nielsen
School of Earth Sciences and Environmental Sustainability, Northern Arizona University, Flagstaff, AZ, USA

C. Roulier
Centro Austral de Investigaciones Científicas (CADIC), Consejo Nacional de Investigaciones Científicas y Técnicas (CONICET), Ushuaia, Tierra del Fuego, Argentina

Instituto de Sociedad, Cultura y Estado (ICSE), Universidad Nacional de Tierra del Fuego (UNTDF), Ushuaia, Tierra del Fuego, Argentina

On the one hand, ecologists have identified biological invasions as one of the three main drivers of modern global environmental change (Vitousek et al. 1997), and even isolated corners of the planet experience this truly global biological phenomenon. In fact, among the biomes of the Southern Cone, it is the remote sub-Antarctic forest ecoregion that has the highest number of invasive exotic terrestrial mammals (in both absolute and relative terms) (Ballari et al. 2016). Furthermore, in the Tierra del Fuego Archipelago, invasive exotic species even dominate the local assemblages of terrestrial mammals and freshwater fish by about 2 to 1 (Anderson et al. 2006a; Valenzuela et al. 2014). Consequently, Tierra del Fuego has become an ideal natural laboratory to study biological invasions (Valenzuela et al. 2014). However, even here scientific publications regarding invasive species' socio-political aspects have been almost non-existent, but paradoxically, researchers and managers of biological invasions in Patagonia actually state that these topics should be their top priorities (Anderson and Valenzuela 2014).

Therefore, not only is there a gap between specialists and the general public regarding the social perceptions and imaginaries of nature and environmental resource management, but there is also some incongruence among specialists regarding what they say is important (i.e., social and practical research) and what they have actually being studying about invasive species in Patagonia (i.e., basic ecology). The mismatch between stated preferences and ultimate actions by invasion biology specialists indicates that academic and management institutions are not articulating current values and concepts about their study objects or nature more generally (Vatn 2005). Because science is the collective production of acceptable truths that change over time, the active articulation of values and science requires continual consideration. Rather than being determined by facts and methodologies alone, scientific communities are constantly (re)organizing the sociocultural norms and values that constitute collective desirable ways of knowing and coexisting (Jasanoff 2004). In the face of these shortcomings in the current science-society linkage regarding invasive beavers, an important aspect for effectively addressing this problem is to find new ways to conceive, study, and manage invasive exotic species. It would behoove both scientists and practitioners to actively engage in the process of this paradigm shift (sensu Kuhn 1970) to incorporate the sociocultural norms and values of this particular time and place.

In this chapter, we tackle portions of this challenge. We present the case of the North American beaver (*Castor canadensis*) in Tierra del Fuego to show how it can be reconceptualized from a biological invasion to a socio-ecological problem. Lessons are drawn from two distinct aspects of the case: (a) the academic approach needed to carry out socio-ecological research on biological invasions and (b) the personal and institutional experience of interdisciplinary teamwork to achieve a new way of studying and managing invasive species. The exploration of this case study should allow for a new synthesis of the ecological and social information about invasive beavers in southern South America, but at the same time it should prove useful beyond invasion biology as part of the socio-ecological paradigm shift taking place in environmental research and practice (e.g., ecological restoration, Suding et al. 2015; conservation biology, Teel et al. 2018).

11.2 The Beaver as a Boundary Object

To consolidate interdisciplinary research and formalize teamwork, we used the invasive North American beaver (*Castor canadensis*) in Tierra del Fuego as a "boundary object" (Star and Griesemer 1989), defined as common objects of study or management (e.g., components of biodiversity) by various communities of practice (e.g., social scientists, ecologists, wildlife biologists, and managers) that are sufficiently malleable to adapt to the requirements of various groups, while maintaining common traits and meaning among these groups. As an analytical tool, boundary objects also serve as the limits to understand the relationship between individual practices and institutional configurations that maintain disciplinary frontiers (Bowker and Star 1999). Therefore, this philosophical concept can be useful to better design interdisciplinary research, as well as align research with policies to successfully address biological invasions. For the purposes of this book chapter, we will divide these efforts into how the beaver serves as a boundary to understand this biological invasion's (a) history, past and present conceptualizations of the beaver and invasive species generally; (b) disciplines, social and biophysical domains of research about the beaver; (c) management, the association between academic and practical issues; and (d) stakeholders, engagement of different social actors on how science and society interact on this topic (Fig. 11.1). Then, we used these categories at the frontier of different realms of understanding to organize, summarize, and assess the state of knowledge regarding *Castor canadensis* in southern Patagonia as part of a critical literature review (sensu Grant and Booth 2009). Peer-reviewed publications and gray literature (e.g., government reports) were used by searching academic databases (e.g., Web of Knowledge and SciELO), checking the citations within publications and the authors' collective experience working on this case since the late 1990s.

11.3 History: Present and Past

As a boundary object, the beaver offers a metaphorical hinge to open the door to both present and past understandings of this species and of nature generally (Archibald et al. 2020). We know that in 1946 twenty beavers were introduced from Canada to Tierra del Fuego by the Argentine Ministry of War (Pietrek and Fasola 2014), which had been presided over by Juan D. Perón from 1944 to 1945. However, this initiative should be understood as a part of broader efforts that were subsequently implemented by Perón as president (1946–1955) to transform the very identity of Argentina from a country that exports food to the rest of the world (i.e., "*Argentina, el granero del mundo*" – Argentina, the world's granary) to a modern, industrial nation (Mateo and Carreras-Doallo 2013). Plus, historically Argentine policies, especially those advanced by presidents like Domingo F. Sarmiento (1868–1874), had promoted European immigration and development initiatives that

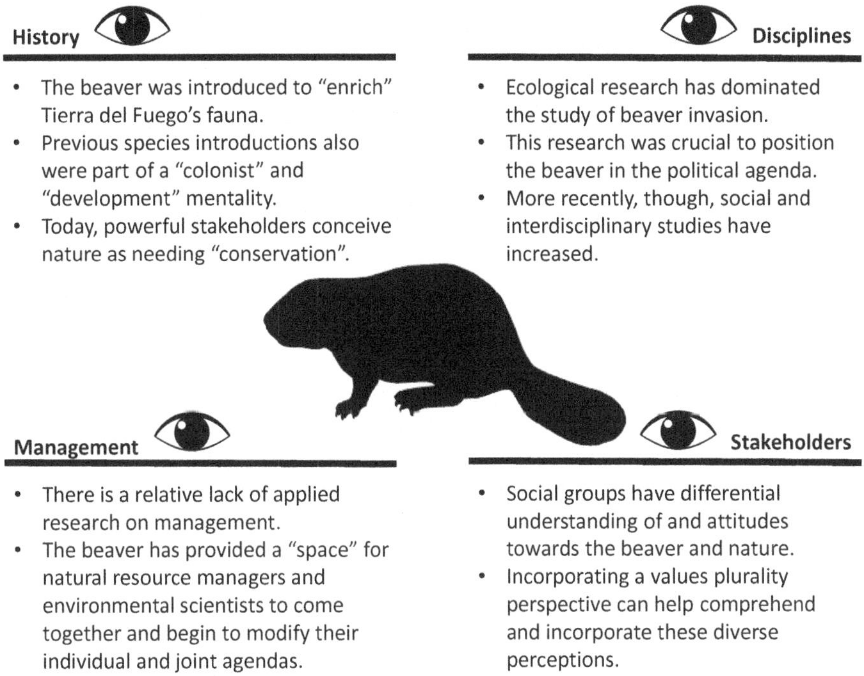

Fig. 11.1 The invasive North American beaver (*Castor canadensis*), introduced to Tierra del Fuego in 1946, was used as a boundary object to analyze this biological invasion from different perspectives that are relevant to different academic and social communities of practice: (**a**) history, past and present conceptualizations of the beaver and invasive species generally; (**b**) disciplines, social and biophysical domains of research about the beaver; (**c**) management, the association between academic and practical issues; and (**d**) stakeholders, engagement of different social actors on how science and society interact on this topic

had tacit or implicit ideas about the superiority of northern European "races" and cultures (Fernández 2017). Indeed, in government propaganda from the time of the beaver's introduction to southern Argentina, emphasis is placed on the release as an effort to "enrich" the Fuegian fauna, and in the same newsreel images of development and purported progress also are shown from in other parts of Patagonia, equating the introduction with agricultural expansion and oil exploration (Anonymous 1946). Consequently, in Argentina, the beaver was protected from hunting until 1980, demonstrating the mentality that believed that this species was a valuable resource. In Blanco and Bárquez (1980), the authors also stated that early technical reports were commissioned about this species precisely to evaluate its process of acclimatization in Tierra del Fuego, indicating that the early concerns about this species were for its survival, rather than considerations on its impacts over the native ecosystem.

However, this development vision began to change in by the 1980s, when the first scientific technical reports appeared in Argentina (Blanco and Bárquez 1980; Marconi and Balabusic 1980) and the first peer-reviewed publication in Chile

(Sielfeld and Venegas 1980), as researchers and managers began to account for the negative impacts of beavers in Fuegian ecosystems. Today, we are arguably in an era where Patagonian nature is conceived by some social groups as "fragile" or "pristine", and wilderness conservation is the outcome of this mentality (Mittermeier et al. 2003; Holmes 2015). In turn, environmental management policies regarding invasive exotic species in large part seek to reconstruct pre-European species assemblages, such as the binational treaty signed between Argentina and Chile in 2009 (Resolución SAyDS 157/10), a vision that portrays nature mostly as a static, ahistorical (and perhaps romantic) entity that can be separated from its sociocultural and biotic history (Robbins and Moore 2013; Lorimer 2015). Nonetheless, this switch from a "development" to a "conservation" social imaginary is not absolute, and presently the beaver's image is also used by some sectors to promote local identity and tourist activities. For example, the beaver is the symbol of the region's premier ski resort (*Cerro Castor* or Beaver Mountain) and is featured in images frequently used for tourism. While there is largely consensus among scientists and managers regarding the fact that invasive species, especially the beaver, are a problem in this part of the world (Zagarola et al. 2014; Anderson et al. 2017), at the same time some politicians continue to promote territorial development based on invasive exotic species, such as a recent effort in Argentina to impose exotic salmon farming in the Beagle Channel, which has shown itself to be hugely environmentally and socially problematic in Chile (Vesco 2018).

In summary, diverse social meanings of the natural world are held by different groups coexisting over historical periods. Society, therefore, is conditioned, but not caused by its natural surroundings, and instead, it is sustained by complex relationships with nature (Castoriadis 2007). In turn, these social imaginaries play a defining role in the specificity of a particular socio-ecological system and allow us to better understand how specific institutions, groups, or historical periods perceived, constructed, and reproduced meanings and significance about species, including invasive ones (Archibald et al. 2020). However, it is important to recognize that these human-nature relationships and divides are not static and therefore cannot be considered as an objective fact that transcends its context. Nonetheless, institutional structures, such as policies, the media, or scientific paradigms, can provide enough continuity that ideas about the natural world solidify for sufficient time to make them seem constant. Using a boundary object like the beaver provides a window to understand the past and present human-nature relationships with invasive exotic species.

11.4 Disciplines: Biophysical and Social Domains

While the earliest peer-reviewed publication about North American beavers in Tierra del Fuego dates from 1980 (Sielfeld and Venegas 1980), it was not until the 1990s that natural scientists began to take significant interest in this topic as a biological invasion as the sub-discipline of invasion biology consolidated (e.g., first English-language university textbook Williamson 1996). Subsequently, there was explosive growth in basic ecological studies of the beaver, which allowed scientists to establish a strong knowledge basis regarding the biophysical domain of this biological invasion. A suite of ecological studies has concentrated on quantifying diverse aspects of the beaver's impacts on (i) expanding across the archipelago at high densities and even colonizing the mainland (Lizarralde 1993; Skewes et al. 2006; Anderson et al. 2006a; Wallem et al. 2007; Sanguinetti et al. 2014); (ii) simplifying the diversity and increasing the productivity of stream food webs, but only at the habitat scale (Anderson and Rosemond 2007, 2010); (iii) increasing carbon and nutrient retention of watersheds (Lizarralde et al. 1996; Anderson et al. 2014a) and increasing the decomposition rate of organic matter, at least for lenga (*Nothofagus pumilio*) leaves (Ulloa et al. 2012); and (iv) affecting riparian vegetation dynamics by reducing tree regeneration and allowing the establishment of exotic grasses (Anderson et al. 2006b; Martínez et al. 2006; Arismendi et al. 2008; Baldini et al. 2008) that persist for decades (Wallem et al. 2010). Additionally, basic studies have been conducted in the earth sciences, including geomorphology of impacted streams (Coronato et al. 2002) and hydrogeomorphic process that affect vegetation patterns (Westbrook et al. 2017). Also, based on these initial studies, ecological models and theory were developed to understand the distribution and ecological impacts of this species (Wallem et al. 2007) and compare it to its native range (Anderson et al. 2009). At the landscape level in southern Patagonia, the beaver's presence was found to be affected by time since colonization and river sinuosity (Davis et al. 2016), as well as vegetation cover, forest type, stream presence, and topography (Henn et al. 2016). At the same time, evidence also indicates that previously thought to be suboptimal habitat, like the Patagonian steppe, does not present a demographic obstacle for beaver expansion to the north (Pietrek and González-Roglich 2015; Pietrek et al. 2016).

Studies of the beaver's social domain appeared later than the natural science publications mentioned above. For example, Santo et al. (2017) showed that while Argentine and Chilean ranchers on Tierra del Fuego have diverse knowledge of the way their land is affected by beavers, the ecosystem services most salient to them included both instrumental values related to production (e.g., water for drinking, animals, and irrigation) but also intrinsic and relational values (e.g., landscape beauty, ecosystem health, balance of nature, and biodiversity). Another integrated study used social surveys to complement dendrochronological analysis used to establish the date of beaver arrival to the South American mainland, which was previously reported for the 1990s (see above) but may even date as early as the 1960s (Graells et al. 2015). From an anthropological perspective, invasive beavers have even been studied recently as a diasporic community (Ogden 2018), and an interdisciplinary collaboration between natural scientists and artists experimented with methods to promote non-anthropocentric forms of communication among researchers and beavers by using scents (Graells et al. 2017). Currently, two doctoral candidates in social sciences (C. Roulier and M. Dicenta, co-authors of this chapter) are using the beaver to explore socio-institutional issues surrounding environmental management and the environmental history of Tierra del Fuego.

Also, given that the media play a central role in focusing attention and reenforcing certain representations in the social imaginary, an analysis was conducted regarding the ways that invasive exotic mammals appear in the Argentine press at national and provincial levels from 2012 to 2015 (Car et al. in press). The starkest finding of this study was that overall biological invasions are not newsworthy enough to be the front-page of any of the studied newspapers at the national or provincial levels. In particular, in Tierra del Fuego, invasive beavers are a problem for natural and rural areas and do appear in the media, but more than 95% of people in southern Patagonia live in cities. Therefore, the absence of beavers as a newspaper cover story demonstrates that even though the topic has been positioned by scientists and managers at the political level, it is still not a high priority or of great public interest. Plus, at a national level, the term "exotic" was found to be a polysemy and also convey positive attributes of nature, such as exuberant and beautiful. Nonetheless, in Tierra del Fuego scientists and government authorities are the ones who speak most frequently to the press about the problems particularly associated with invasive beavers and feral dogs. Even though these sources often spoke from an ecological-scientific perspective, these two species have recognizable social impacts, which demonstrate that when a socio-ecological problem also has territorial or human relevance they can be made more visible and be integrated into public discourse.

11.5 Management: Theory and Practice

After the decision to introduce the beaver in the 1940s, there are no other specific management actions until 1981, when Argentina authorized legal hunting, and 1993, when Chile declared it a harmful species and in so doing authorized its year-round hunting (Anderson et al. 2011). The first such efforts where focused on removing problematic individuals (e.g., those affecting infrastructure, mainly roads) without taking into account the ecological context. In Tierra del Fuego National Park, some unconventional methods were used, such as burning or dynamiting the dams and lodges, and dams were often removed with backhoes, but these actions were largely ineffective (Sanguinetti et al. 2014). By the 1990s, there were a number of projects and studies being conducted at the behest of management agencies in both countries (e.g., Lizarralde and Escobar 1999; Skewes et al. 1999). In 2001, Tierra del Fuego National Park (Argentina) began to work on the creation of the first holistic plan to control beavers developed between scientists, managers, and park rangers (Sanguinetti et al. 2014). This program is the longest-running, continuous management strategy and was designed to include (i) the biological characteristics of the species, (ii) the ecological context of the invaded ecosystem and native species, (iii) management considerations and trapping logistics, (iv) the presence and activities of the national park's visitors, and (v) secured funding through the Argentine National Parks Administration. The park's sustained efforts have successfully controlled this invasive species at the scale of one protected area, reducing the abundance of beavers, avoiding reinvasions, and preventing that new forest areas are impacted (Sanguinetti et al. 2014).

In 2004, government agencies from both countries (e.g., for Argentina the National Parks Administration-*APN* and for Chile the Agriculture and Livestock Service-*SAG*) held the first binational meeting to address the problem of invasive beavers at the Austral Center for Scientific Research (*CADIC-CONICET*). Subsequently, a total of six such events were held with managers and researchers, allowing the two nations to come to agreements on short- and long-term strategies, including implementation of parallel and coordinated incentives and training for trapping to lower beaver densities and reduce impacts, the signing of a binational strategy to control beavers via eradication, and carrying out a technical and economic feasibility study on beaver eradication (Mann 2008). Beginning in 2006, the international NGO Wildlife Conservation Society (WCS-Chile) became involved and was an important catalyst for these binational meetings, bringing international media attention and specialists from the United States and New Zealand as advisors (Ogden and Holmes 2015). In this same year, the provincial government of the Argentine portion of Tierra del Fuego promulgated Law #696/2006 that declared the beaver to be a "harmful species." Later, in 2013, the Argentine National Environment Council (*COFEMA*), called for the deepening research, monitoring, control, and removal of this species by way of increasing the planning and collaboration between the involved jurisdictions (*Resolución COFEMA* #261/2013). The next year the problem was elevated to a national concern, and the beaver was declared an invasive exotic species by the Secretary of the Environment and Sustainable Development (*Resolución SAyDS* #1048/2014).

Based largely on these formal and informal interactions between managers and scientists, a community of practice began to form. This community produced an increasing body of literature regarding the beaver's ecological impacts, which make it possible to position the topic of an invasive species in the political agenda, even though perhaps not a top priority. Over time, however, natural scientists and conservation NGOs had increasing predominance with their traditional ecological/conservation approach. Early on, natural resource managers promoted the seminal studies of invasive beavers in Tierra del Fuego and implemented control efforts aimed at using the beaver for socio-economic benefits. For example, in Chile a regional development fund for the Magallanes Region of Chile (*FONDEMA*) financed a project from 2004 to 2007 to train trappers and pay bounties for pelts, but it also developed thesis projects in the University of Magallanes that dealt with issues involved in the use of the beaver for fur, meat, and handicrafts (Estay 2007; Bahamonde 2007; Paillacar 2007; Caibul 2008). Eventually, seven restaurants in Chile used this as part of the regional wild-caught cuisine, and also over the period of the project the market for furs went from one buyer before the project to five by 2007 (Soto and Cabello 2007). However, beginning with the stronger leadership in 2006 of scientists and NGOs, the goal switched to eradication, and efforts to place a value on the beaver were eschewed (Parkes et al. 2008). Consequently, when

Argentina and Chile signed a binational agreement in 2008, they affirmed their commitment to "restore beaver-degraded ecosystems to their natural condition," by promoting its eradication, and the human dimension of this agreement was relegated to generating scientific information to inform the public and ostensibly keep them from blocking these efforts (see Menvielle et al. (2010) and Malmierca et al. (2011) for general information on the agreement and *Resolución SAyDS* 157/10 (2018) for the text itself).

However, in part due to these interactions, researchers began to modify their own research agenda and conduct more practical management studies, which had originally been commissioned by the local government agencies. Many of these studies, however, remained purely basic research, but the study was justified in the context of management. After signing the binational agreement, for example, there was a need to determine beaver dispersal pathways and processes to plan effective areas needed for eradication (Pietrek et al. 2017), which also can be studied with genetic techniques (Lizarralde et al. 2008; Fasanella et al. 2010). Furthermore, applied ecological assessments were required to understand vegetation regeneration dynamics and determine whether new techniques could increase the survival of seedlings and facilitate riparian forest restoration (Henn et al. 2014). Plus, landscape ecology methods were employed to help determine simple tools that could help prioritize areas for trapping, such as using rivers sinuosity as a highly significant explanatory variable for presence/absence of beavers in watersheds that had intermediate colonization times (Davis et al. 2016).

However, the management of this biological invasion, particularly efforts to eradicate it, requires new knowledge of the socio-ecological system's human dimensions (see integral feasibility assessment in Wallem et al. 2013). In this regard, residents of the Magallanes Region in southern Chile reported a willingness-to-pay of approximately $7.1 million USD per year (2011 values) to restore forest resources affected by beaver invasion, which was based mostly on "heritage value" (48.7%), while option, existence, and use values were between 15% and 19% each (Soto Simeone and Soza-Amigo 2014). Furthermore, this same study found that the lowest socio-economic group was willing-to-pay a greater proportion of its income to achieve this goal. Later, between 2012 and 2015, the only binational study to date (i.e., including both Chilean and Argentine portions of Tierra del Fuego) characterized the perspective of ranchers from both countries regarding potential programs and incentives to participate in beaver eradication and subsequent forest restoration. In general, ranchers agreed with eradication (67%), and it was possible to increase their willingness to participate by increasing payments, increasing the expectation of program success and decreasing requirements for landowner involvement (Santo et al. 2015). Paradoxically, while about 90% of Ushuaia residents who were surveyed in Tierra del Fuego National Park were aware that the beaver has negative ecological impacts, only around 50% supported invasive species eradication programs (Anderson et al. 2016).

From the above literature review, it is evident that there is still a conspicuous absence of research into trapping, considering not just effectiveness and efficiency but also ethical concerns, animal welfare, and responsible life management. In fact, despite the fact that between 2004 and 2016, nearly 12,000 beavers have been trapped in the region, there is still not a standardized trapping protocol between the Argentina and Chile, which is necessary to coordinate best practices, including the sustainability of the projects and to address concerns regarding the humane treatment of nonhuman animals. Moreover, this case shows the need of a paradigm shift that engages with less anthropocentric modes of knowing nature and nonhuman animals. Studying the future and trajectory of these post-beaver landscapes would help move beyond understanding only the economic and biodiversity damages to human outcomes but encompassing the relationships among different species, such as beavers, birds, humans, and trees, with particular modes of socializing and coexisting. Consequently, the need for reconceptualizing this biological invasion is not just an issue of conducting more social research about the beavers impacts on people or how to get society to support eradication but also of delving into other areas that are less familiar to the mostly natural scientists who have traditionally taken an interest in invasive species.

In this sense, while the research agenda still shows a significant bias towards basic ecological questions, the beaver as a boundary object also provided a conceptual space where scientists, managers, and policy-makers have been able to come together to understand one another better and even bring their agendas into better alignment. Indeed, the topic of beaver invasions continues to be discussed in regular meetings between Chile and Argentina's foreign relations officials and environmental authorities, including the yearly Austral Integration Commission, as well as the working group that follows up on the binational agreement. It is important to point out, however, that even though this is a relatively homogenous group, environmental managers and scientists do not always agree on their priorities or perspectives (Anderson et al. 2017).

Clearly, the scientific and political agendas have been in significant dialogue at a binational level thanks to these invasive beavers bringing the two realms into the same room, and these collaborations have led to such tangible products as both countries obtaining several million dollars in funding from the Global Environment Facility (GEF) to conduct pilot eradication projects (GEF 2013). Plus, the Chilean Agriculture and Livestock Service (*SAG*) has continued to promote local capacity-building and trapping on ranches and private property in the Chilean portion of the island, via a the National Fund for Regional Development (*FNDR*) project that ended in 2017. However, the main objective at both the political and discursive levels since the two countries signed the binational agreement in 2008 and obtained GEF funding has been to do away with incentives for individuals to trap beavers, and instead priority has been given to eradication via professional hunters who are paid to remove beavers from an area, rather than facilitating traps or paying bounties to individuals.

11.6 Stakeholders: Science and Society

We know that different stakeholders perceive, relate to, and steward the environment in different ways (Bennett et al. 2016). As such, it is logical that there would be a differential understanding of and attitudes towards beavers. Ranchers, for example, largely perceive the beaver to be a harmful species, but at the same time can hold contradictory values, such as it increasing the landscape's beauty (Anderson et al. 2014b). Similarly, on Navarino Island, the beaver is largely considered to simply be a "typical" species on the island without overly negative connotations (Schuettler et al. 2011). Some social groups even identify with the beaver, as they too considered themselves to be industrious colonists, like the beaver (Berghoefer et al. 2010). In this context, it is important to understand that ecological information and perspectives can be dominant in the academic literature, the media, or public policies.

However, scientific ecology only represents one voice in the science-society dialogue and often is not the primary source for other viewpoints (e.g., animal rights activists who oppose eradication or tourism operators who use the beaver as a symbol for Tierra del Fuego, see also Estévez et al. 2015). Scientists may have high-quality information, arguments grounded in empirical evidence, and in some contexts may even have more opportunities than other social actors for their voices to be heard, as occurs in Argentine Tierra del Fuego; local media report mostly what scientists and authorities say on these issues (Car et al. in press). For example, Anderson et al. (2016) found that about 90% of Ushuaia residents recognize the beaver as a harmful species, but less than half supported removal of invasive species from the island by lethal means. Therefore, a true science-society dialogue is more than just generating information to educate the public (Wynne 1992; Fischer 2009), and instead scientists and managers also need to understand the other values, perspectives, and interests that are at stake (e.g., Haider and Jax 2007; Berghoefer et al. 2008) to not only determine socially acceptable control strategies (Estévez et al. 2015) or to enhance the legitimacy/objectivity of studies by including facts and voices that might escape expert opinions (Harding 1991) but also democratize science and establish ecosystem restoration baselines that go beyond simply ecological or historical criteria (Suding et al. 2015).

11.7 Achieving a Paradigm Shift in Biological Invasions Research

Multiple values, diverse knowledge sources, and varied social representations are embedded in our understanding, research, management, and communication about the biological invasion of beavers in Patagonia. Therefore, conducting more holistic science or participating in more effective actions in dealing with invasive species requires a paradigm shift that enables perspectivism in science or the recognition of not only multiple ways of observing reality but also the very existence of multiple social and natural worlds (Law 2011). However, the field of ecology, which has dominated invasion biology, largely lacks philosophical training (Graham and Dayton 2002), but researchers and managers must understand that science is a social practice (see Díaz 1996). As such, interdisciplinary studies require not just training in various research methods, but also attention must be paid to cultivating the culture and institutions that will allow complex socio-ecological phenomena to be addressed by diverse groups of participants.

This review of the study and management of invasive beavers in Tierra del Fuego was developed from the authors' long-term research, education, and conservation efforts in both the Argentine and Chilean portions of Tierra del Fuego. The overall vision that has guided the team involved in this chapter includes the principles of integration, collaborative networks, and linking basic research with socially relevant applications (Anderson et al. 2011). Plus, to achieve a joint vision, three complementary approaches have been used to promote the consolidation of a socio-ecological paradigm within this group regarding the study and management of biological invasions. Specifically we have (a) developed an integrated socio-ecological conceptual model, (b) utilized philosophical tools to promote dialogue and understanding between disciplinary/personal perspectives, and (c) made the project itself an object of attention, study, and effort (e.g., interviews with participating researchers).

First, to achieve integrated research questions and a common understanding between project participants, we adapted the pulse-press dynamics model (PPD, Collins et al. 2011) by working in multiple sub-projects that occurred over various years in the context of this broader initiative. The resulting integrated conceptual framework (Fig. 11.2) encompasses both the biophysical and social domains of this human-nature system (humans-beavers-Tierra del Fuego), but does not require every researcher (or every sub-project) to engage the entire integrated socio-ecological system simultaneously. Nonetheless, by visualizing and comprehending the relationships of disciplinary questions with the overall problem, it allows for transformative questions, even when they are small or specific areas of inquiry in the iterative way that García (2006) defines interdisciplinarity.

Then, to consolidate the integrated research and teamwork, we applied best practices identified in the literature, which include philosophical tools and concepts for interdisciplinary dialogue (NAS 2005; Eigenbrode et al. 2007). For example, the members of the research team carried out iterative planning and research meetings but also conducted outreach efforts, including developing signs for the national park, among other things. Furthermore, beyond academic research, there have also been iterative events, such as advisory councils, bilateral commissions, and other governmental structures. These spaces have been important for interaction with other stakeholders and to gain trust, mutual cultural and institutional understanding, and vocabulary interchange for creating science-policy platforms and implementing

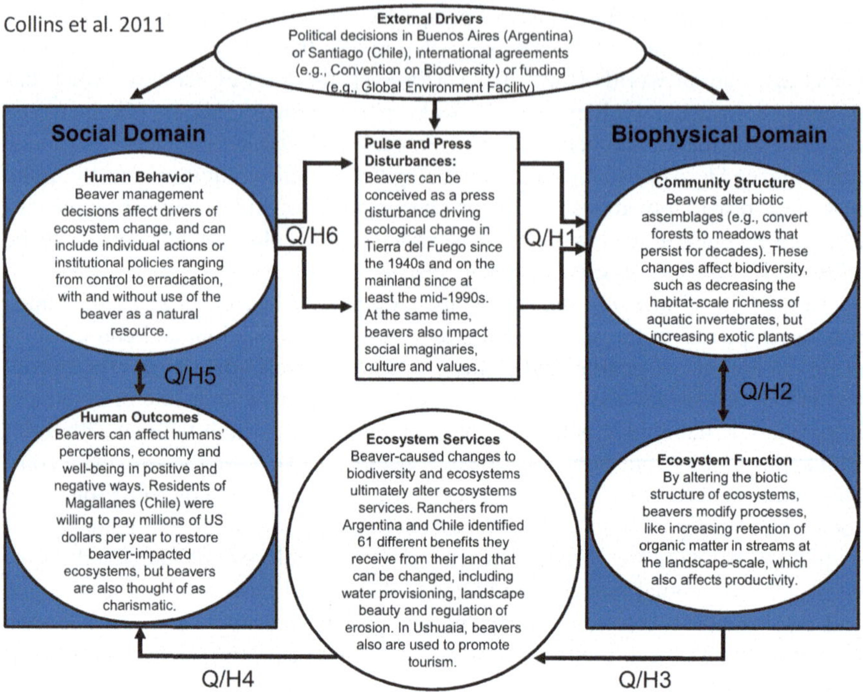

Fig. 11.2 The pulse-press dynamics (PPD) conceptual model was developed by a group of natural and social scientists associated with the US Long-term Ecological Research Network (Collins et al. 2011). To develop interdisciplinary projects, it is useful to employ these integrated frameworks to allow a diverse group of academic disciplines to identify components of the system where each can carry out specific research questions with questions/hypothesis (indicated by Q/H and arrows) that when taken together allow a mechanistic rather than a merely descriptive understanding of the socio-ecological system. At the same time, the integrated vision of the system emerges by the iterative deconstruction and reconstruction of the whole by the team in what García (2006) defines as interdisciplinary and furthermore allows different individuals and disciplines to see the importance and value of their work in the context of the whole, rather than feeling like a junior partner or add-on to achieve superficial quotas for multiple disciplines or participants in a project

co-designed actions. Throughout these activities, the beaver specifically and the concept of biological invasions more broadly became a "bridge" to integrate and coordinate experiences, questions, and results. These include the organization of a participatory symposium around interdisciplinarity and the design and implementation of a team-taught graduate course entitled "Linking human and natural dimensions to understand and conserve socio-ecological systems."

Finally, over the course of these experiences, we have been introspective and self-aware of the transformative work we are attempting to do not only at the academic or applied levels but also as individuals or as institutions. We have conducted surveys and ethnographic interviews among the group participants to reflect on the experience of collaborating, showing at least three dimensions. First, we discovered that the interviews themselves worked as a therapeutic tool for promoting group identity among this epistemic community. While some participants had early doubts regarding their contribution to the project, during and after the interview process, they became aware of the knowledge they had gained/shared about the other's views and about the imprints of some of our core scientific concepts into the group project. The interviews themselves helped identify achievements, actively reflecting on changes, and valuing the group itself. Second, some limits of interdisciplinarity became clear, such as the tendency to identify a person as the representative of an entire discipline, something that might endanger validity, as the rest of the team cannot provide peer accountability. In this line, disciplinary power asymmetries were noticeable as well, between social and natural sciences, and in general projects continue to be led by natural scientists. Of course, however, there is a need for mutual leadership and an enrichment of understanding in both directions. On the other hand, though, despite its benefits, interdisciplinarity requires time that is not always possible to have, as when actions depend on a presidential term. Combined with the integrated model, the interviews showed the capacity to modify our work along the way or to confirm that the conceptual models applied were achieving the results of these diverse disciplines coming to understand the other(s).

At the same time, it is apparent from this diagnosis that not all aspects of interdisciplinarity are intellectual or conceptual; instead they also deal with the social context and the psychology of the individual participant. The success of this collaboration between disciplines and sectors has been possible because it also considered the human beings involved in the socio-ecological system. This phenomenon also became salient during the interviews in two ways. First, through the concept of insecurity used by some participants to describe themselves and the facility to relate to others who also tended to provide their knowledge with self-doubt. Beyond our experience, it has already been argued that insecurity is vital for our contemporary complex environmental problems, which transcend regional, temporal, and disciplinary borders. As Latour has put it, reflexivity does not mean scientific mastery but rather becoming aware that "consciousness does not mean full control" (Latour 2003, p. 36). Second, and related to the idea of insecurity, the importance of the subjects was also reflected when participants acknowledge the respect and affective relations among them. The process of collaboration showed that not only sharing a research problem was effective for teamwork, but also the absence of conflictive personal relations among participants was key to success. In this way, the researchers, managers, and also the humans living in Tierra del Fuego are not just study "objects" for research or management, but they are rather "subjects" with their own individuality, rights, and needs. Specifically, regarding the team involved in this interdisciplinary case study, the researchers understand themselves to affect and be affected by their process of being involved in the study. This important realization is part of successfully working at the interface of different disciplinary (academic) or social realities; it has helped participants feel comfortable in that which is uncomfortable. In fact, with this perspective, natural scientists do not merely educate society or social science about the ills of biological invasions. Rather, multiple forms of understanding nature and invasive species are allowed to coexist as a way to democratize knowledge and to actively include plural perspectives and values at the same level, from a world of biotic interactions, drivers, and feedback loops between biodiversity and ecosystem function to a world of stakeholders, social networks, public policy cycles, and communication strategies.

A final important lesson in this process is to understand these multiple visions as sometimes complementary and others as different views of the world that exceed each other to understand the broader social process and the relationships that are developed between the individuals in the team. Plus, this level of commitment to the project only is possible if there is a willingness to understand the other and modify one's self, which requires patience, confidence, and time. Therefore, it helps when the topic being addressed by the interdisciplinary team is problem-based, rather than emanating from a merely theoretical question that is only relevant to one subgroup. Also, the demands of time and effort make it necessary for there to be institutional support or strong leadership to overcome changes in group membership, funding, or other intrinsic and extrinsic factors that can disrupt the continuity required to achieve transformative work. In all of these cases, it is necessary to have the skills and strategies that go beyond basic research or conceptual methodologies and instead understand the social process of science itself and have the individuals, leaders, institutions, culture, and training necessary to implement new ideas for more effective, situated, and ethical management and research (Kuhn 1970; Díaz 1996).

11.8 Final Remarks

Only recently has "nature" begun to be studied from an interdisciplinary perspective in southern Patagonia, but this is not very different than global trends. The Intergovernmental Science-Policy Platform on Biodiversity and Ecosystem Services (IPBES) was established in 2012 as a way to consolidate and continue previous work by the Millennium Ecosystem Assessment (MA 2005), but specifically it has the mandate to be beyond an academic understanding of human-nature relationships and be more policy relevant, which is achieved by mechanisms for political legitimacy and also the inclusion of diverse knowledge into the assessment of biodiversity and human well-being (Díaz et al. 2015; Pascual et al. 2017). In the case of southern Patagonia, we see examples arising of this shift whereby, for example, watersheds in Ushuaia (Argentina) and Punta Arenas (Chile) were assessed with social surveys (Zagarola et al. 2014) and biological sampling (Zagarola et al. 2017) to determine both socio-cultural values of these ecosystems and also bioindicators of stream health. This approach also showed that "specialists" (scientists and managers) and the general public perceived some issues differently, whereby specialists considered invasive exotic species as a major environmental threat, while the general community did not place a high priority on this driver. At the same time, the study of social imaginaries and media representations can facilitate strategies that impact the public agenda and political policies. Clearly, much work is still to be

Table 11.1 Different paradigms have been used to conceive the human-nature relationship in ecology generally and as it relates to invasion biology specifically. Four ways to consider these relationships are (a) the exclusion of people from ecosystems, (b) the consideration of humans as drivers of disturbance and ecosystem change, (c) the recognition that humans are the recipients of ecosystem goods and services, and (d) the incorporation of a holistic perspective whereby humans simultaneously affect and are affected by nature. In this table, literature regarding invasive beavers in Tierra del Fuego is organized using these categories, illustrating how not only are most studies about the biological invasion of beavers biased towards ecological studies, but they are also within the paradigm of considering humans as drivers of ecosystem disturbance. This paradigmatic limitation of considering humans as a unidirectional effect on nature can help explain the relative lack of success in managing biological invasions, because the complex reality of this socio-ecological problem is not currently being addressed. Recognizing the paradigmatic lenses being used to understand, study, and manage invasive beavers may help expanding research and actions to be more holistic and effective

Human-nature paradigms	Study unit	Research topics regarding biological invasions	Literature on invasive beavers in Tierra del Fuego
Humans omitted from ecosystems	"Natural" ecosystems and biodiversity	Natural history of native species and ecosystems Study and conservation of "pristine wilderness" areas	
Humans as drivers of disturbance and ecosystem change	"Natural" and anthropogenic ecosystems and biodiversity	Invasive species autecology Invasive species impacts Native ecosystem ecology Eradication techniques "Natural" ecosystem restoration	Sielfeld and Venegas (1980), Lizarralde (1993), Lizarralde et al. (2004, 2008), Anderson and Rosemond (2007, 2010), Anderson et al. (2006a, b, 2011, 2014a, b), Baldini et al. (2008), Arismendi et al. (2008), Soto Simone and Soza Amigo (2014), Henn et al. (2014, 2016), Pietrek et al. (2016, 2017), Davis et al. (2016)
Humans as recipients of ecosystems	"Natural" and anthropogenic ecosystems and biodiversity	Ecosystem services Social perceptions, values of nature Social and media representations of nature	Berghoefer et al. (2008, 2010), Schuettler et al. (2011), Anderson et al. (2016), Anderson et al. (2017), Santo et al. (2017), Car et al. (in press)
Humans as co-participants in ecosystems	Systems with "historical" and "novel" biotic, social, & cultural assemblages	Human-centered design of conservation project Community-based management Participatory decision-making Governance	Santo et al. (2015)

done in the integration of social and natural dimensions of environmental issues, but in the case of beavers, advances are being made to update the paradigm (Table 11.1). It is our hope, though, that lessons can be drawn from this example and be applied to future socio-ecological problems to be more effective, efficient, and ethical in both their study and management.

Acknowledgments This work was funded by numerous agencies and grants, including a PIDUNTDF-A project (#02/16), two UNTDF-APN PEININ projects (2015), and two US National Science Foundation grants in the programs on Coupled Nature-Human Systems (*Ecology, Culture and Outcomes: Linking Human Perceptions and Socio-Ecological Thresholds for Ecosystem Restoration – ECO-Link*, GEO-1211877) and International Research Experience for Students (*Patagonian Research Experiences for Students in Sustainability-PRESS,* OISE 1261229). The authors also wish to thank the *Grupo Socio-Eco* at CADIC-CONICET & UNTDF and the many students, colleagues, and research assistants who have collaborated in developing this long-term initiative, especially UNTDF undergraduates Y. Álvarez, E. Arona, and F. Zuliani; NAU students L. Manak, A. Mrotek, K. Slutz, and A. Weber; and ECO-Link colleagues C.J. Donlan, K. Guillozet, A. Santo, and M. Sorice.

References

Anderson CB, Rosemond AD (2007) Ecosystem engineering by invasive exotic beavers reduces in-stream diversity and enhances ecosystem function in Cape Horn, Chile. Oecologia 154:141–153

Anderson CB, Rosemond AD (2010) Beaver invasion alters terrestrial subsidies to subantarctic stream food webs. Hydrobiologia 652:349–361

Anderson CB, Valenzuela AEJ (2014) Do what I say, not what I do. Are we linking research and decision-making about invasive species in Patagonia? Ecol Austral 24:193–202

Anderson CB, Rozzi R, Torres-Mura JC, McGehee SM, Sherriffs MF, Schüttler E, Rosemond AD (2006a) Exotic vertebrate fauna in the remote and pristine sub-Antarctic Cape Horn Archipelago, Chile. Biodivers Conserv 15:3295–3313

Anderson CB, Griffith CR, Rosemond AD, Rozzi R, Dollenz O (2006b) The effects of invasive North American beavers on riparian vegetation communities in Cape Horn, Chile: Do exotic beavers engineer differently in sub-Antarctic ecosystems? Biol Conserv 128:467–474

Anderson CB, Martinez Pastur G, Lencinas MV, Wallem PK, Moorman MC, Rosemond AD (2009) Do introduced North American beavers *Castor canadensis* engineer differently in southern South America? An overview with implications for restoration. Mammal Rev 39:33–52

Anderson CB, Soto N, Cabello JL, Wallem PK, Martínez-Pastur G, Lencinas MV, Antúnez D, Davis E (2011) Building alliances between research and management to better control and mitigate the impacts of an invasive ecosystem engineer: the pioneering example of the North American beaver in the Fuegian Archipelago of Chile and Argentina. In: Francis R (ed) A handbook of global freshwater invasive species. Earthscan Publishing, London, pp 343–355

Anderson CB, Lencinas MV, Valenzuela AEJ, Simononok MP, Wallem PK, Martínez Pastur G (2014a) Ecosystem engineering by an invasive species, the beaver, increases landscape-level ecosystem function but does not affect biodiversity in Tierra del Fuego's freshwater systems. Divers Distrib 20:214–222

Anderson CB, Santo AR, Sorice MG, Guillozet K, Zagarola JP, Valenzuela AEJ, Donlan C (2014b) Más que un ingeniero de ecosistemas: abordando las dimensiones socio-ecológicas del castor. In: XXVII Jornadas Argentinas de Mastozoología. Esquel, Argentina. 4–6 November

Anderson CB, Pizarro JC, Estévez R, Sapoznikow A, Pauchard A, Barbosa O, Moreira-Muñoz A, Valenzuela AEJ (2015) ¿Estamos avanzando hacía una socio-ecología? Reflexiones sobre la integración de las dimensiones 'humanas' en la ecología en el sur de América. Ecol Austral 25:263–272

Anderson CB, Valenzuela AEJ, Van Aert P, Malizia M, Car V, Ader N (2016) Informe final – proyecto PEININ El Parque Nacional Tierra del Fuego como un sistema socio-ecológico. Universidad Nacional de Tierra del Fuego

Anderson CB, Roulier C, Pizarro JC (2017) Perspectivas de actores clave para la gestión y la investigación del acuerdo binacional entre Argentina y Chile para la erradicación del castor norteamericano y la restauración de los ecosistemas afectados. Bosque 38:555–562

Anonymous (1946) Sucesos argentinos. Viaje al Sur. N° 432. Government Newsreel. Buenos Aires

Archibald J, Anderson CB, Dicenta M, Roulier C, Slutz K, Nielsen EA (2020) The relevance of social imaginaries to understand and manage biological invasions in southern Patagonia. Biological Invasions 22:3307–3323

Arismendi I, Szejner P, Lara A, González ME (2008) Impacto del castor en la estructura de bosques ribereños de *Nothofagus pumilio* en Tierra del Fuego, Chile. Bosque 29:146–154

Bahamonde C (2007) Alternativas de elaboración de charqui a partir de carne de *Castor canadensis* (Kuhl 1820, Rodentia). Universidad de Magallanes, Punta Arenas, Chile. Thesis

Baldini A, Oltremari J, Ramírez M (2008) Impacto del castor (*Castor canadensis*, Rodentia) en bosques de lenga (*Nothofagus pumilio*) de Tierra del Fuego, Chile. Bosque 29:162–169

Ballari SA, Anderson CB, Valenzuela AEJ (2016) Understanding trends in biological invasions by introduced mammals in southern South America: a review of research and management. Mammal Rev 46:229–240

Bennett NJ, Roth R, Klain SC, Chan KMA, Clark DA, Cullman D, Epstein G, Nelson MP, Stedman R, Teel TL, Thomas REW, Wyborn C, Curran D, Greenberg A, Sandlos J, Veríssimo D (2016) Mainstreaming the social sciences in conservation. Conserv Biol 31:56–66

Berghoefer U, Rozzi R, Jax K (2008) Local versus global knowledge: Diverse perspectives on nature in the Cape Horn Biosphere Reserve. Environmental Ethics 30:273–294

Berghoefer U, Rozzi R, Jax K (2010) Many eyes on nature: Diverse perspectives in the Cape Horn Biosphere Reserve and their relevance for conservation. Ecol & Society 15: 18. [online] URL: http://www.ecologyandsociety.org/vol15/iss1/art18/

Blanco CA, Bárquez R (1980) La situación del castor (*Castor canadensis*) en la parte argentina de la Isla de Tierra del Fuego. Informe Técnico. Departamento de Investigación y Protección. Dirección Nacional de Fauna Silvestre. Secretaría de Agricultura y Ganadería. Buenos Aires, Argentina

Bowker GC, Star SL (1999) Sorting Things Out: Classification and Its Consequences, Cambridge, MA: MIT Press

Caibul M (2008) Elaboración de embutido tipo paté con carnes silvestres de guanaco, castor y caiquén. Universidad de Magallanes, Punta Arenas, Chile. Thesis

Car V, Ader N, Valenzuela AEJ, Anderson CB (in press) Media representation of invasive introduced mammals: A comparative study of national and provincial trends in Argentina and Tierra del Fuego. In: Invasive exotic mammals of Argentina. Valenzuela AEJ, CB Anderson, S Ballari & R Ojeda (eds.) Sociedad Argentina de Mastozoología. In press

Carpenter SR, Mooney HA, Agard J, Capistrano D, DFries RS, Díaz S, Dietz T, Draiappah AK, Oteng-Yeboah A, Pereira HM, Perrings C, Reid WV, Sarukhan J, Scholes RJ, Whyte A (2009) Science for managing ecosystem services: Beyond the Millennium Ecosystem Assessment. Proceedings of the National Academies of Science of the United States of America 106:1305–1312

Castoriadis C (2007) El imaginario social y la institución. In: La institución imaginaria de la sociedad. Primera Edición. Tusquet Editores, Buenos Aires, Argentina, pp 569–280

Collins SL, Carpenter SR, Swinton SM, Orenstein DE, Childers D, Gragson T (2011) An integrated conceptual framework for long-term social-ecological research. Front Ecol Environ 9:351–357

Coronato A, Escobar J, Mallea C, Lizarralde M (2002) Características geomorfológicas de ríos de montaña colonizados por *Castor canadensis* en Tierra del Fuego. Ecol Austral 13:15–26
Davis E, Valenzuela AEJ, Murcia S, Anderson CB (2016) Habitat selection by introduced North American beavers, *Castor canadensis*, in southern Patagonia. Mastozoologia Neotropical 23:51–61
Díaz E (1996) La ciencia y el imaginario social. Editorial Bilblos, Buenos Aires, Argentina
Díaz S, Demissew S, Carabias J, Joly C, Lonsdale M, Ash N, Larigauderie A, Adhikari JR, Arico S, Báldi A, Bartuska A, Baste IA, Bilgin A, Brondizio E, Chan KMA, Figueroa VE, Duraiappah A, Fischer M, Hill R, Koetz T, Leadley P, Lyver P, Mace GM, Martin-Lopez B, Okumura M, Pacheco D, Pascual U, Pérez ES, Reyers B, Roth E, Saito O, Scholes RJ, Sharma N, Tallis H, Thaman R, Watson R, Yahara T, Hamid ZA, Akosim YA-H, Allahverdiyev R, Amankwah E, Asah TS, Asfaw Z, Bartus G, Brooks AL, Caillaux J, Dalle G, Darnaedi D, Driver A, Erpul G, Escobar-Eyzaguirre P, Failler P, Fouda AMM, Fu B, Gundimeda H, Hashimoto S, Homer F, Lavorel S, Lichtenstein G, Mala WA, Mandivenyi W, Matczak P, Mbizvo C, Mehrdadi M, Metzger JP, Mikissa JB, Moller H, Mooney HA, Mumby P, Nagendra H, Nesshover C, Oteng-Yeboah AA, Pataki G, Roué M, Rubis J, Schultz M, Smith P, Sumaila R, Takeuchi K, Thomas S, Verma M, Yeo-Chang Y, Zlatanova D (2015) The IPBES Conceptual Framework – connecting nature and people. Curr Opin Environ Sustain 14:1–16
Eigenbrode SD, O'Rourke M, Wulfhorst JD, Althoff DM, Goldberg CS, Merrill K, Morse W, Nielsen-Pincus M, Stephens J, Winowiecki L, Bosque-Pérez NA (2007) Employing philosophical dialogue in collaborative science. Bioscience 57:55–64
Estay L (2007) Caracterización física de pieles de castor americano (*Castor canadensis*) provenientes de las islas Tierra del Fuego y Navarino. Universidad de Magallanes, Punta Arenas, Chile. Thesis
Estévez R, Anderson CB, Pizarro JC, Burgman M (2015) Clarifying values, risk perception and attitudes to resolve or avoid social conflicts in invasive species management. Conserv Biol 29:19–30
Fasanella M, Poljak S, Lizarralde M (2010) Invasive North American Beaver (*Castor canadensis*): The distribution of mitocondrial variation across the archipelago of Tierra del Fuego. Mastozoología Neotropical 17:43–52
Fernández A (2017) La ley argentina de inmigración de 1876 y su contexto histórico. Almanack 17:51–85
Fischer F (2009) Democracy and expertise. Oxford University Press, Oxford
García R (2006) Sistemas complejos: conceptos, método y fundamentación epistemológica de la investigación interdisciplinaria. Ediciones Gedisa, Barcelona, Spain
Global Environment Facility (GEF) (2013) Proyecto 83266: Fortalecimiento de los marcos nacionales para la gobernabilidad de las especies exóticas invasoras. http://www.proyectogefeei.cl/proyecto-gefeei/
Graells G, Corcoran D, Aravena JC (2015) Invasion of North American beaver (*Castor canadensis*) in the province of Magallanes, Southern Chile: Comparison between dating sites through interviews with the local community and dendrochronology. Rev Chil Hist Nat 88:3
Graells G, Corcoran D, Marambio C, Gast C (2017) Dear enemy: interspecies communication through artisanal scents. http://www.christygast.com/dear-enemy/
Graham MH, Dayton PK (2002) On the evolution of ecological ideas: paradigms and scientific progress. Ecology 83:1481–1489
Grant MJ, Booth A (2009) A typology of reviews: an analysis of 14 review types and associated methodologies. Health Info Libr J 26:91–108
Haider S, Jax K (2007) The application of environmental ethics in biological conservation: A case study from the southernmost tip of the Americas. Biodivers Conserv 16:2559–2573
Harding S (1991) Whose science? Whose knowledge? Thinking from women's lives. Open University Press, Milton Keynes
Henn JJ, Anderson CB, Kreps G, Lencinas MV, Soler R, Martínez-Pastur G (2014) Determining abiotic and biotic factors that limit transplanted *Nothofagus pumilio* seedling success in abandoned beaver meadows in Tierra del Fuego. Ecol Restor 32:369–378

Henn JJ, Anderson CB, Martínez-Pastur G (2016) Landscape-level impact and habitat factors that explain invasive beaver distribution in Tierra del Fuego. Biol Invasions 18:1679–1688

Holmes G (2015) Markets, nature, neoliberalism, and conservation through private protected areas in southern Chile. Environ & Plan A: Econ & Space 47:850–866

Jasanoff S (2004) States of knowledge: the co-production of science and social order. Routledge, London

Kuhn TS (1970) The structure of scientific revolutions. University of Chicago Press, Chicago

Latour B (2003) Is re-modernization occurring – and if so, how to prove it? A commentary on Ulrich Beck. Theory Culture & Soc 20:35–48

Law J (2011) What's Wrong with a One-World World. http://www.heterogeneities.net/publications/Law2011WhatsWrongWithAOneWorldWorl

Lizarralde M (1993) Current status of the introduced beaver (*Castor canadensis*) population in Tierra del Fuego, Argentina. Ambio 22:351–358

Lizarralde M, Escobar J (1999) Plan de Manejo de la especie *Castor canadensis*. Informe Final. Subsecretaria de Recursos Naturales y Desarrollo Humano, Gobierno de la Provincia de Tierra del Fuego

Lizarralde M, Defferari G, Alvarez S, Escobar J (1996) Effects of beaver (*Castor canadensis*) on the nutrient dynamics of the southern beech forest of Tierra del Fuego (Argentina). Ecol Austral 6:101–105

Lizarralde M, Escobar J, Deferrari G (2004) Invader species of Argentina: a review about beaver (*Castor canadensis*) population situation on Tierra del Fuego ecosystem. Interciencia 29:352–356

Lizarralde M, Gailliet G, Poljak S, Fasanella M, Guilivi C (2008) Assessing genetic variation and population structure of invasive beaver in Tierra del Fuego (Argentina). Biol Invasion 10:613–683

Lorimer J (2015) Wildlife in the Anthropocene: Conservation after Nature. University of Minnesota Press. 264 pp

Malmierca L, Menvielle MF, Ramadori D, Saavedra B, Saunders A, Soto N, Schiavini A (2011) Eradication of beaver (*Castor canadensis*), an ecosystem engineer and threat to southern Patagonia. In: Veitch C, Clout M, Towns D (eds) Island invasives: Eradication and Management. IUCN, Gland, pp 87–90

Mann A (2008) Vertebrados dañinos en Chile: desafíos y perspectivas. Actas del Seminario Taller. Santiago, Chile. Universidad Santo Tomás

Marconi P, Balabusic AM (1980) Distribución y abundancia del castor en Tierra del Fuego con especial referencia a su efecto sobre los ecosistemas. Informe Técnico. Administración de Parques Nacionales, Buenos Aires, Argentina

Martínez-Pastur G, Lencinas MV, Escobar J, Quiroga P, Malmierca L, Lizarralde M (2006) Understorey succession in *Nothofagus* forests in Tierra del Fuego (Argentina) affected by *Castor canadensis*. Appl Veg Sci 9:143–154

Mateo G, Carreras-Doallo XA (2013) La economía social en la Argentina peronista (1946–1955): una mirada desde el discurso oficial. Estudios Interdisciplinarios de América Latina y El Caribe 24:79–104

Menvielle MF, Funes M, Malmierca L, Ramadori D, Saavedra B, Schiavini A, Soto Volkart N (2010) American beaver eradication in the southern tip of South America: main challenges of an ambitious project. Aliens 29:9–16

Millennium Ecosystem Assessment (MA) (2005) Ecosystems and human well-being: Scenarios. In: Carpenter SR, Pingali P, Bennet EM, Zurek MB (eds) Millennium ecosystem assessment. Island Press, Washington, DC

Mittermeier R, Mittermeier C, Brooks TM, Pilgram JD, Konstant WR, da Fonseca GAB, Kormos C (2003) Wilderness and biodiversity conservation. Proc Natl Acad Sci U S A 100:10309–10313

Moss C (2008) Patagonia: a cultural history. Signal Books, Oxford

National Academy of Sciences (NAS) (2005) Facilitating interdisciplinary research. National Academy of Sciences Press, Washington, DC

Ogden LA (2018) The beaver diaspora. Environ Human 10:63–85

Ogden LA, Holmes G (2015) Involucramientos globales del bosque, incluso en el fin del mundo. In: Bustos B, Prieto M, Barton J (eds) Ecología Política en Chile: naturaleza, conocimiento, poder y propiedad, Santiago de Chile. Universidad de Chile

Paillacar O (2007) Rendimiento cárnico de *Castor canadensis* de isla Tierra del Fuego. Universidad de Magallanes, Punta Arenas, Chile. Thesis

Parkes J, Donlan CJ, Campbell K (2008) Control of North American beavers in Tierra del Fuego: feasibility of eradication and alternative options. Informe preparado para el Comité Binacional para la Estrategia de Erradicación de Castores de Patagonia Austral

Pascual U, Balvanera P, Díaz S, Pataki G, Roth E, Stenseke M, Watson R, Basak-Dessane E, Islar M, Keleman E, Maris V, Quaas M, Subramanian SM, Wittmer H, Adlan A, Ahn SE, Al-Hafedh YS, Amankwah E, Asah ST, Berry P, Bilgin A, Breslow SJ, Bullock C, Cáceres D, Daly-Hassen H, Figueroa E, Golden CD, Gómez-Baggethun E, González-Jiménez D, Houdet J, Heune H, Kumar R, Ma K, May PH, Mead A, O'Farrell P, Pandit R, Pengue W, Pichis-Madruga R, Popa F, Preston S, Pacheco-Balanza D, Saarikoski H, Strassburg BB, van den Belt M, Verma M, Wickson F, Yagi N (2017) Valuing nature's contributions to people: the IPBES approach. Curr Opin Environ Sustain 26–27:7–16

Pauchard A, Quiroz CL, García R, Anderson CB, Kalin MT (2011) Invasiones biológicas en América Latina y El Caribe: Tendencias en investigación para la conservación. In: Simonetti JA, Dirzo R (eds) Conservación biológica: perspectivas desde América Latina. Editorial Universitaria, Santiago, Chile, pp 79–94

Pickett STA, Ostfeld RS (1995) The shifting paradigm in ecology. In: Knight RL, Bates SF (eds) A new century for natural resources management. Island Press, Washington, DC

Pietrek AG, Fasola L (2014) Origin and history of the beaver introduction in South America. Mastozoología Neotropical 21:355–359

Pietrek AG, González-Roglich M (2015) Post-establishment changes in habitat selection by an invasive species: beavers in the Patagonian steppe. Biol Invasions 17:3225–3235

Pietrek AG, Escobar JM, Fasola L, Roesler I, Schiavini A (2016) Why invasive Patagonian beavers thrive in unlikely habitats: a demographic perspective. J Mammal 98:283–292

Pietrek AG, Himes Boor GK, Morris WF (2017) How effective are buffer zones in managing invasive beavers in Patagonia? A simulation study. Biodivers Conserv 26:2591–2605

Quiroz CL, Pauchard A, Cavieres LA, Anderson CB (2009) Investigación en invasiones biológicas en Chile: tendencias y desafíos. Revista Chilena de Historia Natural 82:497–505

Resolución SAyDS 157/10 – Acuerdo Argentina y Chile sobre restauración de ecosistemas australes. Accessed 26 Dec 2018. http://argentinambiental.com/legislacion/nacional/resolucion-15710-acuerdo-argentina-chile-restauracion-ecosistemas-australes/

Robbins P, Moore SA (2013) Ecological anxiety disorder: diagnosing the politics of the Anthropocene. Cultural Geographies 30:3–19

Sanguinetti J, Buria L, Malmierca L, Valenzuela AEJ, Núñez C, Pastore H, Chauchard L, Ferreyra N, Massaccesi G, Gallo E, Chehébar C (2014) Manejo de especies exóticas invasoras en Patagonia, Argentina: priorización, logros y desafíos de integración entre ciencia y gestión identificados desde la administración de Parques Nacionales. Ecol Austral 24:183–192

Santo AR, Sorice MG, Donlan CJ, Franck CT, Anderson CB (2015) A human-centered approach to designing invasive species eradication programs on human-inhabited islands. Glob Environ Chang 35:289–298

Santo AR, Guillozet K, Sorice MG, Baird TD, Gray S, Donlan CJ, Anderson CB (2017) Examining private landowners' knowledge systems of an invasive species. Hum Ecol 45:449–462

Sarmiento 1868 is not a citation, but the range 1868-1874 refers to his presidential term, which should be comprehensible from the text

Schuettler E, Rozzi R, Jax K (2011) Towards a societal discourse on invasive species management: A case study of public perceptions of mink and beavers in Cape Horn. J Nat Conserv 19:175–184

Sielfeld W, Venegas C (1980) Poblamiento e impacto ambiental de *Castor canadensis* en Isla Navarino, Chile. Anales del Instituto de la Patagonia 11:247–257

Skewes O, González F, Rubilar M, Quezada R, Olave R, Vargas V, Ávila A (1999) Investigación, aprovechamiento y control del castor en islas Tierra del Fuego y Navarino. Final Report to Servicio Agrícola y Ganadero, XII Región

Skewes O, González F, Olave R, Ávila A, Vargas V, Paulsen P, König H (2006) Abundance and distribution of American beaver, *Castor canadensis* (Kuhl, 1820), in Tierra del Fuego and Navarino Islands, Chile. Eur J Wildl Res 52:292–296

Soto N, Cabello JL (2007) Informe final. Programa control de fauna dañina en la XII[a] Región SAG FONDEMA 2004–2007

Soto Simeone A, Soza-Amigo S (2014) Valoración económica del bosque nativo afectado por la introducción del castor americano en Tierra del Fuego. Bosque 35:229–234

Star SL, Griesemer JR (1989) Institutional ecology, 'translations' and boundary objects: amateurs and professionals in Berkeley's Museum of Vertebrate Zoology. 1907-1939. Soc Stud Sci 19:387–420

Suding K, Higgs E, Palmer M, Callicott JB, Anderson CB, Baker M, Gutrich JJ, Hondula KL, LaFevor MC, Larson BMH, Randall A, Ruhl JB, Schwartz KZS (2015) Committing to ecological restoration: efforts around the globe need legal and policy clarification. Science 348:368–640

Teel T, Anderson CB, Burgman MA, Cinner J, Clark D, Estevez RA, Jones JPG, McClanahan T, Reed MS, Sandbrook C, St. John FAV (2018) Publishing social sciences in *Conservation Biology* to move beyond biology. Conserv Biol 32:6–8

Ulloa E, Anderson CB, Ardón M, Murcia S, Valenzuela AEJ (2012) Organic matter characterization and dynamics in sub-Antarctic streams impacted by invasive beavers. Lat Am J Aquat Res 40:881–892

Valenzuela AEJ, Anderson CB, Fasola L, Cabello JL (2014) Linking invasive exotic species and their ecosystem impacts in Tierra del Fuego to test theory and determine action. Acta Oecol 54:110–118

Vatn A (2005) Institutions and the environment. Edward Elgar, Cheltenham

Vaz AS, Kueffer C, Kull CA, Richardson DM, Schindler S, Muñoz-Pajares AJ, Vicente JR, Martins J, Hui C, Kuhn I, Honrado JP (2017) The progress of interdisciplinarity in invasion science. Ambio 46:428–442

Vesco L (2018) Polémica en Tierra del Fuego por un proyecto de introducir la industria salmonera en el Canal de Beagle. La Nación Published 11 October 2018. Accessed 26 Dec 2018: https://www.lanacion.com.ar/2180761-polemica-tierra-del-fuego-proyecto-introducir-industria

Vitousek PM, Mooney HA, Lubchenco J, Melillo JM (1997) Introduced species: A significant component of human-caused global change. N Z J Ecol 21:1–16

Wallem PK, Jones CG, Marquet PA, Jaksic FM (2007) Identificación de los mecanismos subyacentes a la invasión de *Castor canadensis* (Rodentia) en el archipiélago de Tierra del Fuego, Chile. Rev Chil Hist Nat 80:309–325

Wallem PK, Anderson CB, Martínez-Pastur G, Lencinas MV (2010) Using assembly rules to measure the resilience of riparian plant communities to beaver invasion in subantarctic forests. Biol Invasions 12:325–335

Wallem PK, Soto N, Cabello JL, Castro SA, Jaksic FM (2013) La invasión de *Castor canadensis* en Tierra del Fuego como oportunidad para analizar la valoración del impacto y manejo de vertebrados exóticos invasivos en Chile. In: Jaksic FM, Castro SA (eds) Invasiones Biológicas. Ediciones Universidad Católica de Chile, Santiago, Chile, pp 209–238

Westbrook CJ, Cooper DJ, Anderson CB (2017) Alteration of hydrogeomorphic processes by invasive beavers in southern South America. Sci Total Environ 574:183–190

Williamson M (1996) Biological Invasions. Springer, New York

Wynne B (1992) Misunderstood misunderstanding: social identities and public uptake of science. Public Underst Sci 1:281–304

Zagarola JP, Anderson CB, Veteto J (2014) Perceiving Patagonia – an assessment of social values and perspectives regarding watershed ecosystem services and management in southern South America. Environ Manag 53:769–782

Zagarola JP, Lopez ME, Martínez Pastur G, Anderson CB (2017) The effects of urbanization on streams in Tierra del Fuego. Ecol Austral 27:45–54

Chapter 12
Foreign Carnivore: The Case of American Mink (*Neovison vison*) in South America

L. Fasola, P. Zucolillo, I. Roesler, and José L. Cabello

As in other parts of the globe, the American mink has invaded the southern cone of Latin America successfully. Nowadays it is one of the most widespread and feared invasive vertebrates in southern Argentina and Chile. Unfortunately, these are only 2 countries of a list of over 30 where it has invaded in Europe and Asia (Macdonald and Harrington 2003; Bonesi and Palazon 2007). The wide distribution that it has reached in occupied territories and the difficulties to trap mustelid efficiently discourage the idea of investing on eradication programs (King et al. 2009).

L. Fasola (✉)
ConsejoNacional de InvestigacionesCientíficas y Técnicas (CONICET) – DirecciónTécnica Patagonia Norte, Administración de ParquesNacionales, San Carlos de Bariloche, Río Negro, Argentina

Proyecto Macá Tobiano – Aves Argentinas/Asociación Ornitológica del Plata y AsociaciónAmbiente Sur, Río Gallegos, Argentina

P. Zucolillo
Departamento de Ecología. Facultad de Cs Biológicas, Laboratorio 307, Santiago de Chile, Chile

Center of Applied Ecology and Sustainability (CAPES-UC), Pontificia Universidad Católica de Chile, Santiago, Chile

I. Roesler
Proyecto Macá Tobiano – Aves Argentinas/Asociación Ornitológica del Plata y AsociaciónAmbiente Sur, Río Gallegos, Argentina

EDGE Affiliate-Zoological Society of London, London, UK

Departamento de Ecología, Genética y Evolución – IEGEBA, ConsejoNacional de InvestigacionesCientíficas y Técnicas (CONICET), Facultad de Ciencias Exactas y Naturales, Universidad de Buenos Aires, Pabellón II Ciudad Universitaria, Ciudad Autónoma de Buenos Aires, Buenos Aires, Argentina

J. L. Cabello
Fundación CEQUA, Centro de Estudios del Quaternario, Punta Arenas, Chile

Island Conservation, Santa Cruz, CA, USA

F. M. Jaksic, S. A. Castro, *Biological Invasions in the South American Anthropocene*, https://doi.org/10.1007/978-3-030-56379-0_12

However, recent control programs focused on clear conservation or economic objectives have evidenced that they can deliver auspicious results when immigration is controlled.

12.1 American Mink Introduction to South America

The arrival of this exotic mustelid to both Chilean and Argentinean Patagonia occurred at the same decade. Some establishment dates and location of fur farms, successive events, and numbers of individuals are known and enlighten the tight relation in the history of its introduction in this South American area. The American mink invasion in Patagonia renders a good example of an invader that affects both countries and moves through different and not few geographical connections like valleys, rivers, and lakes across borders.

12.1.1 Argentina

The first mink fur farms were opened in Argentina in the 1930s in Santa Cruz province (Fig. 12.1) (Godoy 1963; Canevari and Vaccaro 2007). Those pioneer farms closed soon after due to low profitability, and no information about their exact location or the number of animals they reared is available. However, the valley of Chico river (at the center of Santa Cruz province) could be the place where those farms were located. During informal interviews, two people from Gobernador Gregores (Fig. 12.1) (in the central valley of Chico river) gave information regarding fur farms in the area. The owner of Estancia (Ea. hereafter) La Angostura (Antonio "Tonchi" Kusanovic) mentioned that a neighboring property (Ea. Las Tunas) reared foxes in the 1960s and that he remembers seeing "small cages" at the premises. Moreover, the owner of Ea. La Lucha (Juan Carlos Fernández) mentioned that foxes were also reared in the area of Ea. TucuTucu (Fig. 12.1). As many of the farms that used to rear foxes also produced mink fur, those mentioned farms were probably the pioneer establishments.

After the first attempt in Santa Cruz, R. García Mata re-initiated the industry by opening in 1935 a farm near Mar del Plata in Buenos Aires province, with over 30,000 individuals (García Mata 1982). The Argentinean Ministry of Rural Affairs boosted the activity in 1953 by offering new entrepreneurs tax exemptions and simplifying the paperwork related to the import of the animals. By 1960, 65 farms were operative in the country. Most of them concentrated in Buenos Aires province (47) and the Patagonian region (13), but some isolated ones were located in other provinces (Godoy 1963). The list includes the main island of the Fuegian archipelago (Isla Grande of Tierra del Fuego-Argentinean side) where there were farms near Río Grande (Jaksic et al. 2002) and in Bahía Aguirre, Mitre Peninsula (Ministry of Economy and Public Works and Services of Tierra del Fuego 2002). Most of the

Argentina: 1)Salta; 2) Mendoza; 3) Córdoba; 4) La Pampa; 5) Buenos Aires; 6) Neuquén; 7) Río Negro; 8) Chubut; 9) Santa Cruz; 10) Tierra del Fuego, Antártida e Islas de Atlántico Sur.
Chile: A) R Maule (VII); B) R BíoBío (VIII); C) R Araucanía (IX); D)R Los Ríos (XIV); E) R Los Lagos (X); F) R Aysén(XI); R Magallanes y Antártida Chilena (XII) separated in Gc) Continental part and Gi) Tierra del Fuego.

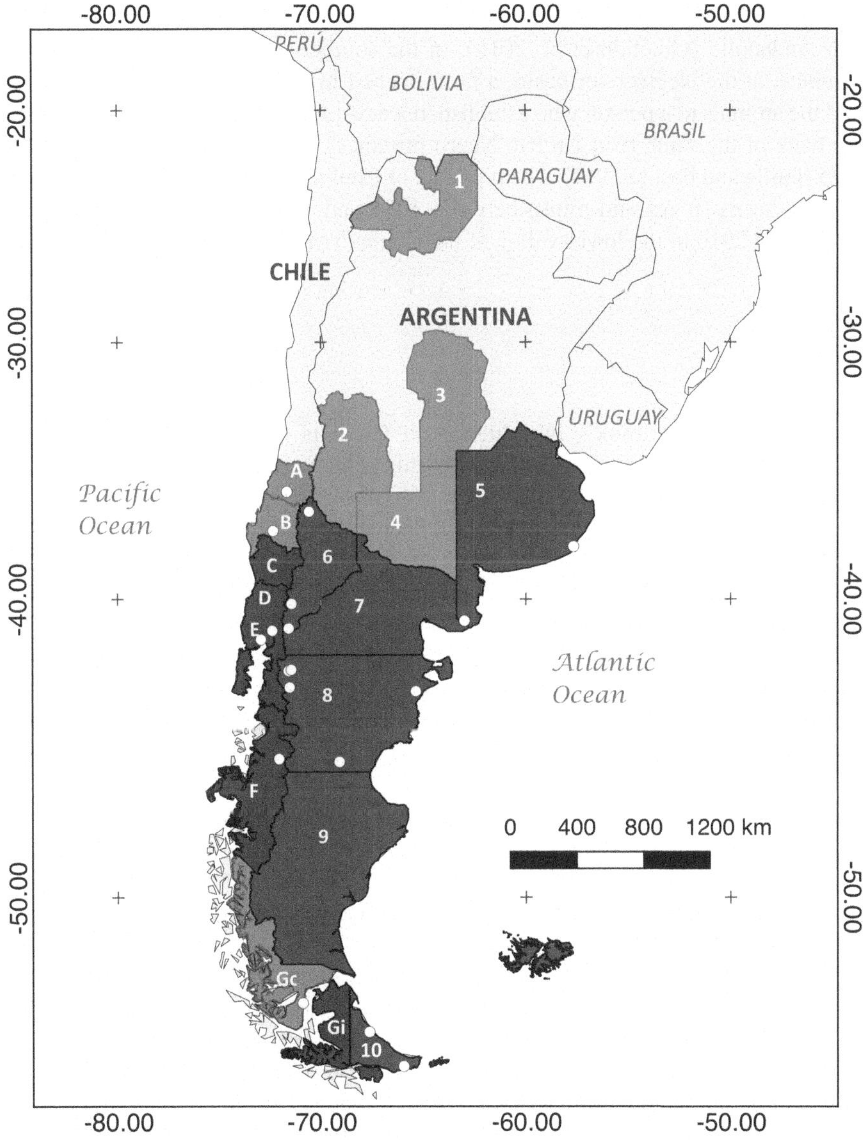

Fig. 12.1 Southern South America**. White points**: location of farms. **Light gray**: provinces (Argentina) and regions (Chile) where farms operated. **Dark gray**: Provinces and Regions were wild populations of American mink are confirmed
Argentina: (1) Salta; (2) Mendoza; (3) Córdoba; (4) La Pampa; (5) Buenos Aires; (6) Neuquén; (7) Río Negro; (8) Chubut; (9) Santa Cruz; (10) Tierra del Fuego, Antártida e Islas de Atlántico Sur
Chile: A) R Maule (VII); B) R Biobío (VIII); C) R Araucanía (IX); D) R Los Ríos (XIV); E) R Los Lagos (X); F) R Aysén (XI); R Magallanes y Antártida Chilena (XII) separated in Gc) Continental part and Gi) Tierra del Fuego

farms were closed by the late 1970s (Jaksic et al. 2002; Pagnoni et al. 1986; Fasola et al. 2011) (Fig. 12.1).

In recent years, new research works presented detailed new data at provincial level that will help understand the present distribution of the species in Argentina. In northern Neuquén province (Fig. 12.1), a farm was operative during the 1970s near Andacollo (Guichón et al. 2016). In the southernmost city of Buenos Aires province, at the Negro river basin, a farm opened in 1965 and closed in 1974 and sold the animals to a person who established a new farm in Viedma city, at the opposite bank of the same river (in Río Negro province). That farm kept working until 1975 (Failla and Fasola 2019). The province of Chubut allowed the establishment of new fur farms (foxes and mink) between 1986 and 1995. One of them remained active until 2010 in the lower valley of the Chubut river (Jones 2011).

12.1.2 Chile

The history of the mink's introduction in Chile is similar and linked to the Argentinean business, being introduced to this neighboring country in independent multiple ways over the 1930s and 1940s. Also, the presence of a geographical barrier as the Andes mountain range has not been an impediment for the movements between both territories, revealing certain degree of permeability (Jaksic et al. 2002; Iriarte et al. 2005). Details about the beginning of this invasion are well documented by Sandoval (1994) and Jaksic et al. (2002) who sought out information about the origin areas and the routes of dispersion, contributing to piece together the history of this invasion. As in Argentina, many fur farms opened up motivated by private investors to promote the economic activity. According to the compiled information, the main establishments were installed in the X, XI, and XII Regions (Fig. 12.1). Specifically, the first import of American mink coming from the United States took place in Punta Arenas (Region XII), undertaken by Edmundo Pisano between 1934 and 1936 and closing in 1950. Part of the breeding stock was sold to a new fur establishment that worked until 1970, when the business failed again and closed. Since then, no wild minks have been detected in that area. But in Aysén (Region XI), near Coyhaique, a fur establishment (property of González and Lagos Inc.) imported American minks from Argentina in 1967 and then more animals were imported by Pavez Castillo Bros & Co. from the same country (Mar del Plata and Sarmiento cities) in 1973. In these cases, when the fur farms went bankrupt, the American minks were partly released into the wild and the rest sold as pelts. In 1971, the presence of feral populations in expansion was warned, evidenced by the increase of complaints to the Agricultural and Livestock Service (SAG) from farmers of the area, due to relentless predation on their poultry. In Los Lagos (Region X) by the 1970s, González and Lagos Inc. of Coyhaique sold animals to future raisers from Puerto Montt city who released part of the stock into the wild when their breeding business failed and reduced the rest to pelts (Sandoval 1994; Jaksic et al. 2002). The feral established populations of this exotic carnivore in Puyehue and Vicente Perez Rosales National Parks may have originated from this fur farm. Nanvas (1987) suggests that this population could be related to another breeding center established in Todos los Santos lake at 1940, but there is no formal reference about his source of information. There have been imports of *N. vison* in other regions, in cities like Los Andes (Región del Biobío) and Longaví (Region del Maule) in 1957 (Bidegain 1963) and Los Angeles (Region Biobío) in 1975, but this breeding centers did not succeed, and they ceased to exist shortly after. To date, there is no record of wild minks in these northern areas (Figs. 12.1 and 12.3).

The reason for the business failure of the American mink fur initiatives in both Argentina and Chile was mainly economic: investment costs exceeded returns. Securing food for captive minks is particularly expensive and represents an important proportion of the fur's final price, whereby fur farms success depended greatly on securing food sources rich in proteins at reasonable prices (García Mata 1982). This is probably the reason why mink farms in the proximity of fish processing plants in Buenos Aires province-Argentina have been profitable since the 1930s and those in Patagonia (Chile and Argentina) were so volatile.

As we can see, economic obstacles in Patagonian farms led owners to take erroneous decisions (i.e., free individuals who were no longer useful when closing the farms). In addition, the few precautions taken when securing the cages on the farms gave rise to multiple escapes of individuals who found in Patagonia an optimal environment to settle (Lizarralde and Escobar 2000; Jaksic et al. 2002) conforming source points of minks. The process of dispersion and establishment of the American mink in the new territory continues today and ignores political boundaries.

In Chile, the species was officially recognized as a harmful species in 1998 in the sixth Article of the Hunting Law Regulations. The regulation allows the American mink hunting throughout the year, without limitation of the number of individuals and thus settling the first governmental measure in pursuit of its control (Agriculture Minister of Chile 1998).

In Argentina, the species has been declared as a detrimental and harmful species by law only in Santa Cruz province (L N° 3353/14) just until 2014. The law expresses the decision of implementing a permanent program to control the species. However, such program has never been regulated. Other provinces (Neuquén province, Río Negro, Chubut, and Tierra del Fuego) recognize American mink as an invader and also include the species in the sport hunting regulation allowing its hunt with no numeric restriction throughout the year. In Buenos Aires province, the species has only recently been confirmed in the wild and is therefore not considered in wildlife regulations yet.

Despite the inclusion of the American mink in hunting regulations in both countries, this single measure seems ineffective for its control, since its expansion and negative effect continue and reveal the need of complementary strategies.

12.2 From Farms to the Wild: Invasion Routes and Current Invasion Fronts

These farms originated different invasion nodes or foci, as the consequence of escapes or intentional releases (Pagnoni et al. 1986; Jaksic et al. 2002). Some wild populations were evident, while the farms that originated them were operative (Pagnoni et al. 1986). Some others became evident decades after the farms were closed (Failla and Fasola 2019; Guichón et al. 2016). Nowadays, American mink presence is impressive all over Patagonia in both Argentina and Chile, thus becoming the exotic carnivore with the greatest distribution in the area occupying the different habitats it finds as it advances through the territory.

In western Argentina, it has a continuous distribution along a straight line of 1300 km from north (S38° 30′ – Neuquén province) to south (S39° 45′ – Santa Cruz province) (Fasola et al. 2011; Fasola and Roesler 2018). But it is not confined to the Andean forest as it has also invaded several rivers that run through the Patagonian steppe (Table 12.1). It is also present in the Fuegian archipelago, widespread not only in the main island (Fasola et al. 2011; Cossa et al. unpublished data; Valenzuela et al. 2012) but also at other several smaller islands (Valenzuela et al. 2013).

Not many invasion routes had been suggested for mink invasion in Patagonia, since there were a high number of fur farms that could have acted as an origin for several foci. However, for the more isolated fur farms or groups of them, hypotheses on invasion routes were suggested. Moreover, some areas were potentially colonized from more than one invasion front, and thus the relative importance of the contribution of each front will be revealed with the use of molecular techniques (see below).

In Argentina, Pagnoni et al. (1986) utilized information gathered from questionnaires distributed by post in western Chubut province to define the distribution of American mink and to hypothesize about invasion routes mainly based on temporal sequences of detections. They concluded that there were two invasion origins: one in Sarmiento city and another in Cholila city (Fig. 12.1). From Sarmiento, animals invaded the big lakes and nearby marshlands that are closed to the urbanized area and, also, dispersed upstream along the Senguer river towards the west of the province reaching Alto Río Senguer area and Fontana and La Plata lakes. They also invaded several tributaries of the Senguer river. Likewise, they suggested that some movements could have head south (Fig. 12.2). The same authors suggested that the population originated from farms in Cholila, moved south invading Los Alerces National Park (NP hereafter), and continued until they reached the localities of Corcovado and Río Pico. Populations spread northwards and might have advanced to the east into the Chubut river basin that drains into the Atlantic. Chehébar et al. (1986) suggest that the colonization process in Nahuel Huapi NP occurred from south to north and that by 1985 minks were only present in the southern half of Nahuel Huapi NP. Peris et al. (2009) report that minks colonized Lanín NP through two different routes. Mink came from the south (from the Nahuel Huapi NP) and also from Chile through Huahum river and Lácar lake. Fasola et al. (2011)

Table 12.1 Hydrological basins in Argentinean and Chilean Patagonia occupied by the species, estimated proportion of the area invaded, national parks affected, ecoregion, and bibliographic source

Province/region	Hydrologic basin[a,e]	Drain	Proportion of occupied area[a]	National parks	Ecoregion	References
Neuquén (ARG)	Neuquén river	Atlantic	Not defined	No	Patagonian steppe	Guichon et al. (2016)
	Huahum river	Pacific	100%	Lanín National Park	Patagonian forest	Fasola et al. (2011), Guichón et al. (2016)
Río Negro (ARG)	Negro river	Atlantic	30%	No	Low Monte	Failla and Fasola, in preparation
Neuquén-Río Negro (ARG)	Limay river	Atlantic	70%	Lanín National Park; Nahuel Huapi National Park	Valdivian temperate forest-Patagonian forest-Patagonian steppe	Chehébar et al. (1986), Fasola et al. (2010), Guichón et al. (2016)
Río Negro-Chubut (ARG)	Manso river and Puelo lake	Pacific	100%	Nahuel Huapi National Park; Lago Puelo National Park	Valdivian temperate forest-Patagonian forest	Fasola et al. (2010)
Chubut (ARG)	Futaleufú river	Pacific	100%	Los Alerces National Park	Valdivian temperate forest-Patagonian forest	Fasola et al. (2010)
	Chubut river	Atlantic	50%	No	Patagonian steppe	Fasola et al. (2010), Jones (2011)
	Carrenleufú and Pico rivers	Pacific	100%	No	Patagonian forest-Patagonian steppe	Fasola et al. (2010)
	Senguer and Chico rivers	Atlantic[a,b]	100%	No	Patagonian steppe	Pagnoni et al. (1986), Fasola et al. (2010)
	Simpson river	Pacific	100%[a,c]	No	Patagonian steppe	

(continued)

Table 12.1 (continued)

Province/ region	Hydrologic basin[a,e]	Drain	Proportion of occupied area[a]	National parks	Ecoregion	References
Santa Cruz (ARG)	Deseado river	Atlantic	50%	Patagonia National Park	Patagonian steppe	Fasola & Roesler (2018)
	Buenos Aires and Pueyrredón lakes	Pacific	100%	Patagonia National Park	Patagonian steppe	Fasola et al. (2010), Fasola & Roesler (2018)
	Mayer river and San Martín lake	Pacific	50%[a,d]	Perito Moreno National Park	Subantarctic Magellanic forest-Patagonian steppe	Fasola & Roesler (2018), Fasola & Roesler (unpublished data)
	Chico river	Atlantic	50%	Perito Moreno National Park	Subantarctic Magellanic forest-Patagonian steppe	Fasola & Roesler (2018), Fasola & Roesler (unpublished data)
	Santa Cruz river	Atlantic	30%	Glaciares National Park	Subantarctic Magellanic forest	Fasola & Roesler (2018)
Tierra del Fuego (ARG)	Fagnano lake	Atlantic	100%	Tierra del Fuego National Park	Subantarctic Magellanic forest	Fasola et al. (2011)
	Basins of Tierra del Fuego	Atlantic/ Beagle Channel	70%	Tierra del Fuego National Park	Magellanic steppe	Fasola et al. (2011), Fasola and Valenzuela (2014), Cossa et al. (in prep.)
Araucanía (CHI)	Imperial river Toltén river	Pacific	20%	Conguillío National Park Villarrica National Park	Valdivian temperate forest-Patagonian forest	CONAF Reports (2014, 2015)

Province/ region	Hydrologic basin[a,e]	Drain	Proportion of occupied area[a]	National parks	Ecoregion	References
Los Ríos (CHI)	Valdivia river Bueno river In between Valdivia R. and Bueno R. Coastal Basin Bueno river	Pacific	100%	Puyehue National Park	Valdivian temperate forest-Patagonian forest	Medina (1997)
Los Lagos (CHI)	Bueno river In between Bueno river and Puelo R. basins and islands Puelo river In between Puelo R. and Yelcho R. In between R. Yelcho and Regional Limits Chiloé Is. and surrounding	Pacific	90%	Puyehue National Park Vicente Perez Rosales National Park Alerce Andino National Park Chiloé National Park Hornopirén National Park Pumalín National Park	Valdivian temperate forest-Patagonian forest	Medina (1997), Ruiz et al. (1996), Quinan and De los Ríos Escalante, (2011), Sepulveda et al. (2011), Vergara (2015), Medina 97, Zucolillo et al. (*In Preparation.*)

(continued)

Table 12.1 (continued)

Province/region	Hydrologic basin[a,e]	Drain	Proportion of occupied area[a]	National parks	Ecoregion	References
Aysén (CHI)	In between Palena R. and Aysén R. Chonos and Guaitecas archipelago Aysén river In between Aysén R., Baker R., and Gral. Martinez Coastal and Is. Baker river Guayeco archipelago	Pacific	100%	Magdalena National Park Queulat National Park San Rafael Lagoon National Park	Patagonian Andean forest Magellanic steppe	Aldridge (1986). CONAF, Rozzi and Sheriffs (2003), Valenzuela and Grau (2005), Bus Leone et al. (2014), Barros et al. (2014), Medina et al. (2015), CONAF reports
Magallanes and Chilean Antartica (CHI)	Tierra del Fuego Southern Channel Beagle Islands and Atlantic territory	Pacific/Magellanic Strait/Beagle Channel	40%	Yendegaia National Park A. de Agostini National Park Torres del Paine National Park *(?)*	Magellanic steppe Subantarctic Magellanic forest	Anderson et al. (2006), Schuttler et al. (2008), Ibarra et al. (2009), Davis et al. (2012), Cabello JL. (2014), Valenzuela et al. (2014), Crego et al. (2015), Cabello et al. (2016), Kusch et al. (2016) (WSC-MMA Report)

[a]Approximate estimation
[b]Under a normal precipitation regime, the basin behaves as endorheic. However, with exceptional precipitation levels it drains into the Atlantic (through the Chico and Chubut rivers)
[c]Even when no formal survey was ever conducted in this area in Argentina, the situation of the same basin in Chile and the proximity to one of the main sources of mink in Argentina support the hypothesis that the area is completely occupied by the species
[d]The area is extremely remote, but based on surveys (Fasola and Roesler 2018), the area indicated here as occupied by the species might be underestimated
[e]Chilean river basin names obtained from "Hydrographic Basins Public Inventory." General Water Department (DGA – Dirección General de Aguas)
*(?)*There have been reliable sightings inside this protected area, but its presence is not officially confirmed with trapped minks yet

suggested that the invasion front that colonized the northwest area of Santa Cruz province is the same as the one that originated in Sarmiento, Chubut, and moved towards the west and south. Fasola and Roesler (2018) finished suggesting that this area could also have received a flow of animals from Coyhaique, Aysén region, Chile (Fig. 12.2). The Perito Moreno NP and Belgrano and Chico river basins, in central-west Santa Cruz province, still have a low occupation of mink. A mink was hunted in Belgrano river, downstream from the Perito Moreno NP in 2002 (Gpque. Mariana Martinez pers. comm.). This was the first record of a wild mink in the basin. From 2013 onwards, minks are often detected, sighted, and trapped in that national park. Therefore, the uncertainty is whether those animals come from unconfirmed farms that could have operated in the basin (i.e., Ea. TucuTucu) or entered from Chile. The occupation of San Martín lake and Desierto lake/de las Vueltas river (northern area of Los Glaciares NP, Santa Cruz river basin) seems to have occurred from the north (Chico river basin). In Tierra del Fuego, the invasion had a single origin in the northern half of the island, and animals might have expanded radially. Radial expansion could be occurring in some of the invasion foci, especially in northern Neuquén and the lowland areas of the Rio Negro river and in the province of Chubut.

Regarding Chile, it has been stated that the exotic mustelid is widely distributed between latitudes 39° and 55° (Medina 1997, Medina Vogel et al. 2015), but occupation patterns may be different depending on the region (Fig. 12.3).

From south where the invasion originated, the landscape is dominated by fjords and channels, and mink distribution seems to be continuous. However, ice fields (North and South Ice Fields) force a broken distribution. Currently, in Magellan's region, *N. vison* is present in the Chilean territory of Tierra del Fuego, where feral populations of this carnivore came from the fur farms on the Argentinean side of the island that intentionally freed them after pelt business failed in 1970 (Sandoval 1994, Lizarralde and Escobar 2000, Fasola et al. 2009), generating dispersion of minks to all direction and towards Chilean lands, lately crossing fjords and channels until reaching the neighboring islands. In 2001, aiming to avoid a plague that threatens the rich native bird life that lives in this area, the presence of the invasive predator was alerted near Puerto Williams. But it was not until 2003 that it was published and informed to the scientific community that minks had reached the Navarino Island (Rozzi and Sheriffs 2003). A few years later, American mink was found at Hoste Island and Picton and Lennox Is., located at the west and southeast of Navarino Island, respectively, becoming the most southern territory where *N. vison* presence is known for certain and then occupying an extensive part of Tierra del Fuego Chilean Territory (Anderson et al. 2006; Cabello 2014; Valenzuela et al. 2014; Crego et al. 2015). Despite Punta Arenas city being one of the first places where the fur business began, currently there is no official record of the establishment of a mink feral population in the continental area of Magellan's region. However, evidence of the presence of American mink was found in some areas near the city (Cabello 2016a, b). Furthermore, American mink sightings were recorded by anglers near the town of Puerto Natales (Rubens river) and the Torres del Paine National Park. One hypothesis explaining mink presence at this latitude sets

Dark and medium grey areas indicate approximate current invaded in Chile and Argentina respectively. Arrows connect sources of animals (mainly farms) and define an average direction. **Black arrows** indicate confirmed routes or obligated colonization directions. **White arrows** indicate hypothesis to explain mink presence at certain areas, mainly where there is a gap in the information of invaded areas, and unconfirmed direction of invasion (where invasion could have occurred in both directions).

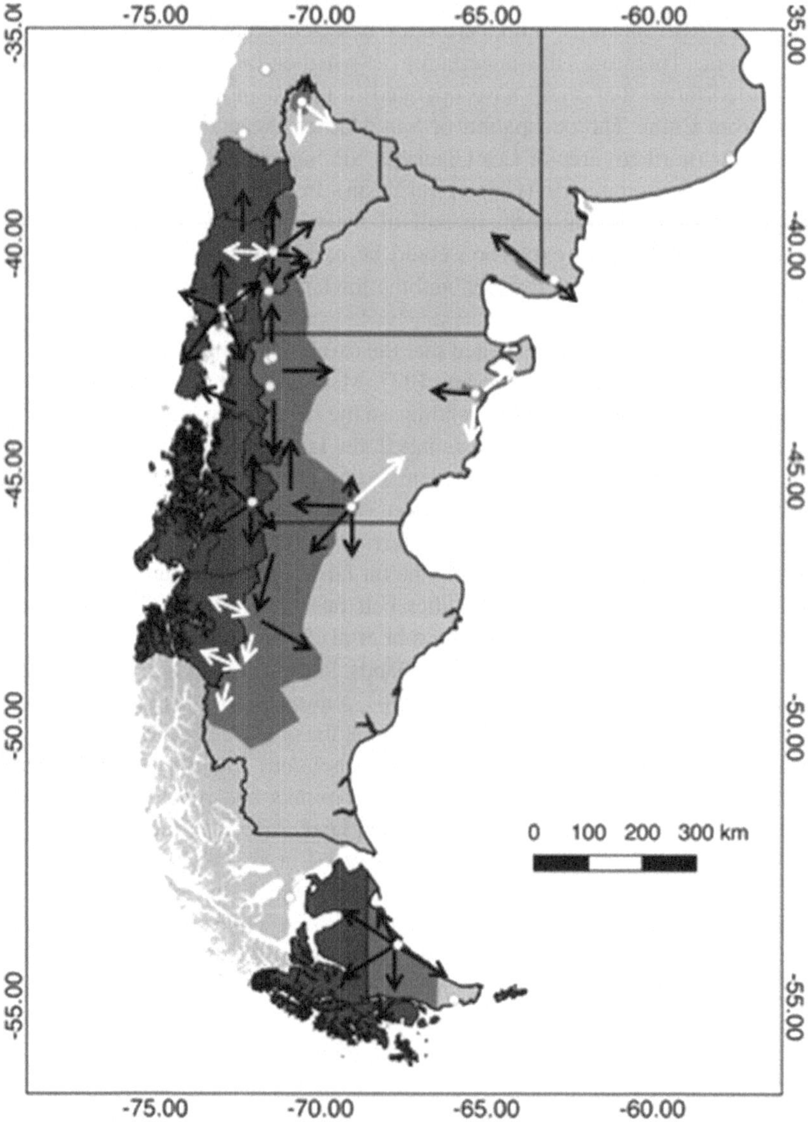

Fig. 12.2 Hypothetical invasion routes that explain current invaded area of American mink in southern South America based on evidence from dates of arrival (or detection), surveys, and literature

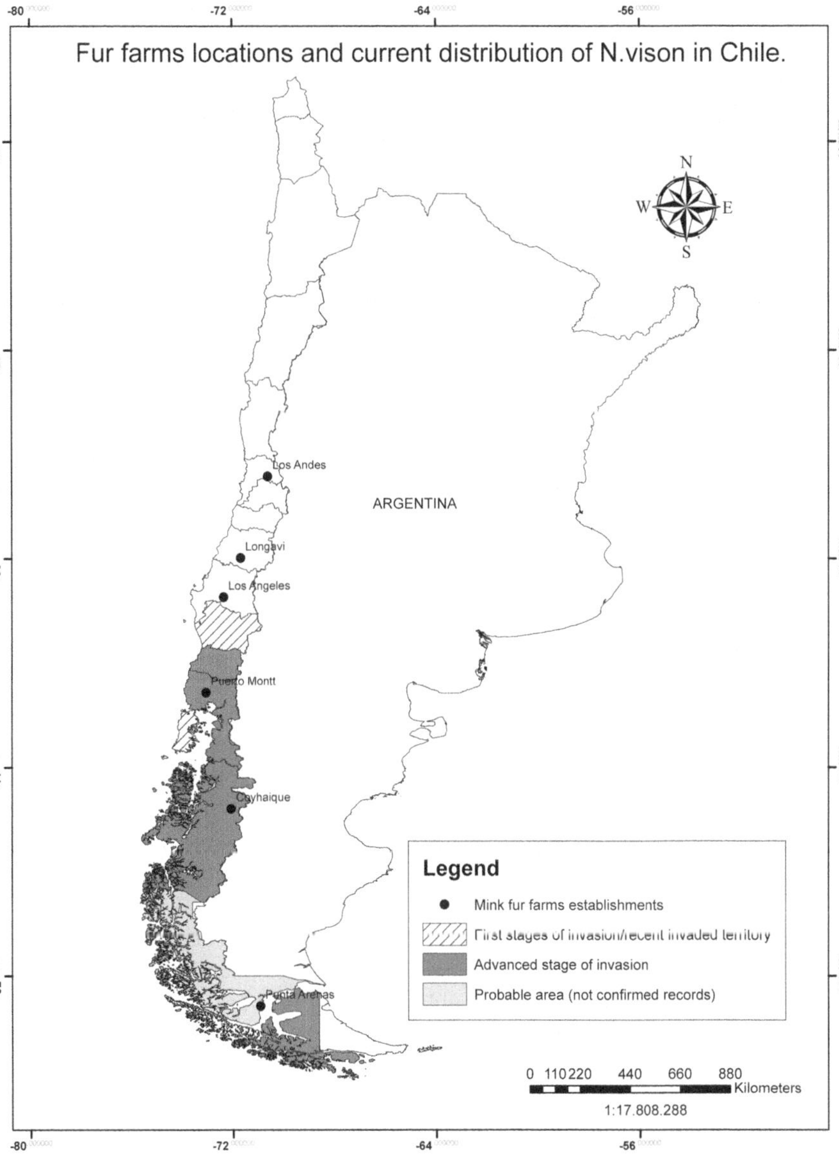

Fig. 12.3 American mink current distributional range in Chile and historical locations of fur farms' establishments

Fig. 12.2 (continued) **Dark and medium gray areas** indicate approximate current invaded in Chile and Argentina, respectively. Arrows connect sources of animals (mainly farms) and define an average direction. **Black arrows** indicate confirmed routes or obligated colonization directions. **White arrows** indicate hypothesis to explain mink presence at certain areas, mainly where there is a gap in the information of invaded areas, and unconfirmed direction of invasion (where invasion could have occurred in both directions) individuals reaching the area from the northwestern area in Chile, trespassing the ice field camp through low-altitude valleys or coming from Argentinean side (Fig. 12.2).

As in Magallanes region, Aysen region landscape are dominated by Patagonian steppe and *Nothofagus sp* forests (Jaksic 1998). Mink doesn't show any difficult invading this habitat. The exotic mustelid is well distributed over most of the area (Bonacic et al. 2010; Medina Vogel et al. 2015) from east to west, probably due to the fact that the first fur farms were installed there and thus the region became a source area of wild mink populations (Figs. 12.2 and 12.3 and Table 12.1).

The same situation described above we find in Los Lagos (X) and Los Ríos (XIV) regions, to the north where the landscape contains diverse ecosystems, such as high-altitude primary and secondary temperate rainforests, steep slopes, active volcanoes, a large hydrographic network that connects the different basins, a large number of wetlands, as well as agricultural ecosystems with high human intervention. Mink has been present in this complex matrix of ecosystems for about 30 years.

It is important to highlight an event related to the expansion of the American mink in X Region: its recent arrival to the island of Chiloé (c. 840,000 ha) (Fig. 12.3 and Table 12.1).

The first record in the area was in 2013. A citizen of Ancud found a dead juvenile male hit by a car on the road, and this incident constitutes the first formal evidence of American mink's presence on this island, where no mink sign had been recorded before (Medina 1997). In fact, the arrival of this carnivore had been warned previously by technicians from CECPAN (Centro de Estudios y Conservación del Patrimonio Natural) who had found evidence of American mink's presence while carrying out an inventory of rodent populations on four islands of the Chonos archipelago, located at south of Chiloé Is. (Valenzuela and Grau 2005). Natural barriers like distance and strong currents have not been sufficient to protect Chiloé from the mink's arrival. Currently, American mink is present mainly in the north of the Chiloé Island and occasionally registered at the central part of the island (Chiloé National Park) and southern seaports (Quellón and San Pedro Is.). The arrival of mink is intimately linked to fisheries activities and the developed boat traffic network associated with this business as well as public transport between continent and island through ferry (Vergara et al. 2015).

The northern confirmed invasion front in Chile is located in the Araucanía region (IX) bordering with Neuquén, an Argentinean province (Fig. 12.3). In 2003 at Villarrica National Park, feces and tracks were registered at different locations, and in 2008 the first mink was captured at the Cerro Ñielol Natural Monument (Temuco). Sightings reported at the Conguillío National Park started the same year. The area is characterized by a wide hydrographic network, native temperate forest, and high bird richness. After confirming the presence of the species within protected areas, CONAF implemented in 2014 a monitoring plan at the Villarrica PN to detect and trap this invader to prevent native birdlife disappearance (Guiñez Lillo 2014; Morales 2015).

In order to reach new areas, mink has shown ability to travel great distances not only following the waterlines of the hydrologic web but also crossing several watershed divides. Also, mink movements across land away from water bodies were recorded in Patagonia. In the province of Santa Cruz, Argentina, mink is known to have crossed distances over 10 km (in a straight line, which may represent much longer distances) away from water bodies when dispersing, in order to reach lakes with high density of waterfowl (Fasola and Roesler unpublished data). Confirmed displacements in open waters are also impressive. In Chile, mink were capable of swimming at least 2 km through open seawaters to reach the shore of the Chiloé Island (Vergara et al. 2015). In freshwater, crossings were confirmed over longer distances to reach opposite shores at big lakes (i.e., Viedma Lake, PN Los Glaciares).

Molecular techniques are being implemented to elucidate genetic structure in wild mink populations at regional or local level (Mora et al. 2018; Malerba et al. 2018). Genetic structuring studies will help researchers understand the past and present invasion routes, fundamental to correct decision-making and development of management actions (Holland 2000; Lecis et al. 2008). This powerful tool has been implemented in several countries where *N. vison* is invasive, primarily to assess the genetic structure to comprehend the invasion process from a dynamic perspective, the dispersion through the landscape, establishment, and connectivity between populations (Lecis et al. 2008; Zalewsky et al. 2009, 2010; Shimatani et al. 2010).

In Patagonia, Mora et al. (2018) analyzed individuals from the Aysén region and recognized three clusters or units significantly different, one of them (Puerto Cisnes) with high mtADN diversity probably resulting from the contribution of multiple farms that operated in the area. The other two clusters demonstrated low or no mtADN diversity evidencing possible "founder effects." Among the three units exist a low dispersion and a restrictive connectivity given more by distance rather than the presence of geographic barriers. Malerba et al. (2018) found that populations within a single control area in the Santa Cruz province are probably originated from different invasion fronts. These pioneer studies highlight how useful this kind of approach is to recognize management units, within and between both countries, to implement effective control strategies over this exotic mustelid in South America.

12.3 How Did American Mink Become One of the Most Successful Vertebrate Invaders in South America?

The key elements that allowed mink to invade a variety of habitats without difficulties and reach the actual distributional range in this new territory are related to some intrinsic characteristics of the species, ecological ones like habitat suitability and interactions within the native recipient communities, exotic species, and other external factors like socio-historical context and history of introduction (year, number, and locations of fur farms). Features, such as trophic plasticity, facultative delayed implantation, hunting skills, and its ability to use almost all types of natural or anthropic structures as refuge, are central for its success in a wide range of climatic and biotic conditions (Enders 1952; Dunstone 1993; Larriviere 1999; Macdonald and Harrington 2003). In addition, the fact that captive breeding was mainly driven by private investments and government policies (in the case of Argentina) had a central role influencing its success and defining the current area of occupancy of the species.

12.3.1 Energetic Sources in Invaded Areas

The American mink is a strict carnivore and opportunistic predator that consumes prey according to their relative availability in the system (Macdonald and Harrington 2003; Melero et al. 2008) (Photo 12.1). It can hunt on land and under water and climb trees (Dunstone 1993). It feeds on a broad variety of food items including terrestrial mammals, fish, birds, crustaceans, amphibians and reptiles at any growth stage such as eggs, chicks and breeding individuals or adults. Prey size goes from an amphipod of a couple of centimeters to a hare of more than a couple of kilograms, and this scheme holds in Patagonia. (*See* Table 12.2.)

Mammals are generally the most consumed prey in Patagonia or the second in importance at coastal areas or some locations in freshwater systems, in both Chile and Argentina (Medina 1997; Previtali et al. 1998; Schütler et al. 2008; Fasola et al.

Photo 12.1 Different types of habitats invaded by American mink in South America. (**a**) Patagonian steppe (Santa Cruz, Argentina)

(**b**) Patagonian Andean forest (Pumalín National Park, Chile),

(**c**) Valdivian forest (Valdivia, Chile)

(**d**) subantarctic Magellanic forest (Tierra del Fuego, Chile)

2011; Valenzuela et al. 2013). In contrast to its native range (Melquist et al. 1981; Ben-David et al. 1997) and Europe (Dunstone 1993) where fish is preferred, mink covers most of its energetic demands with mammalian prey in South America. One of the reasons could be the low diversity and abundance of freshwater fishes in Patagonia. The list of mammal species includes native species like cricetid rodents, coypu, and other occasional species (i.e., an endemic opossum) as well as invasive species like muskrat, murid rodents, beaver, rabbits, and hares. Temporary or locally abundant availability of other prey items accompanies the category described above. For instance, in many freshwater basins of Patagonia, two macro-crustaceans of the genera *Aegla* sp. and *Sammastacus* sp. are so abundant that mink exploits them as a secondary source of energy (Medina 1997; Previtali et al. 1998; Jara Millar 2010; Fasola et al. 2011) and sometimes even as the main source, as it was found in stomach and feces analyses from Los Lagos region, where these genera (*Sammastacus* sp. and *Aegla* sp.) were the most consumed prey, above the other categories, reflecting the high abundance of this crustacean in this habitat (Quinan and De los Ríos Escalante 2011; Garreton Maldonado 2015) (Table 12.2).

Also, minks respond to seasonal peaks in birds' availability. In Patagonia, increases in their abundance during spring and summer reflect an increase of the frequency of this type of prey in mink's diet (Ibarra et al. 2009; Fasola et al. 2011).

Table 12.2 Variety of prey items reported for American mink in South America

Mammals				
		Cricetidae		
			Abrothrix	Ruiz et al. (1996), Schuttler et al. (2008), Jara Millar (2010), Fasola et al. (2011), Valenzuela et al. (2013), Ibarra et al. (2009) Ruiz et al. (1996)
			Geoxus *Loxodontomys*	Fasola et al. (2011)
			Chelemys	Fasola et al. (2011)
			Phyllotis	Fasola et al. (2011)
			Irenomys	Fasola et al. (2011)
			Reithrodon	Fasola et al. (2011)
			Ondatra zibethicus (muskrat)	Schuttler et al. (2008), Valenzuela et al. (2013), Ibarra et al. (2009), Crego et al. (2016)
			Oligoryzomys	Valenzuela et al. (2013), Ruiz et al. (1996), Schuttler et al. (2008), Medina-Vogel et al. (2012), Ibarra et al. (2009)
		Muridae		
			Rattus sp.	Fasola et al. (2011), Valenzuela et al. (2013)
			Mus musculus	Ibarra et al. (2009)
		Leporidae		Schuttler et al. (2008), Fasola et al. (2011), Valenzuela et al. (2013)
		Microbiotheridae		
			Dromiciops gliroides	Ruiz et al. (1996), Fasola et al. (2011)
		Myocastoridae		
		Myocastor coypus		Foester (1973)
		Catoridae		
			Castor canadensis	Valenzuela et al. (2013), Ibarra et al. (2009)
	Unidentified			Previtali et al. (1998), Fasola et al. (2011), Roesler (2016), Medina (1997), Ibarra et al. (2009), Schuttler et al. (2008), Quinan and De los Ríos Escalante (2011), Garreton Maldonado (2015)

(continued)

Table 12.2 (continued)

Mammals					
Birds					
		Podicipedidae			Ruiz et al. (1996), Peris et al. (2009), Schuttler et al. (2008)
			Podilymbus podiceps		Medina (1997)
			Podiceps gallardoi		Roesler (2016)
			Podiceps occipitalis		Roesler (2016)
		Anatidae			Ruiz et al. (1996), Peris et al. (2009), Schuttler et al. (2008)
			Chloephaga picta *Cygnus melancoryphus*		Ibarra et al. (2009)
		Rallidae		Ruiz et al. (1996), Peris et al. (2009), Roesler (2016)	
		Furnariidae			
			Aphrastura spinicauda		Ibarra et al. (2009)
		Troglodytidae			
			Troglodytes musculus		Ibarra et al. (2009)
		Diomedeidae *Thalassarche melanophris*			Kusch et al. (2016)
		Other passerines			Ruiz et al. (1996), Roesler (2016)
	Unidentified				Fasola et al. (2011), Valenzuela et al. (2013), Roesler (2016) Medina-Vogel et al. (2012), Ruiz et al. (1996), Jara Millar (2010), Quinan and De los Ríos Escalante (2011), Garreton Maldonado (2015)
Fish					
	Marine				
		Nototheniidae			Valenzuela et al. (2013)
		Harpagiferidae			Valenzuela et al. (2013)
		Pinguipedidae			

(continued)

Table 12.2 (continued)

Mammals					
			Pinguipes chilensis		Medina-Vogel et al. (2012)
			Prolatilus jugularis		Medina-Vogel et al. (2012)
	Freshwater				
		Percichthyidae			
			Percichthys trucha		Medina (1997)
		Cyprinidae			
			Cyprinus carpio		Medina (1997)
		Percichthyidae			
			Percillia gillissi		Medina (1997)
		Salmonidae			
			Oncorhynchus mykiss		Medina (1997), Ibarra et al. 2009
			Salmo trutta		Medina (1997)
			Salvelinus fontinalis		Ibarra et al. (2009)
		Galaxiidae			
			Aplochiton taeniatus		Ibarra et al. (2009)
			Aplochiton zebra		Ibarra et al. (2009)
			Galaxias maculatus		Ibarra et al. (2009)
			Galaxias platei		Ibarra et al. (2009)
		Exotic			Fasola et al. (2011)
		Native			Fasola et al. (2011)
	Unidentified				Ruiz et al. (1996), Fasola et al. (2011), Valenzuela et al. (2013), Medina-Vogel et al. (2012), Ibarra et al. (2009), Jara Millar (2010), Quinan and De los Ríos Escalante (2011), Garreton Maldonado (2015)
Crustaceans					
		Aeglidae			
			Aegla		Ruiz et al. (1996), Jara Millar (2010), Fasola et al. (2011), Quinan and De los Ríos Escalante (2011), Medina (1997)
		Parastacidae			

(continued)

Table 12.2 (continued)

Mammals				
			Sammastacus	Ruiz et al. (1996), Jara Millar (2010), Fasola et al. (2011), Quinan and De los Ríos Escalante (2011), Medina (1997)
		Varunidae		
			Hemigrapsus crenulatus	Medina-Vogel et al. (2012)
		Bellidae		
			Acanthocyclus hassleri	Medina-Vogel et al. (2012)
	Amphipods			Valenzuela et al. (2013)
	Isopods			Valenzuela et al. (2013)
	Unidentified			Previtali et al. (1998), Valenzuela et al. (2013), Medina-Vogel et al. (2012), Fasola and Roesler (2018), Garreton Maldonado (2015)
Reptiles				
	Unidentified			Ruiz et al. (1996), Fasola et al. (2011), Fasola and Roesler (2018)
Amphibians				
	Unidentified			Quinan and De los Ríos Escalante (2011), Medina-Vogel et al. (2012), Fasola and Roesler (2018)
Insects				
		Aeshnidae		
			Rhionaeschna variegata	Valenzuela et al. (2013)
		Gryllacrididae		Medina (1997)
	Coleoptera			
		Lucanidae		
			Chiasognathus grantii	Medina-Vogel et al. (2012)
	Unidentified			Ruiz et al. (1996), Ibarra et al. (2009), Roesler (2016), Quinan and De los Ríos Escalante (2011)
Mollusk				
	Polyplacophora			
			Chiton	Valenzuela et al. (2013)
	Unidentified			Ruiz et al. (1996), Ibarra et al. (2009)
Annelids				
	Polychaeta			Valenzuela et al. (2013)
Plants				
	Unidentified			Ibarra et al. (2009)

It is important to mention that egg consumption by mink has been repeatedly underestimated in terms of the contribution to their diet because they only consume the soft inside and discard the eggshell, so egg consumption is difficult to assess based on feces and stomach content analyses. Only a couple of published studies made experiments using artificial nests and plasticine eggs to quantify the impact of predation over nesting birds in Tierra del Fuego archipelago (Schuttler et al. 2009; Malley et al. 2011, Liljheström et al. 2013).

Finally, a peculiar mink trophic behavior was described in southern Patagonia. A series of plateaus are located along the Andes on the Argentinean side. On top of these plateaus, hundreds to thousands of lakes and ponds filled by snowmelt water receive thousands of birds during spring and summer. Ducks, coots, geese, plovers, flamingoes, and grebes compose the waterfowl community. Among the mentioned bird groups, there is one in particular which is endemic to those lakes and specially attracts mink: the critically endangered hooded grebe (*Podiceps gallardoi*) (Roesler et al. 2012a, b). Young male minks are the ones accessing those lakes during the dispersion periods, mostly in late summer and early autumn. Despite the fact that other prey items are available all year round (i.e., small mammals, hares, and lizards), mink consumes mainly aquatic birds. When the lakes freeze during winter and birds leave, minks disappear from the area too. Even when the number of waterfowl can reach hundreds or thousands, 90% of the predated individuals are hooded grebes (Fasola and Roesler 2018). Thus, in this area, young mink does not behave as a generalist and opportunistic predator, as it has been described all over its distribution, but becomes a temporal specialist (Fasola and Roesler 2018).

Beyond the peculiarities of each region and with animosity to generalize, the American mink is an opportunist predator capable of exploiting to the maximum available resources, covering the wide range of food according to the environment invaded. It is important to know that besides the availability of each category, prey selection is influenced by search effort, manipulation time, digestibility, and other factors that make it more efficient from an energetic point of view (Bozinovic 1993). It is clear that *N. vison* has a great plasticity that allows individuals to occupy any habitat adapting its alimentary strategy in order to settle, adjusting its diet depending on the moment and time since its arrival to the native community.

12.3.2 Climatic Un-constraints

In its native range, there are 15 American mink subspecies (Dunstone 1993), and their distribution covers most of North America, with the exception of the tundra and the driest areas of the Southern United States (Larriviere 1999). This means that mink can naturally cope with extreme and opposite climates as the Alaskan annual main temperature (−6 °C to 1 °C) and the tropical conditions of the Everglades (22 °C to 25 °C).

In South America, mink has also shown a broad climatic tolerance. Its current distribution covers extreme climates and diverse ecosystems. It is present in part of the Patagonian and Magellanic steppe, almost all of the Patagonian Andean forest, Valdivian forest, and subantarctic Magellanic forest (Photo 12.2a, b, c, d, respectively). Two other invasion foci are developing in northeastern Patagonia on the Atlantic coast, where climate is also dry and temperature in summer can reach values over 30 °C. The range of climatic conditions (extremes in main precipitation and temperature values) that wild populations of mink tolerate in South America is between 131–3455 mm and 3–13.8 °C, respectively. In Buenos Aires province, where the only current fur farms (2) are located (13.4 °C mean annual temperature and 870 mm main annual precipitation), several non-intentional liberations occur regularly, and even some free individuals have been reported, but no wild stable populations had been confirmed yet. This does not mean a lack of potential for colonization of this area but rather that other than climatic constraints may be operating (see *Biological constraints: competitors*).

Photo 12.2 American mink (*Neovison vison*)

12.3.3 What About the Climatic Constraints?

At its native range, Arizona desert presents environmental prohibitive conditions for minks, a combination of dryness and temperature. At some locations in Patagonia, especially in the central and northern area, minks have to face with similar suboptimal environmental conditions. Also, the difference of daylight hours between summer and winter has been identified as a variable that influences American mink's reproductive rates and that may limit the dispersal of young individuals (Garcia Mata 1982). In accordance to the author, this restriction sets a threshold in latitude. Locations of lower latitude than 33° (closer to the equator) are not suitable for mink fur production given that the difference of daylight hours between summer and winter is not enough to secure efficient reproduction (García Mata 1982). Moreover, observations of the photoperiod in Alaska (high latitudes and low temperatures) revealed a tight relationship between seasonal availability of salmon carcasses, timing of lactation, and breeding in female mink, demonstrating the influence of factors other than latitude and day length (Ben David 1997).

Other physical variables might play a key role in promoting wild mink population establishment. For instance, the density of watercourses should be key in defining places where wild mink can colonize successfully and establish populations with different abundances. Some studies carried out in Chile attempt to answer questions related to the potential distribution in order to explain and predict the spread and advance of American mink through forested landscape. Results recognize complexity of hydrographic network and vegetal coverage as the most significant and determinant variables in the process of invasion at least at regional scale (Bonacic 2010; Vergara et al. 2015, Zucolillo et al. in preparation). This sort of relationship should be expanded to include the drier areas of Patagonia that are mainly located in Argentina, where hydrographic networks are poorer and vegetation scarcer.

12.3.4 Biological Constraints: Competitors

Biological communities have different levels of "permeability" towards invasions, and the presence of competitors is a factor that can limit the success of invaders (Elton 1958).

In Spain, the presence of other mustelids has been cited as a factor affecting expansion and restraining the distribution of wild American mink (Ruiz Olmo et al. 1997). There, mink expanded faster in areas where the European otter (*Lutra lutra*) was absent or rare. Also, mink progression was slower in areas with healthy populations of otters or polecats (Ruiz Olmo et al. 1997). This hypothesis is supported by studies conducted in the United Kingdom that suggest that a declining trend in the number of locations occupied by mink might be related to the recovery of otter populations after a reintroduction program (Bonesi et al. 2006).

As a carnivore invader in Patagonia, *N. vison* is immersed in a native community composed of native and exotic carnivores with which will establish relationships within the network. It has been described that carnivore species of the same family show greater resemblance in corporal size and trophic apparatus than species from other families (Glitteman 1985). This should lead to high niche similarity between the members of the same family when competition between members of the same family is expected (i.e., limited resources). Competence is recognized as one of the major mechanisms that shape community structure, and several times, the size influences interactions among mustelids, and direct or indirect interference may occur (Donadio and Buskirk 2006; Bischof et al. 2013). In Patagonia, Mustelidae family is represented by four native species that share at some degree the distributional range, the use of habitat, and prey with the exotic American mink. Two otter species are represented by *Lontra provocax* ("Huillin") and *Lontra felina* ("Chungungo"), and two exclusive terrestrial mustelids represented by *Galictis cuja* ("lesser grisson"/Quique) and the less known *Lyncodon patagonicus* ("Huroncito patagónico"). Most of the existing studies have mainly focused on the couple formed by the southern river otter and the American mink, though some literature has explored relationships between invasive mink and the more terrestrial "lesser grisson."

The southern river otter has suffered a reduction of over an 80% of its original distribution in Argentina (Fasola 2009). In order to understand how mink and river otter adapt to each other's presence, several authors studied their trophic behavior and habitat use (see section *Impact: Endangered competitors in Patagonia*).

The "lesser grison" occurs at low abundance in the vast Patagonia. If the hypothesis stated about the role of potential competitors as a brake effect on American mink spread holds for Patagonia, the disadvantageous population situation of these two species could have acted as a factor favoring mink expansion in Argentinean Patagonia. Nevertheless, the scenario in Patagonian temperate forests, more developed in Chile, is a little different. There, *G. cuja* is frequently sighted at riverside environments (Zucolillo et al. in preparation) where minks inhabit. This study will allow a better understanding about mink and lesser grison co-occupancy.

Within the competitors, there is a wide distributed domestic carnivore that exceeds mink size in several times that potentially constitutes an impediment for the establishment of the exotic mustelid. In Buenos Aires province where mink farms have been functioning since the 1930s, wild populations have never been detected. Given that these farms are immersed in a semi-urban area with industries and family farms for fruit and poultry production, domestic and feral dogs are very common and could be playing a role in preventing the establishment of a wild mink population, or maybe minks in this area are going through an extremely long lag phase (see next). Unlike the situation described above, in which captive animals are venturing into the wild for the first time, there are areas where wild population of mink are already established or in newly invaded areas. In Chiloé Island where mink has recently arrived and on continental areas where mink was established years ago, minks and dogs can coexist and occupy the same habitats as studies with trap cameras reveal (Farías and Svensson 2014; Zucolillo et al. in preparation).

More research about interactions with native carnivores such as foxes (*Lycalopex culpaeus*, *L. fulvipes*, *L. griseus*), kiodkod (*L. guigna*), and *Puma concolor* is needed in order to figure out how the recipient community could be affecting the mink establishment, distribution, and abundances. This topic deserves more attention as research states the importance of larger native predators in controlling smaller invasive predators. In Sweden, population recovery of the native red fox (*Vulpes vulpes*) seems to be causing invasive mink population's decrease (Carlsson et al. 2010).

12.3.5 Biological Constraints: Who Eats Minks?

In North America, minks are predated by avian predators (*Bubo virginianus*, *Buteo* spp.), bobcats, red foxes, coyotes, alligators, and otters (Lariviere 1999), but none of them is thought to control mink population or limit its distribution. Only recovering populations of the white-tailed sea eagle (*Haliaeetus albicilla*) in a Finish archipelago were found to have an effect on mink spatial behavior. Salo et al. (2008) found that swimming distances of females were shorter at places where eagle predation risk was higher, which could reduce population's growth in the long term. However, there are only occasional records of minks predated by culpeo fox. And while red-tailed eagle (*Buteo ventralis*) and black-breasted eagle (*Geranoaetus melanoleucus*) are known to attack skunks (*Conepatus chinga*), there are no records of these avian predators attacking mink.

12.3.6 Time Lags

In the same line, little is known about factors determining time lags on the invasion process of mink in Patagonia. Several types of factors have been identified as influencing time lags as intra-specific interactions at low population densities (Allee effect), lack of genetic diversity (founder effect), and number of introductions and interactions with elements (biotic or abiotic) of the invaded ecosystem (Crook 2011).

The invasion process of mink in different parts of Argentina and Chile was not homogeneous in rates of population increase or geographic spread. In western and central Chubut province (Argentina), the wild populations of mink became evident in less than two decades as well as their expansion process (Pagnoni et al. 1986; Fasola et al. 2011). In the Fuegian archipelago, mink invasion occurred at two different speeds. While wild populations in the main island (Isla Grande of Tierra del Fuego) became evident after decades since the establishment of the fur farms (which is comparable to what was experienced in the province of Chubut), once minks reached and were detected on Navarino Island in 2001 (Rozzi and Sherriffs 2003), its expansion across most of the marine coast and freshwater systems was extremely fast (less than a decade; see Schuttler et al. 2010). In northeastern Patagonia, in the lower valley of the Río Negro river, the increase and spread of American mink were slower. While farms operated between 1960 and the 1970s, wild minks were only occasionally seen in 2011, and now they are known to occupy the last 180 km of that river (Failla and Fasola in preparation). A similar delay is related to a recently invaded area in northwestern Neuquén (Guichón et al. 2016) that could have originated in a farm that operated in the late 1970s (María Laura Guichón pers. comm.). Lastly, in the Pampas region, in Buenos Aires province (northeast from Patagonian region), mink farms have been operative nonstop since the 1930s, and the owner of one of the farms declares that minks often escape from cages (Pedro Massen pers. comm.). Furthermore, at least one massive release event occurred in the Buenos Aires province when all the cages of a fur farm were opened during a flooding and only the 90% of the animals were recovered afterwards (Federico Hozbor comm. pers.).

Unfortunately, no studies have focused on elucidating the factors influencing these differences in time until population's establishment and growth. However, it has been hypothesized that the competitor's release is the cause for the differences between the invasion process between Isla Grande T. del Fuego and Navarino Island (Schuttler et al. 2010).

In the invasion front recently detected in northeastern Patagonia, increasing abundance of the introduced *Cyprinus carpio* was identified as a trigger for mink population growth and expansion after a lag time of decades (Failla and Fasola in prep.). Similar coupled events were evidenced in southern Argentinean Patagonia. Invasive Chinook salmon move upstream from the Atlantic Ocean to the northern head of the Santa Cruz river basin to deposit eggs and die. The start of this phenomenon (R. Sanchez pers. comm.) coincides with the increase in the number of sightings of American mink in the area. In Chilean Patagonia, minks have coexisted with exotic introduced and naturalized salmonids as well as the big cages for fish farming for many years (*Oncorhynchus mykiss*, *O. kisutch*, *S. salar*). High availability of this important energy source such as fish in unlimited quantities may have affected in some way the accelerated establishment of *N. vison* in Chilean Patagonia. Therefore, processes as melting down or facilitation (Simberloff 1999) between these invasive fish and mink may be operating in these systems, but trophic relationships should be investigated in deep to support this hypothesis.

Finally, as it was mentioned before for the Buenos Aires province, interaction with domestic dogs might be preventing or acting as a factor extending the time lag phase of a wild mink population. Either way, the situation merits employing the precautionary principle assuming that any invader has the potential to become abundant (Crook 2011).

12.4 Impacts

Currently this invasion has become quite important due to the ecological and socioeconomic damage that generates at the invaded ecosystems, although this remains without being clearly quantified at least in South America. American mink has been suspected and was found guilty of several impacts on native species, both in Europe and in South America through predation and competition (Macdonald and Harrington 2003). Through predation, mink has impacted negatively in several European bird and mammal species as well as on amphibians (Barreto et al. 1998; Nordström and Korpimäki 2004; Ahola et al. 2006). On the other hand, direct competition between American and European mink seems to be one of the reasons for the severe decline of the latter in some parts of its range (Maran et al. 1998), while anthropogenic factors such as fragmentation, trapping, and even pollution seem to be the cause for local extinction at other areas (Lodé et al. 2001).

In Chile and Argentina, the emergence of feral population of mink was often associated with the decline of some mammal and bird prey species and the abandonment of regular nesting areas for waterfowl. Also, several authors have explored intraguild competition, principally with native species of the mustelid family.

12.4.1 Endangered Prey of Patagonia

Soon after minks were detected in the wild, they were associated with the decline of prey species. Foerster (1973) was the first in reporting mink negative effect on geese reproduction (*Chloephaga* sp.) and coypu (*Myocastor coypus*) populations in Los Alerces National Park in Chubut province and also the first reported attempt in removing animals through trapping.

The effect of mink on birds has been explored and evaluated at two levels. While some authors have focused on reduction on nesting success, others have focused on its effect on populations and even waterfowl diversity and distribution.

Nesting success in relation to mink predation was mainly studied in coastal habitats of the Fuegian archipelago, along the Beagle Channel, where some endemic species were the focus of the studies. There, the kelp geese (*Chloephaga hybrida*) and the flightless steamer duck (*Tachyeres pteneres*) select the islets of the Channel over the coats of the main island (Tierra del Fuego) to reproduce, and one of the reasons for this pattern is the low proportion of islets with permanent presence of mink, among other terrestrial predators (Liljheström et al. 2014). Along the northern coast of Navarino Island, mink predation was found to affect mainly solitary nesting species (as *C. hybrida* and *T. pteneres*), especially those nests that overlap with areas of high abundance of mink and also those nest that are concealed, apparently because birds have low chance of defending the nest when they cannot detect the predator approaching (Schütller et al. 2009). More recently, mink has been associated with the disappearance of chicks from the only inner water colony of black-browed albatross (*Thalassarche melanophris*) from the Almirantazgo gulf in southern Chile (Kusch et al. 2016).

The hooded grebe is positioned 21st among the top 100 priority bird species of the globe (Jetz et al. 2014; EDGE 2016). Roesler et al. (2012a, b) reported an American mink attack over a hooded grebe colony with the impressive result of the 4% of the global population (global population of c. 800 individuals) dead. In that

Photo 12.3 Endangered hooded grebe (*P. gallardoi*) colony attacked by an American mink (Roesler et al. 2012a, b)

event, 33 adult grebes were killed during the incubation period, and this is probably the only time that mink effect on a population could be measured. Actually, the devastating effect that mink can have on a species on the brink of extinction was measured (Photo 12.3).

The habitat of the hooded grebe is an "island" surrounded by desert, with no native ground/aquatic predator; therefore, this grebe lacks completely anti-depredatory strategies to avoid mink attacks. Fasola and Roesler (2018) detected that the mink selected hooded grebe over other species of waterfowl and even other grebe species (silvery grebe *Podiceps occipitalis*) in great numbers.

Mink has also been suggested as the main reason of the massive shrink of the distribution of the austral rail (*Rallus antarcticus*) in Patagonia. In northern Patagonia, the species was last recorded in 1959, and mink was mentioned as the putative reason for its apparent extinction from the region (Fraga 2000). The austral rail was not found again until 1998 in Santa Cruz province-Austral Patagonia (Mazar Barnett et al. 1998) and then found in several locations (Mazar Barnett et al. 2014). However, American mink is still invading Austral Patagonia (Fasola and Roesler 2018) and is thought to be the reason of austral rail absence in the wetlands that have already been colonized (de Miguel et al. 2019).

Torrent duck (*Merganetta armata*) is another charismatic and flagship species of Patagonia, and American mink completely overlaps the duck's distribution there. In northern Argentinean Patagonia, reduction in successful breeding territories and increased abandonment of torrent duck territories were associated with aquatic recreational activities and American mink's predation (Cerón and Trejo 2012).

Finally, at the community level, Pescador et al. (2012) studied species abundance and community composition of waterfowl species in relation to lake characteristics and presence of American mink in northern Patagonia. They found that the number of species was lower in lakes occupied by mink and that some species were found only in lakes free of mink, suggesting that mink could be shaping bird communities in some Patagonian lakes. However, with the information presented by these last authors, it is hard to trace a causality of the mink effect over the avian community. More studies need to be done to understand the real effect, discriminating different groups of birds.

Some of the situations described above support the *tight rope hypothesis* described by Barreto et al. (1998) based on the water vole's (*Arvicola terrestris*) delicate situation. As an example, the hooded grebe faces global climate change in their reproductive distribution as precipitation has diminished and wind has increased making nesting requirements difficult to meet. It also faces other threats as stocked trout that change physicochemical water conditions at lakes and kelp gull inland colonies that increase the predation pressure over grebe nests (Roesler et al. 2012a, b). Thus, the hooded grebe's situation is equivalent to that of the water vole, as climate change and other invasive species have the same impact on populations as the habitat fragmentation does for the vole. Those impacts have pushed the species to critical conditions in which mink could have a rapid and devastating effect because of the combined effect of the threats and not because of the mink's potential to affect the prey species on its own. Again, similar analysis can be done on the austral rail and their habitat transformation for cattle rearing plus mink's predation (de Miguel et al. 2019).

12.4.2 Endangered Competitors in Patagonia

Interspecific competition is considered one of the main forces structuring biological communities (Tilman 1982). Thus, the introduction of a competitor can drive important changes in the invaded system. The magnitude of the effects of this competition will depend on the intensity of the interaction and the competitive skills of the species involved (Keddy 2001).

In Patagonia, there are other species ecologically closed to American mink: southern river otter known as "Huillín" (*Lontra provocax*) and the lesser grison (*Galictis cuja*). Competitive interactions between each of them and the American mink were studied.

American mink is described as a semiaquatic species (Dunstone 1993) that restricts its movements to the first strip of land from the waterline (Harrington and Macdonald 2008) and feeds on both aquatic and terrestrial prey (Macdonald and Harrington 2003). Thus, it can be thought as an ecologically intermediate between the lesser grison and the southern river otter. While the first one is more terrestrial in terms of types of prey and habitat use, the latter show more aquatic habits.

The lesser grison is a South American polecat associated with habitats that offer water and good cover and feeds on different terrestrial vertebrates (Redford and Eisenberg 1992). For this reason, competition with introduced American mink (almost the same weight as the grison) was explored. What was found is that even when *G. cuja* often occupies river valleys, their diet is mostly composed of terrestrial species. On the other hand, minks rely exclusively on aquatic or semiaquatic prey, and so hunting habitat preferences segregate these mustelids in Patagonia (Delibes et al. 2003).

American mink potential competition with the southern river otter was explored, since this otter is classified as Endangered by the IUCN since 2000. In both Chile and Argentina, interaction between the two species was studied in both freshwater and coastal habitats. In freshwater, southern river otter's distribution is almost completely overlapped with the mink's range (Fasola et al. 2009). In Argentina Aued et al. (2003) found a negative relationship between the proportions of occupied sites by otters and mink per water system. Later, Fasola et al. (2009) compared site characteristics preferred by otters and mink in places with and without otters. What they found was that minks tend to select places more similar to those selected by otters within the area where they coexist. Also, by comparing mink's diet between places with and without otters, they found a convergence of mink's diet to otter preferences. They hypothesized that no limiting resources allowed American mink and southern river otter coexistence in freshwater systems.

On coastal areas, different patterns were found. Medina-Vogel et al. (2012) studied mink and otter habitat use, diet, and activity patterns in southern Chilean archipelago and found that minks in the presence of otters tended to select different habitat characteristics and that diet showed some differences suggesting that coexistence was possible due to segregation in these ecological requirements. Moreover, at one location where diet of the two species converged, they found that mink was mainly active during the day, suggesting that in this particular situation otter could be impacting on mink by affecting their activity patterns. In Argentina, on the Beagle Channel, Valenzuela et al. (2012) studied habitat characteristic preferences and diet of both mustelid species. In this study case, results suggested that neither otter nor mink was affected by the presence of the other. Also, their results showed that mink tended to change its diet towards more terrestrial prey items in the presence of otters and, while it continued consuming aquatic prey, the species were different. Thus, trophic segregation at two levels was mediating coexistence in this area. Evidences showed that some degree of competition might be operating at certain areas but that the native southern river otter is the outcompeting species. In recent years, the studies concentrated on otter-mink interaction have turned towards the impact over the endangered *L. provocax* perspective through pathogen transmission and try to explain otter river disappearance from some areas.

Summarizing, American mink seems to be innocuous to either of the two ecologically similar species to itself through competition. More importantly, minks do not seem to be affecting endangered southern river otter through competition (but see *Pathogens and Diseases in American mink*), and probably the latter outcompetes mink when resources become limited. This is parallel to the situations found in

Europe where the same pairs coexist, a native otter and the introduced American mink. As it was mentioned in previous sections, recovering European otter apparently affects mink through interspecific competition, and a decline in the occupation of mink was evidenced (Bonesi et al. 2006).

12.4.3 Family Livelihoods and Tourism

In a different dimension of invasive species impacts, the American mink has become a severe economic problem in the Los Ríos region (Chile) in the last 10 years. Mink predation over poultry is affecting subsistence farming. The economic loss associated with mink attacks has been estimated in 450USD per predation event. Therefore, the local government has initiated a community control program (*Control Comunitario del Visón en la Región de los Ríos*) to reduce American mink predation over wild and domestic animals. When farms get registered to participate in the program for the mink trapping activities, they also enter into a bounty system to incentive trapping. By doing this, they seek to protect local subsistence farming and also preserve autochthonous fauna, not only for environmental conservation but also to protect areas highly important for tourism.

Nature tourism is gaining in importance worldwide, and the Argentinean government at national level is pushing new ecotourism programs. Wildlife observation and photography and particularly bird watching can be negatively affected by mink predation at local levels, and consequently the economic activity may suffer. Currently, 30,000 birdwatchers from all over the world get to Argentina and spend on average 1700 USD per person (Dodyk 2017). In the case of Chile, 30% of the tourists who visit central Chile would be interested in nature tourism (Concha Martinez 2013). Chile is a country that owns 4300 km line coast, islands, and a variety of ecosystems from the densely forested fjords to the driest desert, and birdwatching represents one of the most important activities within nature tourism. Paradoxically, featured places where this activity concentrates are areas where American mink inhabits: north and south Patagonia, Osorno, Pucón, Puerto Varas, Puerto Montt, Valdivia, and Tierra del Fuego.

Mink impacts are complex to value economically since it is a species that essentially affects biodiversity, and therefore its impact is intangible from an economic perspective. Their first attempt to measure the economical and ecological mink impact was made in Argentinean North Patagonia and Chilean Aysén region (Pagnoni et al. 1986; Aldridge 1986), and more recently, the Chilean Environment Ministry (MMA – Ministerio del Medio Ambiente) quantified economic losses caused by several exotic invasive species. The estimated losses due to mink impacts on components of biodiversity reach $ 8,135,449 USD. If the species is not controlled, Chile estimates losses of $ 406,708,227 USD in 20 years in damages to biodiversity. (Cerda et al. 2017)

12.5 Pathogens and Diseases in American Mink

The first veterinarian studies on *N. vison* were developed in Argentina to detect problems in breeding minks and describe causes of dead in captive individuals. The Aleutian disease was the most important cause of death among analyzed captive minks, such as other ones like distemper and tuberculosis (Martino 1991; Martino et al. 2007). Later, prevention of distemper outbreaks in mink farms was addressed (Jar et al. 2010).

With a growing interest in discovering how wild mink could affect native and domestic fauna, a change of perspective has occurred, and veterinary studies are considering an impact on the fauna itself rather than a disease on farm minks. Few authors have addressed the problem of diseases associated with mink in Patagonia. Medina et al. (2010) proposed American mink as an allochthonous reservoir of leptospirosis, toxoplasmosis, and canine distemper. Then, prevalence of *Toxoplasma gondii* (responsible for toxoplasmosis), canine distemper virus, and pathogenic leptospirosis was studied in Chile.

Sepulveda et al. (2011) studied and confirmed *Toxoplasma gondii* prevalence in wild minks in part of X Region (Maullín river). They also found that the proportion of infected animals were higher in groups closer to urban areas and suggested that the high prevalence of *T. gondii* in mink populations could be related to its semi-aquatic habits rather than to the consumption of infected prey. For that reason Sepulveda et al. (2011) suggested that mink could be a sentinel of *T. gondii* prevalence in the environment during culling programs. Ongoing research is focused on the mechanisms behind the spread of toxoplasmosis to otters in relation to the presence of American mink and domestic cats (*Felis catus*) (Barros et al. 2018)

The canine distemper virus (CDV) is an important disease for both domestic and wild animals. CVD was a main concern for fur farmers before vaccines were available. García Mata reported the occurrence of an important outbreak of CVD in Argentinean farms in 1954 that was responsible for important economic losses. The virus could cause the death of up to the 90% of the cubs if they get infected soon after weaning (García Mata 1982). In Chile, Sepulveda et al. (2014) confirmed serological evidence for CVD in wild minks. The authors complemented the study by evaluating CDV in dogs (that resulted serologically positive) and also studied interactions between dogs, minks, and southern river otters since the transmission of the virus is thought to be through direct contact. They probed interaction between dogs and minks in rural areas, while in riparian habitats only minks were found to interact with otters selecting positively their latrines and increasing risk of infection. They concluded that mink could act as bridge host between dogs and other endangered wild carnivores and thus the pathogen transfer risk represents a mechanism through which minks have the potential of impacting on native southern river otter (*Lontra provocax*) as well as other canids frequent in riparian habitats such the endangered Darwin's fox (*L. fulvipes*). As with *T. gondii*, researchers are now studying CDV in otters in relation to different population densities of *N. vison* and domestic dog (*C. l. familiaris*).

Finally, prevalence of pathogenic *Leptospira* spp. was also confirmed in wild mink and resulted higher than values reported elsewhere. These results are important not only for wild species but also for human health since leptospirosis is considered the most widespread zoonosis of the world (Barros et al. 2014).

Unfortunately, similar studies in Argentinean wild populations are only starting. In 2007 was published a scientific note about wild introduced mink in B. Aires, tuberculosis, and probable ways of infection, but only analyzed two minks (Martino et al. 2007).

12.6 Control of American Mink in South America

Mink control in South America has been approached from several aspects, and information in methodologies, planning strategies, and actual experiences has been generated.

12.6.1 Trapping Methods and Efficiency Assessment

During the 1980s, the most used trapping methods to capture minks were fox traps/snares, hunting dogs, and armory (Pagnoni et al. 1986). At present, live cage traps and spring traps (lethal) are used. The choice depends on the extension of the trapping area, logistics, and by-catch avoidance.

One of the first studies to test American mink trapping methods was carried out by Cabello et al. (2011) and Davis et al. (2012), who compared trapping effectiveness among different settings for lethal traps (conibear) and testing effectiveness of baits in Navarino Island, where mink is the only terrestrial predator and dead trapping is an option. Results stated that traps inside restricted entrance cubby sets and baited with fresh fish are more suitable for mink control in this type of habitat.

In different regions of Chilean Patagonia, where the assemblage of predators is more complex and constrains the type of traps that could be deployed, Medina et al. (2015) tested live trapping efficiency. They used a fixed allocation of traps and evaluate the effect of removal on the estimated population through modeling. They provided information on a trapping strategy (trap effort and length of trapping period) that could be applied to reduce a 70% of mink population. They also provide information of the number of trap operators required and waterline length that can be feasibly covered. This type of studies represents a priority to supply wildlife managers.

Fasola and Valenzuela (2014) proposed an exercise to select priority areas for mink control from entire Patagonia. Given the current distribution of mink in Argentina, its eradication is no longer feasible, and thus, they proposed a method for the prioritization of areas based on conservation values and mink suitability habitat. They also suggested an initial strategy to design a trapping plan and evaluate implementation costs and efforts.

12.6.2 Ultimate Experiences

In Austral Argentinean Patagonia, a mink control program is part of the strategy to preserve the critically endangered hooded grebe (Roesler et al. 2012a, b). Fasola and Roesler (2018) presented a strategy with control goals that vary across the area as distance from hooded grebe colonies increases and mink density changes. They presented a trapping method that is an adaptation of the floating method presented by Reynolds et al. (2004) and the use of dead traps reported by Davis et al. (2012). The new trapping method was found adequate to deal with the logistic constraints associated with trap in distant and remote streams of the Patagonian steppe and avoid by-catch. Finally, the strategy complements trapping with the action of conservation agents called "colony guardians" who monitor continuously the grebe's colonies and would hunt any mink trespassing trapping lines (Roesler et al. 2016). The program has completed the sixth season, and results are still auspicious with zero hooded grebe predated since the actions' implementation (Fasola and Roesler unpublished data).

In Chile, two agencies take action on the matter. One of them is CONAF (Corporación Nacional Forestal) takes responsibility inside protected areas, and SAG (Servicio Agrícola y Ganadero) has inherence over some invasive species outside protected areas.

One of the efforts of regional scope in Chile is financed with FNDR (National Funds of Regional Development) and executed by Regional SAG from Los Ríos region. They implemented a plan that involves community assistance. Local farmers deploy most of the live trapping effort. They are mainly poultry producers directly affected by mink predation and receive basic training before entering the program. A group of veterinaries assists the farmers when they report a trapped animal and release the captured animal in case of a non-target species or take the animal for humane killing when the trapped individual is an American mink. This program is also based on a bounty system. To date, the number of mink removed by the program from the region is over 2000 individuals (E. Raffo pers. comm). Additionally in this region, SAG signed an agreement with CONADI (Corporación Nacional de Desarrollo Indígena) in the current year, to identify and control this invader in the area with a budget of \$ 5.000.000 CLP.

A special situation has place in the recent invasion front in Chiloé Island (Los Lagos region). The arrival of mink there has been monitored from the beginning by "early warning plan" that involves community participation through a complaint system. The plan is supported by workshops designed to teach how to identify the invading mustelid and to proceed in order to inform the authorities responsible for trapping mink (SAG – CECPAN ONG). Multiple catches have been made in the north of the island, preventing a rapid advance to other basins, thanks to the cooperative work (Vergara et al. 2015).

All these initiatives, in both Chile and Argentina, combine education actions to disseminate information orientated to different social actors concerned with this invasion (e.g., rural communities, citizens, schools, forest rangers). The final purpose is to raise awareness about the problem of living with mink and how to proceed and highlight the importance of working collectively to face the plague. These initiatives are generally carried out in coordination between CONAF, SAG, NGOs, and university research centers in Chile and mainly National Parks Administration and NGOs in Argentina.

Other local initiatives for controlling mink have different objectives, and not many of them have produced a publication. However, these experiences have been spread in workshops, meetings, or technical reports. In that sense, and with the objective of generating a network of experts, technicians, forest rangers, and wildlife managers from Chile and Argentina, the "Collaborative Network for American Mink Control" (*Red de Colaboración para el Control del visón Americano – RCCV*) was created, and the first meeting was celebrated past July 2018. Network of experts was conceived as a way to detect failure in procedures, manage needs, share experiences, unify protocols, agree about actions, favor joint projects, and access funding. Also, the network seeks to define or select priority areas for urgent mink control as a short-term goal. The exchange of experiences and details shared by the different members is the main strength of this network.

One thing that was noticed in the RCCV meeting was that different environments and/or communities and/or available resources impose different requirements that model or adjust the strategy for the control activities that should be developed. It was highlighted the lack of trapping protocols and good practices guidelines. Therefore, as another product of the meeting, a Good Practice Guide is being edited (Cabello unpublished data). This can be the beginning for the elaboration and contribution to a regulation on the subject.

12.7 Conclusions and Prospects for Further Research Applied to Mink Management in South America

The real challenge imposed by American mink invasion in South America and other parts of the globe is that control efforts should be applied in areas that are open to immigration. Much of the research needed must be focused on determining barriers and distances that would reduce or avoid immigration form sources of individuals.

Not less important, pre-defined trapping designs would be desirable. This involves defining the appropriate trap to different situations, baits and olfactive attractants, allocation of traps, trapping efforts, and trapping periods. However, these are difficult things to define in advance. In the past meeting of the RCCV, it was clear that every trapping experience went through an adaptation phase during which distance between traps, bait, attractant, and trapping session length should be adjusted to local reality. This also had to do with differences in habitats, assemblage of carnivores, and also mink abundance. Fasola and Roesler (2016) showed that trapping efficiency decayed after 1 year of control in Patagonia and that trapping season needed to be extended in order to reach the same level of trapping probability of the first season. While decay in animal abundance is a partial indicator of success, it also imposes a new challenge. Detection depends on animal abundance (Mackenzie 2005), and since every method has a detection probability (i.e., individuals avoiding traps), control strategy should be adapted continuously. Fasola and Roesler (unpublished data) had reached a plateau for removing mink with the methods explained in their work on mink control for the conservation of the hooded grebe (Fasola and Roesler 2016). Intensifying trap density and extending trapping season would be prohibitively expensive, and thus, incorporating sniffer dogs and winter trapping sessions to deal with low abundances of mink after years of control will be the next step (Fasola & Roesler pers. comm.).

Technology could be a solution at some situations. There are methods that would decrease the time of technicians in the field as remote alarm systems in traps (Cabello 2016a, b) and novel self re-setting traps (such as Good Nature brand). In this sense, research focused on trapping efficiency in order to adjust trapping effort (e.g., improving trapping rates) or to modify removal method would be an extremely valuable supply to managers.

Another research line that should be deepened is Success Indicators for Control Strategies. One of the longest and biggest mink culling programs in the Cairngorm Mountains, Northeastern Scotland, led by citizen conservationists (Bryce et al. 2011) found, along with the reduction in captures (as an indicator of less animals), that the male/female ratio was male biased after the first year of control as immigrant males tend to move longer and faster (Oliver et al. 2016). They found that with large-scale culling, male numbers dominate in re-colonized areas and this reduces the potential population re-growth. As an example, this biased sex ratio was also found after the first year of control in Santa Cruz (Fasola and Roesler 2016) and on subsequent years (Fasola and Roesler unpublished data) probably indicating that the spatial scale of this program may be adequate.

Oliver et al. (2016) introduce the idea of generating "attractive dispersal sinks" where immigrants are attracted but where demographic rates are kept low with culling. Melero et al. (2018) formalize this idea under the name of "ecological traps" and suggest that targeting control into areas to create this sort of patches could be a "promising tool" to overcome immigration into culled sectors where mink should be kept at very low densities or eradicated. This concept should be explored for Patagonian cases, as it may be a reasonable solution for control programs.

Until now, management to control mink has been approached only with actions targeted on the species. However, the ecology of communities could also support management of this invasive species. Top predators have in several contexts the potential of structuring communities, especially in aquatic environments (Sergio et al. 2008). The recovery of Eurasian otters in Europe was often associated with a decline in mink occupation meaning that favoring otters in the area could regulate mink. In Patagonia, American mink coexist with southern river otter in several catchments (Medina et al. 1997; Fasola et al. 2009; Medina et al. 2012). Then, an integrated approach could involve actions to recover otter populations together with more traditional control activities as trapping or hunting. This approach of community-level management of *N. vison* was also suggested by Crego et al. 2016 for Navarino Is. (Chile) in relation to other invasive species regulation (*C. canadensis*) and muskrat (*O. zibethicus*).

To conclude, invasion of American mink situation in South America is closely related to its introduction history, control management actions applied, and the intrinsic characteristics of the communities that allowed its successful establishment. Control of this invasive carnivore is economically, ecologically, and sociologically required and convenient. In Chilean and Argentinean Patagonia, we face different scenarios of this invasive carnivore invasion process (different habitats, community structures, food resources, invasion rates, invasion stages, knowledge of local reality, etc.). This complexity makes hard to think about eradication as a feasible management measure and more localized goals might be achievable. By having different situations, solutions will need to be adjusted according to the existing knowledge, economical budget, and conservation urgencies (Simberloff 2003; Myers 2000).

In Patagonia there is already useful information to begin with the prioritization of the control areas and trapping methodologies to devise a control strategy (Fasola and Valenzuela 2014; Vergara et al. 2015; Medina et al. 2015; Fasola and Roesler 2016; Mora et al. 2018). Invasive fronts represent an opportunity to hold advance of this harmful mustelid, and populations are not established yet. Detection and early warning are actions recommended at initial stages like the ones we observe in Chile (Chiloé Is., Temuco, and continental Magallanes) and Argentina (Neuquén, Buenos Aires, and part of Santa Cruz province). Highly recommended to increase early detection capacity at these areas by developing and using new technologies (i.e., camera traps) and mink tracker/sniffer dogs and investing in improving detection skills in the nearby human communities. The aim is to forge ahead of mink establishment and spread through the quick obtain of information to supply managers and prevent this ecological nightmare from continuing to expand in South America at the risk of losing native species.

References

Agriculture Minister of Chile (1998) Reglamento de la ley de Caza N°4601, sustituida por la Ley N°19.473. 56 pp https://bcn.cl/2ewpa

Ahola M, Nordström M, Banks PB, Laanetu N, Korpimäki E (2006) Alien mink predation induces prolonged declines in archipelago amphibians. Royal Soc 273(1591):1261–1265

Aldridge D (1986) El vison: Un problema ecológico y económico para Aysén. CONAF Report

Anderson CB, Rozzi R, Torres-Mura JC, McGehee SM, Sherriffs MF, Schuttler E, Rosemond AD (2006) Exotic vertebrate fauna in the remote and pristine sub-Antarctic Cape Horn Archipelago, Chile. Biodivers Conserv 15:3295–3313

Aued MB, Chéhebar C, Porro G, Macdonald DW, Cassini MH (2003) Environmental correlates of distribution of southern river otters (Lontra provocax) at different ecological scales. Oryx 37:413–421

Banks P, Nordström M, Ahola M, Salo P, Fey R, Korpimäki E (2008) Impacts of alien mink predation on island vertebrate communities of the Baltic Sea Archipelago: review of a long term experimental study. Boreal Environ Res 13:3–16

Barreto GR, Rushton SP, Strachan R, Macdonald DW (1998) The role of habitat and mink predation in determining the status and distribution of declining populations of water voles in England. Anim Conserv 1:129–137

Barros M, Saenz L, Lapierre L, Nuñez C, Medina Vogel G (2014) High prevalence of pathogenic Leptospira in alien American mink (Neovison vison) in Patagonia. Rev Chil Hist Nat 87:1–5

Barros M, Cabezón O, Dubey JP, Almeria S, Ribas MP, Escobar LE, Ramos B, Medina VG (2018) Toxoplasma gondii infection in wild mustelids and cats across an urban-rural gradient. PLoS One 13(6):e0199085

Ben-david M, Hanley TA, Klein DR, Schell DM (1997) Seasonal changes in diets of coastal and riverine mink: the role of spawning Pacific salmon. Can J Zool 75:803–811

Bidegain J (1963) Crianza del visón y posibilidades de su explotación en Chile. DVM Thesis, Universidad de Chile. Santiago

Bischof R, Haamed S, Ali H, Kabir M, Younas M, Shah KA, Din JU, Nawaz MA (2013) Using time-to-event analysis to complement hierarchical methods when assessing determinants of photographic detectability during camera trapping. Methods in Ecology and Evolution

Bonacic C, Orhens O, Hernandez F (2010) Estudio poblacional y orientado al control de la especie invasora Neovison vison (Schreber,1777), en el sur de Chile.Fauna Australis

Bonesi L, Palazon S (2007) The American mink in Europe: status, impact and control. Biol Conserv 134:470–483

Bonesi L, Strachan R, McDonald DW (2006) Why are fewer signs of mink in England? Considering multiple hypotheses. Biol Conserv 130:268–277

Bozinovic F (1993) Fisiologia de la alimentacion y digestion en vertebrados: Modelos y Teorias. Rev Chil Hist Nat 66:375–382

Bryce R, Oliver MK, Davies L, Gray H (2011) Turning back the tide of American mink invasion at an unprecedented scale through community participation and adaptive management. Biol Conserv 144:575–583

Bus Leone P, Cerda J, Sala S, Reid B (2014) Mink (Neovison vison) as a natural vector in the dispersal of diatom Didymosphenia geminata. Diatom Res 29:259–266

Cabello JL (2014) Estudio Socio-ecológico de las especies de vertebrados invasores en la Region de Magallanes y Antártica Chilena. Master Thesis, Universidad de Magallanes

Cabello JL (2016a) Informe final Prospección, Control y Capacitación sobre Castor en Magallanes y Ultima Esperanza. LP 1164-14 LP16. SAG

Cabello JL (2016b) Sistema de Monitoreo Remoto para trampas de Especies Invasoras.CORFO 16PIRE-60744. PatagoniaWildlife. Punta Arenas

Cabello JL, Soto-Volkart N, Soto-Mayorga C (2011) Informe del plan de Control del Visón Americano en Isla Navarino. Servicio Agrícola y Ganadero. Magallanes y Antartica Chilena

Canevari M, Vaccaro O (2007) Guía de mamíferos del sur de América del Sur. Editorial L.O.L.A., 413 págs. Buenos Aires

Carlsson NOL, Jeschke JM, Holmqvist N, Kindberg J (2010) Long term data on invaders: when the fox is away, the mink will play. Biol Invasions 12:633–641

Cerda C, Skewes O, Cruz G, Tapia P, Araos A, Baeriswyl F, Critician P (2017) Valoracion economica del impacto de siete species exoticas invasoras sobre los sectores productivos y la biodiversidad en Chile. Proyecto GEF/MMA/PNUD Especies Exoticas Invasoras. 27p

Ceron G, Trejo A (2012) Torrent Duck (Merganetta armata) population trend in northwestern Patagonia, Argentina. Ornitologia Neotropical 23:407–415

Chehébar C (1985) A survey of the southern river otter Lutra provocax Thomas in Nahuel Huapi National Park. Argentina. Biol Conserv 32:299–307

Chehébar C, Gallur A, Giannico G, Gottelli M, Yorio P (1986) A survey of the southern river otter Lutra provocax in Lanín Puelo and Los Alerces National Parks, Argentina and evaluation of its conservation status. Biol Conserv 38:293–304

Concha MA (2013) Blackbird Chile birdwatching. Seminario Para optar al Grado de Ingeniero Comercial, Mención Administración. Facultad de Economia y Negocios. Universidad de Chile

Crego RD, Jimenez JE, Rozzi R (2015) Expansión de la invasión del Visón Norteamericano (Neovison vison) en la Reserva de la Biosfera de Cabo de Hornos, Chile. Anales del Instituto de la Patagonia 43:157–162

Crego RD, Jiménez JE, Rozzi R (2016) A synergistic trio of invasive mammals? Facilitative interactions among beavers, muskrats, and mink at the southern end of the Americas. Biol Invasions 18:1923–1938

Crook J A. (2011) Lag times. In: Encyclopedia of biological invasions. Simberloff D & Rejmanék M (Eds). University of California Press, Berkeley. 404–410pp

Davis EF, Anderson CB, Valenzuela AEJ, Cabello JL, Soto N (2012) American mink (Neovison vison) trapping in the Cape Horn Biosphere Reserve: enhancing current trap systems to control an invasive predator. Ann Zool Fennici 49:18–22

Delibes M, Travaini A, Zapata SC, Palomares F (2003) Alien mammals and the trophic position of lesser grison (Galictis cuja) in Argentinean Patagonia. Can J Zool 81:157–162

Dodyk L (2017) Excursiones pelágicas en Argentina. Aves Argentinas-Revista de Naturaleza y Conservación 49:36–38

Donadio E, Buskirk SW (2006) Diet, morphology and interspecific killing in Carnivora. The Am Narut 167(4):524–535

Dunstone N (1993) The mink. T. &. A.D. Poyser, London

EDGE (2016) Evolutionary distinct and globally endangered: top 100 bird species. Zoological Society of London. Downloaded from http://www.edgeofexistence.org/birds/top_100.php on 10/12/2016

Elton CS (1958) The ecology of invasions by animals and plants. Chapman and Hall, London

Enders RK (1952) Reproduction in the mink Mustela vison. Proc Am Philos Soc 96:691–755

Failla M and Fasola L (2019) Visón americano: un nuevo invasor del Río Negro, Patagonia Argentina. Mastozoología Neotropical 26: 482-486

Farías AA, Svensson GL (2014) Ecoregional vulnerability assessment for the functional richness of south American carnivores (Mammalia: Carnivora). J Mamm Evol 21(4):437–450

Fasola L (2009). Distribución, alimentación e interacciones de dos mustélidos semiacuáticos en los bosques andino patagónicos: el Huillín (*Lontra provocax*), nativo, y el visón Americano (*Mustela vison*), introducido. Tesis doctoral. Universidad de Buenos Aires. 96 pp

Fasola L, Roesler I (2016) Invasive predator control program in austral Patagonia for endangered bird conservation. Eu J Wildlife Res. https://doi.org/10.1007/s10344-016-1032-y

Fasola L, Roesler I (2018) A familiar face with a novel behavior raises challenges for conservation: American mink in arid Patagonia and a critically endangered bird. Biol Conserv 218:217–222. https://doi.org/10.1016/j.biocon.2017.12.031

Fasola L, Valenzuela A (2014) Invasive carnivores in Patagonia: defining priorities for their management using the American mink (Neovison vison) as study case. Ecol Austral 24:183–192

Fasola L, Chehébar C, Macdonald DW et al. (2009) Do alien North American mink compete for resources with native South American river otter in Argentinean Patagonia? Journal of Zoology, 277, 187–195

Fasola L, Muzio J, Chehébar C, Porro G, Cassini MH (2011) Range expansion and prey use of American mink in Argentinean Patagonia: dilemmas for conservation. Eu J Wildlife Res 57:283–294

Foerster R (1973). Estudio integral sobre la presencia de visones (Mustela vison) en el ParqueNacional los Alerces y zona de in uencia. Technical report, Centro de Documentación, Administración de ParquesNacionales, Argentina

Fraga R (2000) Introduced feral mink *Mustela vison* in Patagonia: A plausible cause of population declines in the Austral rail *Rallus antarcticus*? Cotinga 13:71–72

Ministry of Economy and Public Works and Services of Tierra del Fuego (2002) Proyecto de creación de un área protegida en el extremo sudoriental de la Isla Grande de Tierra del Fuego, Republica Argentina. Ministerio de Economía y Servicios públicos, Secretaría de Planeamiento y desarrollo, Subsecretaría de Planeamiento, Ushuaia, Tierra del Fuego. 100pp

García MR (1982) El visón. Su cria en cautividad. Editorial Hemisferio Sur, Buenos Aires, Argentina. Pp226

Garretón Maldonado EF (2015) Estudio de la dieta del vison (Neovison vison) en el Monumento Natural Lahuen Ñadi, por medio del análisis de contenido gastrointestinal y de materia fecal. Memoria de Titulo para obtención del Titulo Profesional MV. Universidad San Sebastian. Puerto Montt

Gerell R (1971) Population studies on mink Mustela vison in southern Sweden. Oikos 8:83–109

Glitteman JL (1985) Carnivore body size: ecological and taxonomic correlates. Oecologia 67(4):540–554

Godoy J (1963) Fauna Silvestre. Vol I, II, VII. Technical Report. Consejo Federal de Inversiones, Buenos Aires, Argentina

Guichón ML, Monteverde M, Piudo L, Sanguinetti J, Di Martino CV (2016) Mamíferosintroducidos en la provincia de Neuquén: Estado actual y prioridades de manejo. MastozoologíaNeotropical 23:255–255

Guiñez-Lillo B (2014) Diagnóstico de la presencia, mapeo y propuesta de un plan de control para la erradicación de Neovison vison en el Parque Nacional Villarrica, Región de Araucanía, Chile. CONAF Report

Harrington LA, Macdonald D (2008) Spatial and temporal relationships between invasive American mink and native European polecats in the southern United Kingdom. J Mammal 89:991–1000

Holland BS (2000) Genetics of marine bioinvasions. Hydrobiologia 420(1):63–71

Ibarra JT, Fasola L, Macdonald DW, Rozzi R, Bonacic C (2009) Invasive American mink Mustela vison in wetlands of the Cape Horn Biosphere Reserve, southern Chile: what are they eating? Oryx 43:87–90

Iriarte A, Lobos G, Jaksic F (2005) Invasive vertebrates species in Chile and their control and monitoring by governmental agencies. Rev Chil Hist Nat 78:143–154

Jaksic FM (1998) Vertebrate invaders and their ecological impacts in Chile. Biodivers Conserv 7:1427–1445

Jaksic FM, Iriarte JA, Jiménez JE et al (2002) Invaders without frontiers: cross-border invasions of exotic mammals. Biol Invasions 4:157–173

Jar AM, Ramayo LG, Stempler A, Goldman LH, Mundo SL (2010) Observaciones sobre el plan de vacunación contra el distemper en visones (Mustela vison) en un criadero de Argentina. Rev Argent Microbiol 42:189–192

Jara Millar RF (2010) Estudio dietario y parasitológico del visón Neovison vison en la cuenca del Río Maullín, Región de Los Lagos, Chile. Memoria de Título para optar al Título De Médico Veterinario. Universidad Austral de Chile, Valdivia

Jetz W, Thomas GH, Joy JB et al (2014) Global distribution and conservation of evolutionary distinctness in birds. Curr Biol 24:919–930

Jones AC (2011) Distribución del visón (Mustela vison) en la provincia del Chubut y posiblescon-flictos con actividadeshumanas. Undergraduate Thesis, National University of Patagonia

Keddy PA (2001) Competition. Population and Community Biology Series. Kluwer Academic Publishers

King C, McDonald RM, Martin RD, Dennis T (2009) Why is eradication of invasive mustelids so difficult? Biol Conserv 142:806–816

Kusch A, Caceres Murrie A, Chacon M, Terán D, Vila A, Uhart M, Munzenmayer R, Dougnac (2016) Monitoreo de la colonia de albatros ceja negra y Control del vison americano. Islote Albatros- Seno Almirantazgo. Tierra del Fuego. MMA-WCS

Langenscheidt M (2015) Monitoreo de Huillín (Lontra provocax) en el Parque Nacional Chiloé y sus sectores aledaños. CONAF Report

Lariviere S (1999) Mustela vison. Mamm Species 608:1–9

Lecis R, Ferrando A, Ruiz Olmo J, Mañas S, Domingo-Roura X (2008) Population genetic structure and distribution of introduced American mink (Neovison vison) in Spain, based on microsatellite variation. Conserv Genet 9:1149–1161

Liljesthröm M, Schiavini A, Sáenz Samaniego RA, Fasola L, Raya Rey A (2013). Kelp Geese (*Chloephaga hybrida*) and the Flightless Steamer Ducks (*Tachyeres pteneres*) in the Beagle Channel: the importance of islands in providing safe nesting habitat. *The Wilson Journal of Ornithology*, 125:583–591

Liljesthröm M, Fasola L, Valenzuela AEJ, Raya Rey A, Schiavini A (2014) Nest predators of flightless steamer-ducks (Tachyeres pteneres) and flying steamer-ducks (Tachyeres patachonicus). Waterbirds 37:210–214

Lizarralde MS, Escobar JM (2000) Exotic mammals in Tierra del Fuego. Ciencia Hoy 10:52–63

Lode T, Cormier JP, Le Jacques D (2001) Decline in endangered species as an indication of anthropic pressures: the case of European mink Mustela lutreola western population. Environ Manag 28:727–735

Macchi PJ, Vigliano PH (2014) Salmonid introduction in Patagonia: the ghost of past, present and future management. Ecol Austral 24:162–172. Asociacion Argentina de Ecología

Macdonald DW, Harrington LA (2003) The American mink: the triumph and tragedy of adaptation out of context. NZ J Zool 30:421–441

Mackenzie DI (2005) What are the issues with presence—absence data for wildlife managers? Special section: the value and utility of presence – absence data to wildlife monitoring and research. J Wildlife Manag 69:849–860

Malerba MS, Fasola L, Roesler I, Pereda MP, de Miguel A, Martín L & Mahler B (2018). Variabilidadgenética de visones americanos *Neovison vison* asilvestrados en la provincia de Santa Cruz: ¿se cumple la paradoja genética de las especies invasoras?.Mastozoología Neotropical. doi: https://doi.org/10.31687/saremMN.18.25.2.0.23

Malley B, Anderson C, Stodola K, Rosemond A (2011) Identifying native or exotic predators of ground nesting songbirds in subantarctic forests in southern Chile. Anales Instituto Patagonia 39(1):51–57

Maran T, Macdonald DW, Kruuk H, Sidorovich NV, Rozhnow VV (1998) The continuing decline of the European mink *Mustela lutreola*: evidence for the intraguild aggression hypothesis. In: Dunstone N, Gorman ML (eds) Behaviour and ecology of riparian mammals. Cambridge University Press, Cambridge

Martino P, Villar JA (1991) A note on diseases of mink. J Veterinary Med (Zentralblatt für Veterinaermedizin) B 38:227–230

Martino P, Gatti M, Bautista E, Stanchi N (2007) Case report: tuberculosis in introduced American Mink Mustela vison. *Small Carnivore Conserv* 36:46–47

Mazar Barnett J, della Seta M, Imberti S, Pugnali G (1998) Notes on the rediscovery of the Austral Rail Rallus antarcticus in Santa Cruz, Argentina. Cotinga 10:96–101

Mazar Barnett J, Imberti S, Roesler I (2014) Distribution and habitat use of the Austral Rail (Rallus antarcticus) and perspectives on its conservation. Bird Conserv Int 24:114–125

Medina G (1997) A comparison of the diet and distribution of Southern river otter (Lutra provocax) and mink (Mustela vison) in Southern Chile. J Zool (Lond) 242:291–297

Medina Vogel G, Barros M, Monsalve R, Pons D (2015) Assessment of the efficiency in trapping North American mink (Neovison vison) for population control in Patagonia. Rev Chil Hist Nat 88(1):1–12

Medina-Vogel G (2010) Ecology of emerging infectious diseases and wild species conservation. Arch Med Vet 42:11–24

Medina-Vogel G, Barros M, Organ JF, Bonesi L (2012) Coexistence between the southern river otter and the alien invasive North American mink in marine habitats of southern Chile. J Zool (Lond). https://doi.org/10.1111/jzo.1201

Melero Y, Palazón S, Bonesi L, Gosálbez J (2008) Feeding habits of three sympatric mammals in NE Spain: the American mink, the spot ted genet, and the Eurasian otter. Acta Theriol 53:263–273

Melero Y, Cornulier T, Oliver MK, Lambin X (2018) Ecological traps for large-scale invasive species control: predicting settling rules by recolonising American mink post-culling. J Appl Ecol 55:1769–1779

Melquist WE, Whitman JS, Hornocker MG (1981) Resource partitioning and coexistence of sympatric mink and river otter populations. In: Chapman J, Pursley D (eds) Worldwide furbearer conference, vol 1. Worldwide Furbearer Conference Inc., Frostburg, pp 187–220

Mora M, Medina Vogel G, Sepúlveda M, Vergara D, Alvarez VR, Vianna J (2018) Genetic structure of introduced American mink (Neovison vison) in Patagonia: colonization insights and implications for control and managements strategies. Wildl Res 45(4):344–356
Morales MC (2015) Informe de Avance Plan de Control del Visón Americano en el sector Puesco del Parque Nacional Villarrica. Región de la Araucanía. CONAF Report
Myers JH, Simberloff D, Kuris AM, Carey JR (2000) Eradication revisited: dealing with exotic species. Trends Ecol Evol 15(8):316–320
Nanvas J (1987) Los vertebrados exóticos introducidos en Argentina. Revista de Museo Argentino de Ciencias Naturales "Bernardino Rivadavia" Zoologia. Argentina 14:7–38
Nordström M, Korpimäki E (2004) Effects of island isolation and feral mink removal on bird communities on small islands in the Baltic Sea. J Anim Ecol 73:424–433
Oliver MK, Piertney SB, Zalewski A, Melero & Lambin X. (2016) The compensatory potential of increased immigration following intensive American mink population control is diluted by male-biased dispersal. Biol Invasion Vol 18(10):3047–3061
Pagnoni G, Garrido J, Marin M (1986) Impacto económico y ambiental del visón. Mustela vison (Schreber 1877) en el norte de la Patagonia. Technical report, CENPAT- CONICET, Dirección de Fauna de la Provincia de Chubut. 20pp
Peris S, Sanguinetti J, Pescador M (2009) Have Patagonian waterfowl been affected by the introduction of the American mink Mustela vison? Fauna & Flora International, Oryx 43:648–654
Pescador M, Díaz S, Peris S (2012) Abundances of waterbird species on lakes in Argentine Patagonia as a function of season, lake size and the presence of mink. Hydrobiologia 697:111–125
Previtali A, Cassini MH, Macdonald DW (1998) Habitat use and diet of the American mink (Mustela vison) in Argentinian Patagonia. J Zool 246:443–486
Quinan E, De Los Ríos Escalante P (2011) Presence of decapods in the diet of an introduced mink (Neovison vison (Schreber, 1777)) population in a Chilean North Patagonian lake. Crustaceana 84(1):125–127
Redford KH, Eisenberg JF (1992) Mammals of the Neotropics: the Southern Cone. Chile, Argentina, Uruguay, Paraguay, vol 2. The University of Chicago Press, Chicago, pp 1–430
Reynolds JC, Short MJ, Leigh RJ (2004) Development of population control strategies for mink *Mustela vison*, using floating rafts as monitors and trap sites. Biol Conserv 120:533–543
Roesler I (2016) Conservación del Macá Tobiano (Podiceps gallardoi): factores que afectan la viabilidad de sus poblaciones. Universidad de Buenos Aires, Argentina (PhD thesis, 274 pp.)
Roesler I, Imberti S, Casañas H et al (2012a) A new threat for the globally Endangered Hooded Grebe *Podiceps gallardoi*: the American mink Neovison vison. Bird Conserv Int 1988:1–6
Roesler I, Imberti S, Casañas H et al (2012b) Hooded Grebe Podiceps gallardoi population decreased by eighty per cent in the last twenty-five years. Bird Conserv Int 22:371–382
Roesler I, Fasola L, Casañas H et al (2016) Colony guardian programme improves recruitment in the critically endangered hooded grebe *Podiceps gallardoi* in Austral Patagonia. Argentina Conserv Evid 13:62–66
Rozzi R, Sherriffs MF (2003) El visón (Mustela vison Schreber, Carnivora: Mustelidae), unnuevomamíferoexóticopara la Isla Navarino. Annales del Instituto de la Patagonia 31:97–104
Ruiz J, Schlatter R, Bucher D (1996) Estudio de la situación del vison (Mustela vison, Schreber 1777) y su impacto sobre las comunidades autóctonas de la X Región, como aporte a la protección y recuperación de las áreas silvestres protegidas del estado. CONAF, X Región, Pto Montt. Chile
Ruiz-Olmo J, Palazon S, Bueno F (1997) Distribution, status and colonization of the American mink *Mustela vison* in Spain. J Wildlife Res 2:30–36
Salo P, Nordstrom M, Thompson RL, Korpimaki E (2008) Risk induced by a native predator reduces alien mink movements. J Anim Ecol 77:1092–1098
Sandoval RJ (1994) Estudio ecológico del visón asilvestrado (Mustela vison, Schreber) en la XI Región. DVM Thesis, Universidad Austral de Chile
Sargeant AB, Swanson GA, Doty HA (1973) Selective predation by mink Mustela vison, on waterfowl. Am Midi Nat 89:208–214

Schüttler E, Cárcamo J, Rozzi R (2008) Diet of the American mink Mustela vison and its potential impact on the native fauna of Navarino Island, Cape Horn Biosphere Reserve, Chile. Revista Chilena Historia Natural 81:585–598

Schüttler E, Klenke R, McGehee S et al (2009) Vulnerability of ground-nesting waterbirds to predation by invasive American mink in the Cape Horn Biosphere Reserve, Chile. Biol Conserv 142:1450–1460

Schüttler E, Ibarra JT, Gruber B, Rozzi R, Jax K (2010) Abundance and habitat preferences of the southern- most population of mink: implications for managing a recent island invasion. Biodivers Conserv 19:725–743

Sepúlveda MA, Muñóz-Zanzi C, Rosenfeld C, Jara R, Pelican KM, Hill D (2011) Toxoplasma gondii in feral American minks at the Maullín river, Chile. Vet Parasitol 175:60–65

Sepúlveda MA, Singer RA, Silva-Rodríguez EA, Eguren A, Stowhas P, Pelican K (2014) Invasive American mink: linking pathogen risk between domestic and endangered carnivores. Ecohealth 11:409–419

Sergio F, Caro T, Brown D, Clucas B, Hunter J, Ketchum J, McHugh K, Hiraldo F (2008) Top predators as conservation tools: ecological rationale, assumptions and efficacy. Annu Rev Ecol Evol Syst 39:1–19

Shimatani Y, Fukue Y, Kishimoto R, Masuda R (2010) Genetic variation and population structure of the feral American mink (Neovison vison) in Nagano, Japan, revealed by microsatellite analysis. Mammal Study 35(1):1–7

Simberloff D (2003) How much information in population biology is needed to manage introduced species? Conserv Biol 17(1):83–92

Simberloff D, Von Holle B (1999) Positive interactions of non indigenous species: invasional meltdown? Biol Invasions 1:21–32

Tilman D (1982) Resource competition and community structure. Monogr. Pop. Biol. 17. Princeton University Press, Princeton, p 296

Valenzuela J, Grau J (2005) Occurrence of American mink on the Chonos Archipelago of southern Chile. Oryx 39(1):12–16

Valenzuela A, Raya Rey A, Fasola L, Schiavini A (2012) Understanding the inter-specific dynamics of two coexisting predators in Tierra del Fuego Archipelago: the native southern river otter and the exotic American mink. Biol. Invasions.

Valenzuela AEJ, Raja Rey A, Fasola L, Sáenz Samaniego RA, Schiavini A (2013) Trophic ecology of a top predator colonizing the southern extreme of South America: feeding habits of invasive American mink (Neovison vison) in Tierra del Fuego. Mamm Biol 78:104–110

Valenzuela AEJ, Anderson C, Fasola L, Cabello J (2014) Linking invasive exotic species and their ecosystem impacts in Tierra del Fuego to test theory and determine action. *Acta Oecologica*, 54:110–118

Valenzuela AE, Sepúlveda MA, Cabello JL, Anderson CB (2016) El visón americano en Patagonia: un análisis histórico y socioecológico de la investigación y el manejo. Mastozoología Neotropical 23:289–304

Vergara G, Valenzuela J, Parragué-Mignone C, Langenscheidt (2015) Records and current state of American mink (Neovison vison) in the Chiloé Archipelago, Chile. Boletín del Patrimonio Natural de Chiloé 1(1):2–13

Zalewski A, Piertney S, Zalewska H, Lambin X (2009) Landscape barriers reduce gene flow in an invasive carnivore: geographical and local genetic structure of American mink in Scotland. Mol Ecol 18:1601–1615

Zalewski A, Michalska-Parda A, Bartoszewicz M, Kozakiewicz M, Brzezinski M (2010) Multiple introductions determine the genetic structure of an invasive species population: American mink Neovison vison in Poland. Biol Conserv 143:1355–1363

Chapter 13
Taxonomic Homogenization of the Freshwater Fish Fauna in Chile: Analyzing the Ichthyogeographic Provinces

Pablo Rojas, Sergio A. Castro, Irma Vila, Evelyn Habit, and Fabián M. Jaksic

13.1 Taxonomic Homogenization of the Freshwater Fish Faunas

It is widely recognized that anthropogenic activity modifies habitat conditions and species composition in freshwater communities (Abell et al. 2008). The conservation status of the freshwater communities, and particularly of the freshwater fish fauna, is worrying because a massive compositional change in diversity is undergoing (Mack et al. 2000; Revenga et al. 2005; Dudgeon et al. 2006). The processes underlying this phenomenon can be reduced to mechanisms that promote the addition of new species (invasion or exotic species introduction) and the extinction or extirpation of the native ones (McKinney and Lockwood 1999; Olden and Poff 2003, 2004). As consequence, both invasions and extinctions are expected to

P. Rojas
Laboratorio de Ecología y Biodiversidad, Universidad de Santiago de Chile, Santiago, Chile

Laboratorio de Limnología, Departamento de Ciencias Ecológicas, Universidad de Chile, Santiago, Chile

S. A. Castro (✉)
Laboratorio de Ecología y Biodiversidad, Universidad de Santiago de Chile, Santiago, Chile
e-mail: sergio.castro@usach.cl

I. Vila
Laboratorio de Limnología, Departamento de Ciencias Ecológicas, Universidad de Chile, Santiago, Chile

E. Habit
Departamento de Sistemas Acuáticos, Facultad de Ciencias Ambientales y Centro de Ciencias Ambientales EULA, Universidad de Concepción, Concepción, Chile

F. M. Jaksic
Center of Applied Ecology and Sustainability, Departamento de Ecología, Facultad de Ciencias Biológicas, Pontificia Universidad Católica de Chile, Santiago, Chile

F. M. Jaksic, S. A. Castro, *Biological Invasions in the South American Anthropocene*, https://doi.org/10.1007/978-3-030-56379-0_13

contribute to taxonomic homogenization (McKinney and Lockwood 1999; Olden and Rooney 2006). The taxonomic homogenization is a process that can be defined as a temporal (or spatial) increase of the similarity (or spatial turnover) among communities that initially were more dissimilar (Olden et al. 2011; Toussaint et al. 2014). Currently, it is considered that the taxonomic homogenization is the first step of a more complex and complete process that includes ecological and evolutionary facets at different levels and scales (Olden et al. 2010, 2011).

Numerous studies have analyzed freshwater fish assemblages and their compositional changes at basin scales. Most studies have evidenced the occurrence of taxonomic homogenization, and currently it is accepted that these communities obey to a global homogenization trend (McKinney and Lockwood 1999; Rahel 2002; Olden and Poff 2004; Taylor 2004; Clavero and García-Berthou 2006; Marchetti et al. 2006; Olden et al. 2008). Indeed, the main biogeographical realms (i.e., Afrotropical, Australian, Nearctic, Neotropical, Oriental, and Palearctic) show that compositional similarity has increased among basins (Leprieur et al. 2009; Villéger et al. 2011, 2014, 2015; Toussaint et al. 2014, 2016). However, these biogeographical realms are not uniform, showing regions that differ in their geological and evolutionary history (Ruggiero and Ezcurra 2003). Considering that basins derive from a long evolutionary history (Lévêque et al. 2008; Nelson et al. 2016), the taxonomic homogenization constitutes a disruption (perhaps irreversible) to this ichthyogeographical order (Olden 2006). Thus, a central question here is how the taxonomic homogenization affects these regions and what compositional scenarios are to be expected in the future (Olden 2006).

Chile is considered a biogeographic island within the Neotropical realm because natural geographical barriers match its geopolitical borders (Dyer 2000). Andean Range, Pacific Ocean, and Atacama Desert impose environmental limits along ca. 4500 km of latitudinal extension (Arratia 1997; Dyer 2000). From a zoogeographical point of view, the South American ichthyofauna includes the Neotropical and Andean regions (Vila and Quezada-Romegialli 2018). The Titicaca province belongs to the Neotropical region, while the provinces Atacama, Chilean, Valdivian Lakes, and Patagonia belong to the Andean region (see nomenclature in Vila and Quezada-Romegialli 2018). These ichthyogeographic regions derive from evolutionary processes associated with the emergence of the Andes range, 180 million years ago (Charrier et al. 2005; Vila et al. 2006); therefore, the Chilean freshwater fish fauna has experienced an isolated evolution, separately from other South American basins (Arratia 1982).

Currently, the complete Chilean freshwater fish fauna is composed by 42 species plus a small number of additional taxa that remain to be described (Vila and Quezada-Romegialli 2018). In general terms, this species richness increases with latitude (up to 40° S) from 1 to 18 species per basin, and then decreases gradually to 6–11 species per basin at 58 °S (Habit et al. 2006; Vila and Quezada-Romegialli 2018). In spite of the reduced species richness, this fauna is highly endemic, reaching approximately 80% of endemism (Vila and Quezada-Romegialli 2018). Additionally, these species are considered relict because they come from ancient lineages evolved inside Neotropical and Austral regions (Arratia 1982; Vila et al.

1999; Dyer 2000; Habit et al. 2006; Vila and Habit 2015; Vila and Quezada-Romegialli 2018). On the other hand, 28 exotic species naturalized in different Chilean watersheds have been described (Iriarte et al. 2005; Castro et al. 2014). These species were introduced attending to economic, productive, and ornamental purposes (Basulto 2003; Camus and Jaksic 2009).

Analyzing the fish composition of the Mediterranean-climate regions of the world, Marr et al. (2013) found that Chilean basins located in the Mediterranean region (30° – 33° S) are homogenized with respect to their climatic analogues. Castro et al. (2014) found an incipient trend of homogenization among basins belonging to different Chilean biogeographical regions, and Vargas et al. (2015) documented taxonomic homogenization among basins. These studies, however, use metrics that overestimate the taxonomic homogenization because they are affected by both spatial turnover and nestedness (Baselga 2010). The spatial turnover implies the replacement of some species by others as a consequence of environmental sorting or spatial and historical constraints (Qian et al. 2005), while species nestedness occurs when the biotas of sites containing smaller numbers of species are subsets of the biotas at richer sites (Ulrich and Gotelli 2007).

The present chapter analyzes the taxonomic homogenization of the Chilean freshwater fish fauna, taking advantage that the geopolitical limits coincide with biogeographical ones (Vila and Pardo 2008). Specifically, we studied how taxonomic homogenization occurs by comparing the spatial turnover between basins that belong to different ichthyogeographical provinces. Thus, we hypothesized that if the ichthyogeographical regions of Chile are undergoing a process of taxonomic homogenization, the spatial turnover values should have decreased significantly when basins are compared from historic to the present time.

13.2 Basins and Fishes

Chilean basins are embedded in five ichthyogeographical provinces: Titicaca (TIT; 8° – 23° S), Atacama (ATA; 18° – 26° S), Chilean (CHI; 27° – 38°S), Valdivian Lakes (VAL; 38° – 43° S), and the Patagonia region (PAT; 44° – 55° S) (Fig. 13.1; Vila and Quezada-Romegialli 2018). Specifically, the basins studied were as follows: Chungará (Chu; TIT), Lauca (Lau; TIT), Isluga (Isl; TIT), Ascotán (Asc; TIT), Camarones (Cam; ATA), Loa (Loa; ATA), Copiapó (Cop; CHI), Huasco (Hua; CHI), Elqui (Elq; CHI), Limarí (Lim; CHI), Choapa (Cho; CHI), Aconcagua (Aco; CHI), Maipo (Mai; CHI), Rapel (Rap; CHI), Mataquito (Mat; CHI), Maule (Mau; CHI), Itata (Ita; CHI), Andalién (And; CHI), Biobío (Bio; CHI), Imperial (Imp; CHI), Toltén (Tol; VAL), Valdivia (Val; VAL), Bueno (Bue; VAL), Maullín (Mal; VAL), Palena (Pal; PAT), Cisnes (Cis; PAT), Aysen (Ays; PAT), Baker (Bak; PAT), and Serrano (Ser; PAT) (Table 13.1). Altogether these basins cover 40% of the continental area (Fig. 13.1).

Through an exhaustive bibliographic review covering 1928–2018, we gathered distributional information for native and exotic freshwater fishes in Chile. To

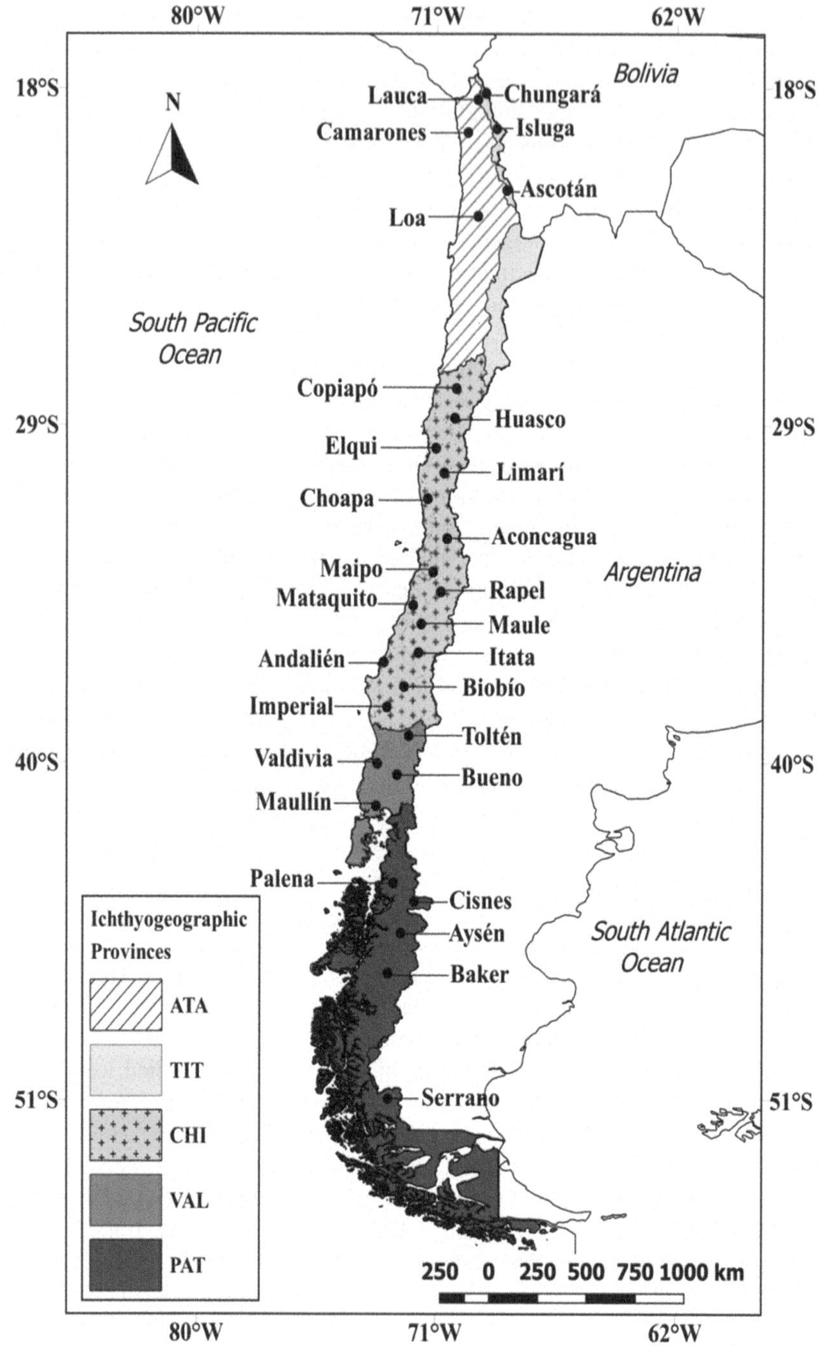

Fig. 13.1 Geographic distribution of the 29 basins of the five ichthyogeographical regions of Chile

Table 13.1 Historical and current fish composition for 29 watersheds belonging to five ichthyogeographical provinces of Chile. For each region, the presence of native and exotic species is indicated by 1 = Present; 0 = Absent; 1/0 = Extinct species. TIT = Titicaca province; ATA = Atacama province; CHI = Chilean province; VAL = Valdivian Lagos rovince; PAT = Patagonia. Chungará (Chu), Lauca (Lau), Camarones (Cam), Isluga (Isl), Ascotán (Asc), Loa (Loa), Copiapó (Cop), Huasco (Hua), Elqui (Elq), Limarí (Lim), Choapa (Cho), Aconcagua (Aco), Maipo (Mai), Rapel (Rap), Mataquito (Mat), Maule (Mau), Itata (Ita), Andalién (And), Biobío (Bio), Imperial (Imp), Toltén (Tol), Valdivia (Val), Bueno (Bue), Maullín (Mal), Palena (Pal), Cisnes (Cis), Aysen (Ays), Baker (Bak), and Serrano (Ser)

Species	ALT				ATA		CHI														VAL				PAT				
	Chu	Lau	Isl	Asc	Cam	Loa	Cop	Hua	Elq	Lim	Cho	Aco	Mai	Rap	Mat	Mau	Ita	And	Bio	Imp	Tol	Val	Bue	Mal	Pal	Cis	Ays	Bak	Ser
Native species																													
Aplochiton marinus[a] (Eigenmann 1928)	0	0	0	0	0	0	0	0	0	0	0	0	0	0	0	0	0	0	0	0	0	1	0	0	0	0	0	1	0
Aplochiton taeniatus[a] (Jenyns 1842)	0	0	0	0	0	0	0	0	0	0	0	0	0	0	0	0	0	0	0	0	1	1	1	1	1	1	1	1	1
Aplochiton zebra[a] (Jenyns 1842)	0	0	0	0	0	0	0	0	0	0	0	0	0	0	0	0	0	0	1	1	1	1	1	1	1	1	1	1	1
Basilichthys cf. *semotilus*[a] (Cope 1874)	0	0	0	0	1	1	0	0	0	0	0	0	0	0	0	0	0	0	0	0	0	0	0	0	0	0	0	0	0
Basilichthys microlepidotus[a] (Jenyns 1841)	0	0	0	0	0	0	1	1	1	1	1	1	1	1	1	1	1	1	1	1	1	1	1	1	0	0	0	0	0
Brachygalaxias bullocki[a] (Regan 1908)	0	0	0	0	0	0	0	0	0	0	0	0	0	0	0	0	1	1	1	1	1	1	1	1	0	0	0	0	0
Brachygalaxias gothei[a] (Busse 1982)	0	0	0	0	0	0	0	0	0	0	0	0	0	0	0	1	0	0	0	0	0	0	0	0	0	0	0	0	0
Bullockia maldonadoi[a] (Eigenmann 1928)	0	0	0	0	0	0	0	0	0	0	0	0	0	0	0	0	1	1	1	1	0	0	0	0	0	0	0	0	0

(continued)

Table 13.1 (continued)

Species	ALT				ATA		CHI														VAL				PAT				
	Chu	Lau	Isl	Asc	Cam	Loa	Cop	Hua	Elq	Lim	Cho	Aco	Mai	Rap	Mat	Mau	Ita	And	Bio	Imp	Tol	Val	Bue	Mal	Pal	Cis	Ays	Bak	Ser
Caragola lapicida[a] (Gray 1851)	0	0	0	0	0	0	0	0	0	0	0	1	1	1	1	1	1	1	1	1	1	1	1	1	1	1	1	1	1
Cheirodon australe[a] (Eigenmann 1928)	0	0	0	0	0	0	0	0	0	0	0	0	0	0	0	0	0	0	0	0	0	1	1	1	0	0	0	0	0
Cheirodon galusdae[a] (Eigenmann 1928)	0	0	0	0	0	0	0	0	0	0	0	0	0	0	1	1	1	1	1	1	0	0	0	0	0	0	0	0	0
Cheirodon kiliani[a] (Campos 1982)	0	0	0	0	0	0	0	0	0	0	0	0	0	0	0	0	0	0	0	0	0	1	0	0	0	0	0	0	0
Cheirodon pisciculus[a] (Girard 1855)	0	0	0	0	0	0	0	1	1	1	1	1	1	1	0	0	0	0	0	0	0	0	0	0	0	0	0	0	0
Diplomystes camposensis[a] (Arratia 1987)	0	0	0	0	0	0	0	0	0	0	0	0	0	0	0	0	0	0	0	0	1	1	0	0	0	0	0	0	0
Diplomystes chilensis[a] (Molina 1782)	0	0	0	0	0	0	0	0	0	0	0	0	1/0	0	0	0	0	0	0	0	0	0	0	0	0	0	0	0	0
Diplomystes incognitus[a] (Arratia and Quezada-Romegialli 2017)	0	0	0	0	0	0	0	0	0	0	0	0	0	1	1	1	1	0	0	0	0	0	0	0	0	0	0	0	0
Diplomystes nahuelbutaensis[a] (Arratia 1987)	0	0	0	0	0	0	0	0	0	0	0	0	0	0	0	0	0	0	1	1	0	0	0	0	0	0	0	0	0

Species	ALT				ATA		CHI														VAL				PAT				
	Chu	Lau	Isl	Asc	Cam	Loa	Cop	Hua	Elq	Lim	Cho	Aco	Mai	Rap	Mat	Mau	Ita	And	Bio	Imp	Tol	Val	Bue	Mal	Pal	Cis	Ays	Bak	Ser
Galaxias globiceps[a] (Eigenmann 1928)	0	0	0	0	0	0	0	0	0	0	0	0	0	0	0	0	0	0	0	0	0	0	0	1	0	0	0	0	0
Galaxias maculatus (Jenyns 1842)	0	0	0	0	0	0	0	1	1	1	1	1	1	1	1	1	1	1	1	1	1	1	1	1	1	1	1	1	1
Galaxias platei (Steindachner 1898)	0	0	0	0	0	0	0	0	0	0	0	0	0	0	0	0	0	0	0	0	1	1	1	1	1	1	1	1	1
Geotria australis (Gray 1851)	0	0	0	0	0	0	0	0	0	0	0	1	1	1	1	1	1	1	1	1	1	1	1	1	1	1	1	1	1
Hatcheria macraei (Girard 1855)	0	0	0	0	0	0	0	0	0	0	0	0	0	0	0	0	0	0	0	1	1	1	1	0	1	1	1	1	0
Nematogenys inermis[a] (Guichenot 1848)	0	0	0	0	0	0	0	0	0	0	0	0	1	1	1	1	1	1	1	1	0	0	0	0	0	0	0	0	0
Odontesthes brevianalis[a] (Günther 1880)	0	0	0	0	0	0	0	0	1	1	1	1	1	1	1	1	1	1	1	1	1	1	1	1	0	0	0	0	0
Odontesthes hatcheri (Eigenmann 1909)	0	0	0	0	0	0	0	0	0	0	0	0	0	0	0	0	0	0	0	0	0	0	0	0	0	0	1	1	0
Odontesthes mauleanum[a] (Steindachner 1896)	0	0	0	0	0	0	0	0	0	0	0	1	1	1	1	1	1	1	1	1	1	1	1	1	0	0	0	0	0
Olivaichthys viedmensis (Mac Donagh 1931)	0	0	0	0	0	0	0	0	0	0	0	0	0	0	0	0	0	0	0	0	0	0	0	0	0	0	0	1	0
Orestias ascotanensis[a] (Parenti 1984)	0	0	0	1	0	0	0	0	0	0	0	0	0	0	0	0	0	0	0	0	0	0	0	0	0	0	0	0	0

(continued)

Table 13.1 (continued)

Species	ALT				ATA		CHI														VAL				PAT				
	Chu	Lau	Isl	Asc	Cam	Loa	Cop	Hua	Elq	Lim	Cho	Aco	Mai	Rap	Mat	Mau	Ita	And	Bio	Imp	Tol	Val	Bue	Mal	Pal	Cis	Ays	Bak	Ser
Orestias cf. *agassii*[a] (Valenciennes 1946)	0	0	1	0	0	0	0	0	0	0	0	0	0	0	0	0	0	0	0	0	0	0	0	0	0	0	0	0	0
Orestias chungarensis[a] (Vila and Pinto 1986)	1	0	0	0	0	0	0	0	0	0	0	0	0	0	0	0	0	0	0	0	0	0	0	0	0	0	0	0	0
Orestias laucaensis[a] (Arratia 1982)	0	1	0	0	0	0	0	0	0	0	0	0	0	0	0	0	0	0	0	0	0	0	0	0	0	0	0	0	0
Orestias parinacotensis[a] (Arratia 1982)	0	1	0	0	0	0	0	0	0	0	0	0	0	0	0	0	0	0	0	0	0	0	0	0	0	0	0	0	0
Percichthys melanops (Girard 1855)	0	0	0	0	0	0	0	0	0	0	0	1	1	1	1	1	1	1	1	0	0	0	0	0	0	0	0	0	0
Percichthys trucha (Valenciennes 1833)	0	0	0	0	0	0	0	0	0	0	0	1	1	1	1	1	1	1	1	1	1	1	1	1	1	0	0	1	0
Percilia gillissi[a] (Girard 1855)	0	0	0	0	0	0	0	0	0	0	0	1	1	1	1	1	1	1	1	1	1	1	1	0	0	0	0	0	0
Percilia irwini[a] (Eigenmann 1928)	0	0	0	0	0	0	0	0	0	0	0	0	0	0	0	0	0	1	1	0	0	0	0	0	0	0	0	0	0
Trichomycterus areolatus[a] (Valenciennes 1833)	0	0	0	0	0	0	0	1	1	1	1	1	1	1	1	1	1	1	1	1	1	1	1	1	0	0	0	0	0

Species	ALT				ATA		CHI														VAL				PAT				
	Chu	Lau	Isl	Asc	Cam	Loa	Cop	Hua	Elq	Lim	Cho	Aco	Mai	Rap	Mat	Mau	Ita	And	Bio	Imp	Tol	Val	Bue	Mal	Pal	Cis	Ays	Bak	Ser
Trichomycterus chiltoni[a] (Eigenmann 1927)	0	0	0	0	0	0	0	0	0	0	0	0	0	0	0	0	0	0	1	0	0	0	0	0	0	0	0	0	0
Trichomycterus chungaraensis[a] (Arratia 1983)	1	0	0	0	0	0	0	0	0	0	0	0	0	0	0	0	0	0	0	0	0	0	0	0	0	0	0	0	0
Trichomycterus laucaensis[a] (Arratia 1983)	0	1	0	0	0	0	0	0	0	0	0	0	0	0	0	0	0	0	0	0	0	0	0	0	0	0	0	0	0
Trichomycterus rivulatus (Valenciennes 1840)	0	1	1	0	0	0	0	0	0	0	0	0	0	0	0	0	0	0	0	0	0	0	0	0	0	0	0	0	0
Exotic species																													
Acipenser baeri (Brandt 1869)	0	0	0	0	0	0	0	0	0	0	0	0	1	0	0	0	0	0	0	0	0	0	0	0	0	0	1	0	0
Acipenser transmontanus (Richardson 1836)	0	0	0	0	0	0	0	0	0	0	0	0	1	0	0	0	0	0	0	0	0	0	0	0	0	0	0	0	0
Ameiurus melas (Rafinesque 1820)	0	0	0	0	0	0	0	0	0	0	0	0	0	0	0	1	1	0	0	0	0	0	0	0	0	0	0	0	0
Ameiurus nebulosus (Lesueur 1819)	0	0	0	0	0	0	0	0	0	0	0	0	1	1	1	1	1	0	0	0	0	0	0	0	0	0	0	0	0
Australoherus facetus (Jenyns 1842)	0	0	0	0	0	0	0	0	1	1	1	1	1	1	0	0	1	1	1	0	0	0	0	0	0	0	0	0	0

(continued)

Table 13.1 (continued)

Species	ALT				ATA		CHI														VAL				PAT				
	Chu	Lau	Isl	Asc	Cam	Loa	Cop	Hua	Elq	Lim	Cho	Aco	Mai	Rap	Mat	Mau	Ita	And	Bio	Imp	Tol	Val	Bue	Mal	Pal	Cis	Ays	Bak	Ser
Carassius auratus (Linnaeus 1758)	0	0	0	0	0	0	0	0	0	0	0	0	1	1	1	1	1	1	1	0	0	0	0	0	0	0	0	0	0
Carassius carassius (Linnaeus 1758)	0	0	0	0	0	0	0	0	0	0	0	0	1	0	0	0	0	0	1	0	0	0	0	0	0	0	0	0	0
Cheirodon interruptus (Jenyns 1842)	0	0	0	0	0	0	0	1	1	1	1	1	1	0	0	0	0	0	0	0	0	0	0	0	0	0	0	0	0
Cnesterodon decemmaculatus (Jenyns 1842)	0	0	0	0	0	0	0	0	0	0	0	1	1	0	0	0	0	0	0	0	0	0	0	0	0	0	0	0	0
Coregonus clupeaformis (Mitchill 1818)	0	0	0	0	0	0	0	0	0	0	0	0	0	0	0	0	0	1	1	0	0	0	0	0	0	0	0	0	0
Ctenopharyngodon idella (Cuvier and Valenciennes 1884)	0	0	0	0	0	0	0	0	0	0	0	0	0	0	0	0	0	0	1	0	0	0	0	0	0	0	0	0	0
Cyprinus carpio (Linnaeus 1758)	0	0	0	0	0	0	0	0	1	1	1	1	1	1	1	1	1	1	1	1	1	1	1	0	0	0	0	0	0
Gambusia affinis (Baird and Girard 1853)	0	0	0	0	1	1	1	1	1	1	1	1	1	1	1	1	1	1	1	1	1	1	1	1	0	0	0	0	0
Gambusia holbrooki (Girard 1859)	0	0	0	0	1	1	1	1	1	1	1	1	1	1	1	1	1	1	1	1	1	1	1	1	0	0	0	0	0
Ictalurus punctatus (Rafinesque 1818)	0	0	0	0	0	0	0	0	0	0	0	0	0	0	1	1	0	0	0	0	0	0	0	0	0	0	0	0	0

Species	ALT				ATA		CHI														VAL				PAT				
	Chu	Lau	Isl	Asc	Cam	Loa	Cop	Hua	Elq	Lim	Cho	Aco	Mai	Rap	Mat	Mau	Ita	And	Bio	Imp	Tol	Val	Bue	Mal	Pal	Cis	Ays	Bak	Ser
Jenynsia multidentata (Jenyns 1841)	0	0	0	0	0	0	0	0	0	0	0	0	1	0	0	0	0	0	0	0	0	0	0	0	0	0	0	0	0
Odontesthes bonariensis (Valenciennes 1835)	0	0	0	0	0	0	0	0	0	1	1	1	1	1	1	1	1	1	1	1	1	1	1	0	0	0	0	0	0
Oncorhynchus gorbuscha (Walbaum 1792)	0	0	0	0	0	0	0	0	0	0	0	0	0	0	0	0	0	0	0	0	0	0	0	0	0	1	1	0	0
Oncorhynchus keta (Walbaum 1792)	0	0	0	0	0	0	0	0	0	0	0	0	0	0	0	0	0	0	0	0	0	0	0	0	0	0	1	0	0
Oncorhynchus kisutch (Walbaum 1792)	0	0	0	0	0	0	0	0	0	0	0	0	0	0	0	0	0	0	0	0	0	0	0	1	1	1	1	1	1
Oncorhynchus masou (Brevoort 1856)	0	0	0	0	0	0	0	0	0	0	0	0	0	0	0	0	0	0	0	0	0	0	0	0	0	0	0	1	0
Oncorhynchus mykiss (Walbaum 1792)	1	1	0	0	0	1	1	1	1	1	1	1	1	1	1	1	1	1	1	1	1	1	1	1	1	1	1	1	1
Oncorhynchus nerka (Walbaum 1792)	0	0	0	0	0	0	0	0	0	0	0	0	0	0	0	0	0	0	0	0	0	0	0	0	0	0	1	0	0
Oncorhynchus tshawytscha (Walbaum 1792)	0	0	0	0	0	0	0	0	0	0	0	0	0	0	0	0	0	0	0	1	1	1	1	1	1	1	1	1	1

(continued)

Table 13.1 (continued)

Species	ALT				ATA		CHI														VAL				PAT				
	Chu	Lau	Isl	Asc	Cam	Loa	Cop	Hua	Elq	Lim	Cho	Aco	Mai	Rap	Mat	Mau	Ita	And	Bio	Imp	Tol	Val	Bue	Mal	Pal	Cis	Ays	Bak	Ser
Salmo salar (Linnaeus 1758)	0	0	0	0	0	0	0	0	0	0	0	0	0	0	0	0	0	0	0	0	0	1	1	1	1	1	1	1	1
Salmo trutta (Linnaeus 1758)	0	0	0	0	0	1	1	1	1	1	1	1	1	1	1	1	1	1	1	1	1	1	1	1	1	1	1	1	1
Salvelinus fontinalis (Mitchill 1814)	0	0	0	0	0	0	0	0	0	0	0	1	0	0	0	0	0	0	0	0	0	0	0	1	1	1	1	1	1
Tinca tinca (Linnaeus 1758)	0	0	0	0	0	0	0	0	0	0	0	1	1	0	0	0	0	0	1	0	0	0	0	0	0	0	0	0	0
Total native species	2	4	2	1	1	1	1	4	5	5	5	11	13	13	13	14	15	15	18	16	15	18	15	14	8	7	8	11	6
Total exotic species	1	1	0	0	2	4	4	5	7	8	8	11	16	9	9	10	10	9	12	7	7	8	8	8	6	7	10	7	6
Total species (natives + exotics)	3	5	2	1	3	5	5	9	12	13	13	22	29	22	22	24	25	24	30	23	22	26	23	22	14	14	18	18	12
Average number of native species/basin inside the province	2.3				1.0		10.6														16.0				8.0				
Average number of exotic species/basin inside the province	0.5				3.0		8.9														7.8				7.2				
Average number of total species/basin inside the province	2.8				4.0		19.5														23.3				15.2				
Native species richness per province	8				1		23														19				11				
Exotic species richness per province	1				4		22														10				11				

[a]Endemic species

analyze the information, we elaborated two species × basin matrices. The first one considered only native species in historical times (pre-European distribution), whereas the second included current native and exotic species (post-European distribution). For analytical purposes, both matrices were coded with values 1 and 0 for the presence and absence of species, respectively.

For the first matrix, named the pre-European matrix, only the historical records of the most probable native species in each province – before the introduction of exotic species – were included. In the post-European matrix, we recorded the current species distribution, including the exotic naturalized species but excluding native species recorded as extinct. Thus, we obtained the most up-to-date and probable distribution of the native fish species in each ichthyogeographical province. Species whose taxonomic status is being debated due to a possible synonym or hybridization were not considered.

13.3 Quantifying Species Turnover

We assessed the species turnover separately for both the pre-European and post-European matrices. Turnover was calculated using Simpson's index: $\beta_{sim} = \min\{b,c\}/(a + \min\{b,c\})$ (Baselga 2010). This index ranges between 0 and 1 for null and maximum turnover; *a* represents the number of shared species between two areas (basins); *b* is the number of unique species that occur in one basin; and *c* is the number of unique species that occur in another basin (Koleff et al. 2003). In this way, we obtained values for $\beta_{sim,pre\text{-}European}$ and $\beta_{sim,post\text{-}European}$, and then the differential was calculated as $\Delta\beta_{sim} = \beta_{sim,post\text{-}European} - \beta_{sim,pre\text{-}European}$. The differential distributions obtained were analyzed by the Wilcoxon median test of paired samples.

13.4 Spatial Turnover Among Ichthyogeographic Provinces

For the five ichthyogeographical provinces, we recorded a total of 69 freshwater fish species, of which 41 were native and 28 were exotic species (Table 13.1). Among native species, 32 (78%) corresponded to endemic species (species with an asterisk in Table 13.1). Currently, the province with the lowest fish richness is Atacama (Table 13.1), with a total of five species (one native and four exotics), and the highest richness is recorded at the Chilean province (Table 13.1), with a total of 45 taxa (23 natives and 22 exotics; Table 13.1).

The historical indices ($\beta_{sim,pre\text{-}European}$) recorded minimum and maximum values between 0.00 and 1.00 (Fig. 13.2, while the current values ($\beta_{sim,post\text{-}European}$) ranges were 0.22 and 1.00 (Fig. 13.2). Thus, differential values ($\Delta\beta_{sim}$) were between 0.62 and 0.58 and showed a median significantly different from zero (Wilcoxon test; $Z = 3.4$; $n = 292$; $P < 0.05$). These results indicate that species turnover has decreased

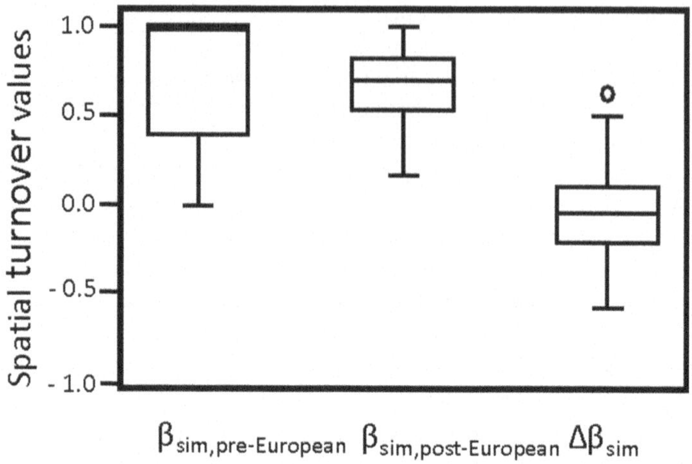

Fig. 13.2 Box plot of the values of the β_{sim} current versus historic indices and their corresponding differentials ($\Delta\beta_{sim}$). Specifically, box show median ±1 standard deviation and whiskers show minimum to maximum values; circle is an outlier value

significantly from historical (pre-European) to current (post-European) times, thus denoting taxonomic homogenization for the Chilean ichthyogeographic provinces.

According to our data, taxonomic homogenization of the Chilean fish fauna is promoted by exotic species, because they are more widely distributed than native ones (Habit et al. 2010), thus reducing the spatial turnover between the ichthyogeographic provinces. In Chile, although the introduction of exotic species began in the nineteenth century, serious efforts have been carried out in the last 60 years (Basulto 2003; Camus and Jaksic 2009). Among exotic fishes, the most widely distributed species were *Gambusia affinis*, *Gambusia holbrooki*, *Oncorhynchus mykiss* and *Salmo trutta*, all of them occupying ≥20 basins (see Table 13.1). For native fishes, only the native *Galaxias maculatus* was distributed in ≥20 basins (Table 13.1).

Theoretical studies (Olden and Poff 2003) indicate that extinction of native species can promote taxonomic homogenization. However, although extinctions at local scales (rivers or tributaries) have been reported in Chile (e.g., *Galaxias maculatus*, *Aplochiton taeniatus* and *Brachygalaxias bullocki*; Habit et al. 2010), only one case of extinction has been confirmed (*Diplomystes chilensis*; Arratia and Quezada-Romegialli 2017). Nevertheless, it is important noting that even if under our scale of analysis (i.e., basins within ichthyogeographic provinces) the extinction of species seems to have small importance, it is also likely that its magnitude will increase in the future as consequence of the potential extinction of taxa that are currently considered as with conservation concerns (see below).

Other homogenizing mechanisms considered in the literature (Leprieur et al. 2011), such as the case of species introduced by translocation, have little numerical importance in Chile. Species translocated correspond to species introduced within their native biogeographical zone in basins where they did not historically occur. Only five species (*Australoheros facetus*, *Cheirodon interruptus*, *Cnesterodon decemmaculatus*, *Jenynsia multidentata* and *Odontestes bonariensis*) that have been introduced in Chile came from other Neotropical basins (Vila and Pardo 2008; Vila and Quezada-Romegialli 2018). However, there are no records of translocated species among Chilean basins both before and after European colonization (Basulto 2003; Camus and Jaksic 2009). The null translocation rate for native fishes is related to their lack of attractiveness for consumption, recreational fishing, ornamental purposes, or aquaculture (Vila and Pardo 2008). In contrast, this mechanism (i.e., species translocation) has been shown to be important in other Neotropical basins, where up to 61% of the exotic species introduced into a given basin correspond to translocated species from other basins of the same region (Villéger et al. 2011).

Recent studies indicate that basins located in the Neotropical realm transit towards taxonomic homogenization (Villéger et al. 2011), but our results indicate that Chilean ichthyogeographical provinces show different susceptibilities, magnitudes, and rates. Indeed, at least two conditions support this situation. First, the number of native species in the Chilean basins is lower than in other basins from the Neotropical realm (Vila and Quezada-Romegialli 2018); on average terms, Chilean basins have 8.9 native species per basin (calculated from Table 13.1), while the Neotropical basins reach 25 native species per basin (Villéger et al. 2011; Toussaint et al. 2016). Second, the average number of exotic species per basin in Chile (6.9 exotic species per basin; Table 13.1) is higher than those recorded for the other Neotropical basins (2.0 species per basin; Leprieur et al. 2009; Villéger et al. 2011). Thus, under these conditions, Chilean basins show a higher rate of taxonomic homogenization than its Neotropical counterparts.

13.5 Prospects

An important challenge is determining which future trend is expected for the taxonomic homogenization detected in Chile. Approximately, 90% of the native freshwater ichthyofauna in Chile (40 species) are of conservation concern (Campos et al. 1998). Basically, they have been classified as vulnerable and endangered species (MMA 2018). Thus, the rate of extinction of native species will likely increase in the future. Complementarily, the situation of exotic species is more complex to predict. On one hand, Chile has a legal framework that intends to reduce the introduction of new fish species, making it unlikely that any new exotic species reach naturalization. On the other hand, because most of exotic species were introduced only recently in Chile (<60 years; Basulto 2003; Camus and Jaksic 2009), it is expected that some of them (i.e., *Carassius carassius*, *Oncorhynchus masou*, *Oncorhynchus nerka*, *Tinca tinca*) will be able to spread their distributional range, occupying a number of basins. Currently this can be facilitated, for instance, by considering the meager governmental capacity to control the furtive release of species, or accidental escapes (Camus and Jaksic 2009). In this last case, for example, Sepulveda et al. (2009, 2013) has estimated >1 million of salmonid escape per year from farms in Chile.

Then, native species loss (some of them endemic) as well as the spreading of currently extant exotic species will likely contribute to the reduction of species turnover and deepen the taxonomic homogenization reported here. In this context, it is imperative to study the putative effects of exotic species on other dimensions beyond biological diversity. Exotic species imply not only taxonomic homogenization but also functional homogenization of the basins (Olden et al. 2010; Villéger et al. 2014). For instance, trout, salmon, and carp (Habit et al. 2006, 2015; Vila and Habit 2015) are generally larger than native species and show generalist trophic habits in mature stages, preying on native species (Vila et al. 2006; Habit et al. 2015) or competing with them for habitat and other resources (Soto et al. 2006; Penaluna et al. 2009; Correa et al. 2012; Elgueta et al. 2013; Habit et al. 2015). Available evidence suggests that these effects affect the abundance and historical distribution of native species (Habit et al. 2010; Arismendi et al. 2009), suggesting a future decrease in the functional uniqueness of ecosystems, deteriorating both their functioning and stability over time (Villéger et al. 2014).

Summarizing, the ichthyogeographical provinces in Chile are in a taxonomic homogenization process of their freshwater fish fauna, shared with other basins inside the Neotropical realm. This process is being driven mainly by the presence of exotic species that are distributed over a wide geographical range, breaking the biogeographic barriers historically established in Chile. Therefore, actions must be taken to avoid the extinction of native species and the propagation of exotic ones in order to preserve the biosingularity of the ichthyogeographical provinces of continental Chile.

References

Abell B, Thieme M, Revenga C, Bryer M, Kottelat M, Bogutskaya N, Coad B, Mandrak N, Contreras S, Bussing W, Stiassny M, Skelton P, Allen G, Unmack P, Naseka A, Ng R, Sindorf N, Robertson J, Armijo E, Higgins J, Heibel T, Wikramanayake E, Olson D, López H, Reis R, Lundberg J, Sabaj M, Petry P (2008) Freshwater ecoregions of the world: a new map of biogeographic units for freshwater biodiversity conservation. Bioscience 8:403–414

Arismendi I, Soto D, Penaluna B, Jara C, Leal C, León-Muñoz J (2009) Aquaculture, non-native salmonid invasions, and associated declines of native fishes in lakes of the northern Chilean Patagonia. Freshw Biol 54:1135–1147

Arratia G (1982) Peces del altiplano chileno. In: Veloso A, Bustos E (eds) El hombre y los ecosistemas de montaña. pp. 93–133. Oficina regional de Ciencia y Tecnología de la Unesco para América Latina y el Caribe. Montevideo, Uruguay

Arratia G (1997) Brazilian and Austral freshwater fish faunas of South America: A contrast. In: Ulrich H (ed) Tropical biodiversity and systematics. Proceedings of the international Symposium on Biodiversity and Systematics in Tropical Ecosystems, Museum Alexander Koenig, Bonn

Arratia G, Quezada-Romegialli C (2017) Understanding morphological variability in a taxonomic context in Chilean Diplomystids (Teleostei: Siluriformes), including the description of a new species. PeerJ 5:e2991

Baselga A (2010) Partitioning the turnover and nestedness components of beta diversity. Glob Ecol Biogeogr 19:134–143

Basulto S (2003) El largo viaje de los salmones, una crónica olvidada. Editorial Maval, Santiago
Campos CH, Dazarola G, Dyer B, Fuentes L, Gavilán JF, Huaiquín L, Martínez G, Meléndez R, Pequeño G, Ponce F, Ruiz VH, Sielfeld W, Soto D, Vega R, Vila I (1998) Categorías de Conservación de Peces Nativos de Aguas Continentales de Chile. Boletín del Museo Nacional de Historia Natural 47:101–122
Camus P, Jaksic F (2009) Piscicultura en Chile: entre la productividad y el deterioro Ambiental. 1856–2008. Santiago de Chile. Ediciones Universidad Católica de Chile
Castro SA, Delpiano C, Vila I Jaksic FM (2014) Invasión de peces dulceacuícolas en Chile y efectos sobre las regiones biogeográficas. In: Jaksic FM & SA Castro. Invasiones biológicas en Chile: Causas globales e impactos locales. Ediciones Universidad Católica de Chile, 415–435
Charrier R, Bustamante M, Comte D, Elgueta S, Flynn JJ, Iturra N, Muñoz N, Pardo M, Thiele R, Wyss AR (2005) The Abanico extensional basin: regional extension, chronology of tectonic inversion, and relation to shallow seismic activity and Andean uplift. N Jb Geol Paläont 236:43–77
Clavero M, Garcia-Berthou E (2006) Homogenization dynamics and introduction routes of invasive freshwater fish in the Iberian Peninsula. Ecol Appl 16:2313–2324
Correa C, Bravo AP, Hendry AP (2012) Reciprocal trophic niche shifts in native and invasive fish: salmonids and galaxiids in Patagonian lakes. Freshw Biol 57:1769–1781
Dudgeon D, Arthington AH, Gessner MO, Kawabata Z-I, Knowler DJ, Lévêque C, Naiman RJ, Prieur-Richard A-H, Soto D, Stiassny MLJ, Sullivan CA (2006) Freshwater biodiversity: Importance, threats, status and conservation challenges. Biol Rev 81:163–182
Dyer B (2000) Systematic review and biogeography of the freshwater fishes of Chile. Estudios Oceanológicos 19:77–98
Elgueta A, Gonzalez J, Ruzzante DE, Walde SJ, Habit E (2013) Trophic interference by *Salmo trutta* on *Aplochiton zebra* and *Aplochiton taeniatus* in southern Patagonian lakes. J Fish Biol 82:430–443
Habit E, Dyer B, Vila I (2006) Estado de conocimiento de los peces dulceacuícolas de Chile. Gayana 70:100–113
Habit E, Piedra P, Ruzzante D, Walde S, Belk M, Cussac V, Gonzalez J, Colin N (2010) Changes in the distribution of native fishes in response to introduced species and other anthropogenic effects. Glob Ecol Biogeogr 19:697–710
Habit E, González J, Ortiz-Sandoval J, Elgueta A, Sobenes C (2015) Efectos de la invasión de peces en ríos y lagos de Chile. Ecosistemas 24:43–51
Iriarte J, Lobos G, Jaksic FM (2005) Invasive vertebrate species in Chile and their control and monitoring by governmental agencies. Rev Chil Hist Nat 78:143–154
Koleff P, Gaston KJ, Lennon J (2003) Measuring beta diversity for presence-absence data. J Anim Ecol 72:367–382
Leprieur F, Olden JD, Lek S, Brosse S (2009) Contrasting patterns and mechanisms of spatial turnover for native and exotic freshwater fish in Europe. J Biogeogr 36:1899–1912
Leprieur F, Tedesco PA, Hugueny B, Beauchard O, Dürr HH, Brosse S, Oberdorff T (2011) Partitioning global patterns of freshwater fish beta diversity reveals contrasting signatures of past climate changes. Ecol Lett 14:325–334
Lévêque C, Oberdorff T, Paugy D, Stiassny MLJ, Tedesco PA (2008) Global diversity of fish (Pisces) in freshwater. Hydrobiologia 595:545–567
Mack RN, Simberloff D, Lonsdale WM, Evans H, Clout M, Bazzaz FA (2000) Biotic invasions: causes, epidemiology, global consequences, and control. Ecol Appl 10:689–710
Marchetti MP, Lockwood JL, Light T (2006) Effects of urbanization on California's fish diversity: differentiation, homogenization and the influence of spatial scale. Biol Conserv 127:310–318
Marr SM, Olden JD, Leprieur F, Arismendi I, Caleta M, Morgan DI, Nocita A, Šandra R, Trkan AS, García-Berthou E (2013) A global assessment of freshwater fish introductions in Mediterranean-climate regions. Hydrobiologia 719:317–329
McKinney M, Lockwood JL (1999) Biotic homogenization: a few winners replacing many losers in the next mass extinction. Trends Ecol Evol 14:450–453

Ministerio del Medio Ambiente (MMA) (2018). At: http://especies.mma.gob.cl/CNMWeb/Web/WebCiudadana/Default.aspx. Accessed 26 Sept 2018

Nelson JS, Grande TC, Wilson M (2016) Fishes of the world. John Wiley & Sons, Inc., Hoboken

Olden JD (2006) Biotic homogenization: a new research agenda for conservation biogeography. J Biogeogr 33:2027–2039

Olden JD, Poff NL (2003) Toward a mechanistic understanding and prediction of biotic homogenization. Am Nat 162:442–460

Olden JD, Poff NL (2004) Ecological processes driving biotic homogenization: testing a mechanistic model using fish faunas. Ecology 85:1867–1875

Olden JD, Rooney TP (2006) On defining and quantifying biotic homogenization. Glob Ecol Biogeogr 15:113–120

Olden JD, Kennard MJ, Pusey BJ (2008) Species invasions and the changing biogeography of Australian freshwater fishes. Glob Ecol Biogeogr 17:25–37

Olden JD, Kennard MJ, Leprieur F, Tedesco PA, Winemiller KO, García-Berthou E (2010) Conservation biogeography of freshwater fishes: recent progress and future challenges. Divers Distrib 16:496–513

Olden JD, Kennard MJ, Lawler JJ, Poff NL (2011) Challenges and opportunities for implementing managing relocation to mitigate the threat of climate change to freshwater biodiversity. Conserv Biol 25:40–47

Penaluna B, Arismendi I, Soto D (2009) Evidence of interactive segregation between introduced trout and native fishes in northern Patagonian Rivers, Chile. Trans Am Fish Soc 138:839–845

Qian H, Ricklefs RE, White PS (2005) Beta diversity of angiosperms in temperate floras of eastern Asia and eastern North America. Ecol Lett 8:15–22

Rahel FJ (2002) Homogenization of freshwater faunas. Annu Rev Ecol Syst 33:291–315

Revenga C, Campbell I, Abell R, de Villiers P, Bryer M (2005) Prospects for monitoring freshwater ecosystems towards the 2010 targets. Philos Trans R Soc B 360:397–413

Ruggiero A, Ezcurra C (2003) Regiones y transiciones biogeográficas: complementariedad de los análisis en biogeografía histórica y ecológica. In: Morrone JJ, Llorente-Bousquets J (eds) Una perspectiva latinoamericana de la biogeografía. Universidad Autónoma de México, Mexico, pp 141–154

Sepúlveda M, Farías F, Soto E (2009) Salmon escapes in Chile. WWF Chile, Valdivia

Sepulveda M, Arismendi I, Soto D, Jara F, Farias F (2013) Escaped farmed salmon and trout in Chile: incidence, impacts, and the need for an ecosystem view. Aquac Environ Interact 4:273–283

Soto D, Arismendi I, González J, Sanzana J, Jara F, Jara C, Guzmán E, Lara A (2006) Southern Chile, trout and salmon country: invasion patterns and threats for native species. Rev Chil Hist Nat 79:97–117

Taylor EB (2004) An analysis of homogenization and differentiation of Canadian freshwater fish faunas with an emphasis on British Columbia. Can J Fish Aquat Sci 61:68–79

Toussaint A, Beauchard O, Oberdorff T, Brosse S, Villéger S (2014) Historical assemblage distinctiveness and the introduction of widespread non-native species explain worldwide changes in freshwater fish taxonomic dissimilarity. Glob Ecol Biogeogr 23:574–584

Toussaint A, Charpin N, Brosse S, Villéger S (2016) Global functional diversity of freshwater fish is concentrated in the Neotropics while functional vulnerability is widespread. Sci Rep 6:22125

Ulrich W, Gotelli NJ (2007) Null model analysis of species nestedness patterns. Ecology 88:1824–1831

Vargas PV, Arismendi I, Gómez-Uchida D (2015) Evaluating taxonomic homogenization of freshwater fish assemblages in Chile. Rev Chil Hist Nat 88:1–10

Vila I, Habit E (2015) Current situation of the fish fauna in the Mediterranean region of Andean river systems in Chile. FiSHMED Fishes in Mediterr Environ 2015(002):19p

Vila I, Pardo R (2008) Peces límnicos. In: Biodiversidad de Chile, Patrimonio y desafíos, Conama (eds) pp. 302–307. Ocho Libros. Santiago, Chile

Vila I, Quezada-Romegialli C (2018) Peces límnicos. In: Biodiversidad de Chile, Patrimonio y desafíos. Ministerio del Medio Ambiente (eds) Ocho Libros, Santiago, Chile

Vila I, Fuentes L, Contreras M (1999) Peces límnicos de Chile. Boletín del Museo de Historia Natural 48:61–75

Vila I, Pardo R, Dyer B, Habit E (2006) Peces límnicos: diversidad, origen y estado de conservación. In: Vila I, Veloso A, Schlatter R, Ramírez C (eds) Macrófitas y vertebrados de los sistemas límnicos de Chile. Editorial Universitaria, Santiago, Chile, pp 73–102

Villéger S, Blanchet S, Beauchard O, Oberdorff T, Brosse S (2011) Homogenization patterns of the world's freshwater fish faunas. Proc Natl Acad Sci U S A 108:18003–18008

Villéger S, Grenouillet G, Brosse S (2014) Functional homogenization exceeds taxonomic homogenization among European fish assemblages. Glob Ecol Biogeogr 23:1450–1460

Villéger S, Blanchet S, Beauchard O, Oberdoff T, Brosse S (2015) From current distinctiveness to future homogenization of the world's freshwater fish faunas. Divers Distrib 21:223–235

References

Adams KL, Wendel JF (2005) Polyploidy and genome evolution in plants. Curr Opin Plant Biol 8:135–141

Aguilera N, Guedes LM, Becerra J, Villaseñor-Parada C, González L, Hernández V (2015) Allelopathic effect of the invasive *Acacia dealbata* link (Fabaceae) on two native plant species in south-central Chile. Gayana Botánica 72:231–239

Aguilera N, Guedes LM, Becerra J, González L (2017) Is autotoxicity responsible for inhibition growth of new conspecific seedlings under the canopy of the invasive *Acacia dealbata* link? Gayana Botánica 74:1–14

Alexander JM, Lembrechts J, Cavieres LA, Daehler C, Haider S, Kueffer C, Liu G, McDougall K, Milbau A, Pauchard A, Rew LJ, Seipel T (2016) Plant invasions into mountains and alpine ecosystems: current status and future challenges. Alp Bot 126:89–103

Alpert P, Bone E, Holzapfel C (2000) Invasiveness, invasibility and the role of environmental stress in the spread of non-native plants. Perspect Plant Ecol Evol System 3:52–66

Amori G, Lapini L (1997) Le specie di mammiferi introdotte in Italia: il quadro della situazione attuale. Supplement alle Ricerche di Biologia della Selvaggina 27:249–267

An M, Pratley JE, Haig T (1997) Phytotoxicity of vulpia residues: I. Investigation of aqueous extracts. J Chem Ecol 23:1979–1995

Anderson CB, Rosemond AD (2007) Ecosystem engineering by invasive exotic beavers reduces in-stream diversity and enhances ecosystem function in Cape Horn, Chile. Oecologia 154:141–153

Anderson CB, Martínez-Pastur G, Lencinas MV, Wallem PK, Moorman MC, Rosemond AD (2009) Do introduced North American beavers *Castor canadensis* engineer different in southern South America? An overview with implications for restoration. Mammal Rev 39:33–52

Anderson MJ, Crist O, Chase JM, Vellend M, Inouye BD, Freestone AL, Sanders NJ, Cornell HV, Comita LS, Davies KF, Harrison SP, Kraft NJB, Stegen JC, Swenson NG (2011) Navigating the multiple meanings of β-diversity: a roadmap for the practicing ecologist. Ecol Lett 14:19–28

Andow DA, Kareiva PM, Levin SA, Okubo A (1990) Spread of invading organisms. Landsc Ecol 4:177–188

Andreu J, Vilà M, Hulme PE (2009) An assessment of stake-holder perceptions and management of noxious alien plants in Spain. Environ Manag 43:1244–1255

Arianoutsou M, Delipetrou P, Vilà M, Dimitrakopoulos PG, Celesti-Grapow L, Wardell-Johnson G, Henderson L, Fuentes N, Ugarte-Mendes E, Rundel PW (2013) Comparative patterns of plant invasions in the Mediterranean biome. PLoS One 8:e79174

Arim M, Abades SR, Neill PE, Lima M, Marquet PA (2006) Spread dynamics of invasive species. Proc Nat Acad Sci U S A 103:374–378

F. M. Jaksic, S. A. Castro, *Biological Invasions in the South American Anthropocene*, https://doi.org/10.1007/978-3-030-56379-0

Arismendi I, Szejnerc P, Larab A, González ME (2008) Impacto del castor en la estructura de bosques ribereños de *Nothofagus pumilio* en Tierra del Fuego, Chile. Bosque 29:146–154

Armesto JJ, Manuschevich D, Mora A, Smith-Ramírez C, Rozzi R, Abarzúa AM, Marquet PA (2010) From the Holocene to the Anthropocene: a historical framework for land cover change in southwestern South America in the past 15,000 years. Land Use Policy 27:148–160

Arnold ML (1997) Natural hybridization and evolution. Oxford University Press, New York

Arnott DL, Vanni MJ (1996) Nitrogen and phosphorus recycling by the zebra mussel (*Dreissena polymorpha*) in the western basin of Lake Erie. Can J Fish Aquat Sci 53:646–659

Aronson MF, La Sorte FA, Nilon CH, Katti M, Goddard MA, Lepczyk CA, Dobbs C (2014) A global analysis of the impacts of urbanization on bird and plant diversity reveals key anthropogenic drivers. Proc R Soc Lond B 281:20133330

Arroyo MTK, Marticorena C, Matthei O, Cavieres L (2000) Plant invasions in Chile: present patterns and future predictions. In: Mooney HA, Hobbs RJ (eds) Invasive species in a changing world. Island Press, Washington, DC, pp 385–421

Austin-Alchon S (2003) A pest in the land: new world epidemics in a global perspective. University of New Mexico Press, USA

Avise J (2004) Molecular markers, natural history, and evolution, Sinauer, MA

Ayala FJ (1982) Population and evolutionary genetics: a primer. Benjamin/Cummings, Menlo Park

Badano EI, Villarroel E, Bustamante RO, Marquet P, Cavieres LA (2007) Ecosystem engineering facilitates invasions by exotic plants in high-Andean ecosystems. J Ecol 95:682–688

Badano EI, Bustamante RO, Villarroel E, Marquet PA, Cavieres LA (2015) Facilitation by nurse plants regulates community invasibility in harsh environments. J Veg Sci 26:756–767

Bailey JP, Child LE (1996) A survey of the distribution of *Fallopia* × *bohemica* (Chrtek & Crtková) J Bailey (Polygonaceae) in the British Isles. Watsonia 21:187–198

Baiser B, Olden JD, Record S, Lockwood JL, McKinney ML (2014) Pattern and process of biotic homogenization in the New Pangaea. Proc R Soc Lond B 279:4772–4777

Baker HG (1974) The evolution of weeds. Annu Rev Ecol Syst 5:1–24

Ballester A, Vieitez E (1979) The allelopathic potential of *Erica australis* L. and *E. arborea* L. Bot Gaz 140:433–436

Barnes SS, Matisoo-Smith E, Hunt TL (2006) Ancient DNA of the Pacific rat (*Rattus exulans*) from Rapa Nui (Easter Island). J Archaeol Sci 33:1536–1540

Baselga A (2010) Partitioning the turnover and nestedness components of beta diversity. Glob Ecol Biogeogr 19:134–143

Baselga A, Jiménez-Valverde A, Niccolini G (2007) A multiple-site similarity measure independent of richness. Biol Lett 3:642–645

Becerra PI, Bustamante RO (2011) Effect of a native tree on seedling establishment of two exotic invasive species in a semiarid ecosystem. Biol Invasions 13:2763–2773

Becker T, Dietz H, Billeter R, Buschmann H, Edwards P (2005) Altitudinal distribution of plant species in the Swiss Alps. Perspect Plant Ecol Evol System 7:173–183

Belmonte J, Vilà M (2004) Atmospheric invasion of non-native pollen in the Mediterranean region. Am J Bot 91:1243–1250

Benson AJ (1999) Documenting over a century of aquatic introductions in the United States. In: Claudi R, Leach JH (eds) Nonindigenous freshwater organisms: vectors, biology, and impacts. Lewis Publishers, Washington, DC, pp 1–31

Berryman AA (1999) Principles of population dynamics and their application. Stanley Thornes, Cheltenham

Binggeli P (1994) Misuse of terminology and anthropomorphic concepts in the description of introduced species. Bull Br Ecol Soc 25:10–13

Blair RB (2001) Birds and butterflies along urban gradients in two ecoregions of the United States: is urbanization creating a homogeneous fauna? In: Lockwood JL, McKinney ML (eds) Biotic homogenization. Kluwer Academic/Plenum, New York, pp 33–56

Blair RB, Johnson EM (2008) Suburban habitats and their role for birds in the urban-rural habitat network: points of local invasion and extinction? Landsc Ecol 23:1157–1169

Bond WJ, Keeley JE (2005) Fire as a global 'herbivore': the ecology and evolution of flammable ecosystems. Trends Ecol Evol 20:387–394

Bonino N, Cossios D, Menegheti J (2008) Dispersión de la liebre europea (*Lepus europaeus*) en Sudamérica. Comunicación Técnica Recursos Naturales 152:1–9

Born W, Rauschmayer F, Brauer I (2005) Economic evaluation of biological invasions: a survey. Ecol Econ 55:321–336

Borsa P, Rolland V, Daguin-Thiébaut C (2012) Genetics and taxonomy of Chilean smooth-shelled mussels, *Mytilus* spp. (Bivalvia: Mytilidae). C R Biol 335:51–61

Brooks ML, D'Antonio CM, Richardson DM, Grace JB, Keeley JE (2004) Effects of invasive plants on fire regimes. Bioscience 54:677–688

Brown JH, Lomolino MV (1998) Biogeography. Sinauer, Sunderland

Brown BJ, Mitchell RJ, Graham SA (2002) Competition for pollination between an invasive species (*Purple loosestrife*) and a native congener. Ecology 83:2328–2336

Buchmann SL (1983) Buzz pollination in angiosperms. In: Dones, Little RJ (eds) Handbook of experimental pollination biology. Scientific and Academic Editions, New York, pp 73–113

Bustamante RO, Castor C (1998) The decline of an endangered temperate ecosystem: the ruil (*Nothofagus alessandrii*) forest in central Chile. Biodivers Conserv 7:1607–1626

Byers JE, Noonburg EG (2003) Scale dependent effect of biotic resistance to biological invasion. Ecology 84:1428–1433

Cadotte MW (2006) Darwin to Elton: early ecology and the problem of invasive species. In: Cadotte MW, McMahon SM, Fukami T (eds) Conceptual ecology and invasion biology: reciprocal approaches to nature. Springer, The Netherlands, pp 15–33

Callaway RM (2002) The detection of neighbors by plants. Trends Ecol Evol 17:104–105

Callaway RM, Ridenour WM (2004) Novel weapons: invasive success and the evolution of increased competitive ability. Front Ecol 2:436–443

Callaway RM, Brooker RW, Choler P, Kikvidze Z, Lortie CJ, Michalet R, Paolini L, Pugnaire FI, Newingham B, Aschehoug ET, Armas C, Kikodze D, Cook BJ (2002) Positive interactions among alpine plants increase with stress. Nature 417:844–848

Calvete C (2005) Modelling the effect of population dynamics on the impact of rabbit hemorrhagic disease. Conserv Biol 20:1232–1241

Campell K, Donlan CJ (2005) Feral goat eradications on islands. Biol Conserv 19:1362–1374

Campos JA, Herrera M, Biurrun I, Loidi J (2004) The role of alien plants in the natural coastal vegetation in central-northern Spain. Biodivers Conserv 13:2275–2293

Camps-Rabadà J (2008) Los primeros conejos llegaron a América iniciado noviembre de 1493. Información Veterinaria 9:21–24

Camus P (2006) Ambiente, bosques y gestión forestal en Chile (1541–2005). Lom Ediciones, Santiago

Camus P, Jaksic F (2009) Piscicultura en Chile: entre la productividad y el deterioro ambiental (1856–2008). Ediciones de la Pontificia Universidad Católica de Chile, Santiago

Camus P, Castro S, Jaksic F (2008) El conejo europeo en Chile: historia de una invasión biológica. Historia 41:305–339

Canals M, Torres-Contreras H, Castro SA, Jaksic FM (2014) Dinámica poblacional de la expansion de enfermedades infecciosas como modelo de invasiones biologicas. In: Jaksic FM, Castro SA (eds) Invasiones biológicas en Chile. Ediciones Universidad Católica de Chile, pp 459–485

Carlton JT (1985) Transoceanic and interoceanic dispersal of coastal marine organims: the biology of ballast water. Oceanogr Mar Biol Annu Rev 23:313–371

Carlton JT (1996) Biological invasions and cryptogenic species. Ecology 77:1653–1655

Carvallo GO, Castro SA (2017) Invasions but not extinctions change phylogenetic diversity of angiosperm assemblage on southeastern Pacific Oceanic islands. PLoS One 12:e0182105

Carvallo GO, Teillier S, Castro SA, Figueroa JA (2014) The phylogenetic properties of native- and exotic-dominated plant communities. Austral Ecol 39:304–312

Casado MA, Acosta-Gallo B, Sánchez-Jardón L, Martín-Forés I, Castro I, Ovalle C, del Pozo A, de Miguel JM (2015) Interactive effects of source and recipient habitats on plant invasions: distribution of exotic species in Chile. Divers Distrib 21:609–619

Case TJ, Gilpin ME (1974) Interference competition and niche theroy. Proc Natl Acad Sci U S A 71:3073–3077

Cassey P, Blackburn TM, Russell GJ, Jones KE, Lockwood JL (2004) Influences on the transport and establishment of exotic bird species: An analysis of the parrots (Psittaciformes) of the world. Glob Chang Biol 10:417–426

Cassey P, Lockwood JL, Blackburn TM, Olden JD (2007) Spatial scale and evolutionary history determine the degree of taxonomic homogenization across island bird assemblages. Divers Distrib 13:458–466

Castilla JC, Collins AG, Meyer CP, Guíñez R, Lindberg DR (2002) Recent introduction of the dominant tunicate, *Pyura praeputialis* (Urochordata, Pyuridae) to Antofagasta, Chile. Mol Ecol 11:1579–1584

Castilla JC, Uribe M, Bahamonde N, Clarke M, Desqueyroux-Faúndez R, Kong I, Moyano H, Rozbaczylo N, Santelices B, Valdovinos C, Zavala P (2005) Down under the southeastern Pacific: marine non-indigenous species in Chile. Biol Invasions 7:213–232

Castro SA, Jaksic FM (2008a) How general are global trends in biotic homogenization? Floristic tracking in Chile, South America. Glob Ecol Biogeogr 17:524–531

Castro SA, Jaksic FM (2008b) Role of non-established plants in determining biotic homogenization patterns in Pacific oceanic islands. Biol Invasions 10:1299–1309

Castro SA, Figueroa JA, Muñoz-Schick M, Jaksic FM (2005) Minimum residence time, biogeographic origin, and life cycle as determinants of geographic distribution of naturalized plants in continental Chile. Divers Distrib 11:183–191

Castro SA, Muñoz-Schick M, Jaksic FM (2007) Transit toward floristic homogenization on oceanic islands in the southeastern Pacific. J Biogeogr 34:213–222

Castro SA, Bozinovic F, Jaksic FM (2008) Ecological efficiency and legitimacy in seed dispersal of an endemic shrub (*Lithrea caustica*) by the European rabbit (*Oryctolagus cuniculus*) in Central Chile. J Arid Environ 72:1164–1173

Castro SA, Badano E, Guzman D, Cavieres L (2009) Biological invasion of a refuge habitat: *Anthriscus caucalis* (Apiaceae) decreases diversity, evenness, and survival of native herbs in the Chilean matorral. Biol Invasions 12:1295–1303

Castro SA, Daehler CC, Silva L, Torres-Santana CW, Reyes-Betancort JA, Atkinson R, Jaramillo P, Guezou A, Jaksic FM (2010) Floristic homogenization as a teleconnected trend in oceanic islands. Divers Distrib 16:902–910

Castro SA, Escobedo VM, Aranda J, Carvallo GO (2014) Evaluating Darwin's naturalization hypothesis in experimental plant assemblages: phylogenetic relationships do not determine colonization success. PLoS One 9:e105535

Cavieres LA, Quiroz CL, Molina-Montenegro MA, Muñoz AA, Pauchard A (2005) Nurse effect of the native cushion plant *Azorella monantha* on the invasive non-native *Taraxacum officinale* in the high-Andes of central Chile. Perspect Plant Ecol Evol System 7:217–226

Cavieres LA, Badano EI, Sierra-Almeida A, Molina-Montenegro MA (2007) Microclimatic modifications of cushion plants and their consequences for seedling survival of native and non-native herbaceous species in the high Andes of central Chile. Arct Antarct Alp Res 39:229–236

Cavieres LA, Quiroz CL, Molina-Montenegro MA (2008) Facilitation of the non-native *Taraxacum officinale* by native nurse cushion species in the high-Andes of central Chile: are there differences between nurses? Funct Ecol 22:148–156

Cavieres LA, Sanhueza AK, Torres-Mellado G, Casanova-Katny A (2017) Competition between native Antarctic vascular plants and invasive *Poa annua* changes with temperature and soil nitrogen availability. Biol Invasions 20:1597–1610

Celesti-Grapow L, Pyšek P, Jarošík V, Blasi C (2006) Determinants of native and alien species richness in the urban flora of Rome. Divers Distrib 12:490–501

Celesti-Grapow L, Capotorti G, Del Vico E, Lattanzi E, Tilia A, Blasi C (2013) The vascular flora of Rome. Plant Biosyst 147:1059–1087
Chace JF, Walsh JJ (2006) Urban effects on native avifauna: a review. Landsc Urban Plan 74:46–69
Chacón E, Saborio G (2006) Análisis taxonómico de las especies de plantas introducidas en Costa Rica. Lankesteriana 6:139–147
Chapin FS III, Zavaleta ES, Eviner VT, Naylor RL, Vitousek PM, Reynolds HL, Hooper DU, Lavorel S, Sala OE, Hobbie SE, Mack MC, Díaz S (2000) Consequences of changing biodiversity. Nature 405:234–242
Chase JM, Kraft NJB, Smith KG, Vellend M, Inouye BD (2011) Using null models to disentangle variation in community dissimilarity from variation in α-diversity. Ecosphere 2:1–11
Chesson P (2000) Mechanisms of maintenance of species diversity. Annu Rev Ecol Syst 31:343–366
Chew MK, Laubichler MD (2003) Natural enemies: metaphor or misconception? Science 301:52–54
Choler P, Michalet R, Callaway RM (2001) Facilitation and competition on gradients in alpine plant communities. Ecology 82:3295–3308
Chown SL, Lee JE, Hughes KA, Barnes J, Barrett PJ, Bergstrom DM, Frenot Y (2012) Challenges to the future conservation of the Antarctic. Science 337:158–159
Christian CE (2001) Consequences of a biological invasion reveal the importance of mutualism for plant communities. Nature 413:635–639
Christy MT, Savidge JA, Rodda GH (2007) Multiple pathways for invasion of anurans on a Pacific island. Divers Distrib 13:598–607
Clavero M, García-Berthou E (2006) Homogenization dynamics and introduction routes of invasive freshwater fish in the Iberian Peninsula. Ecol Appl 16:2313–2324
Clement EJ, Foster MC (1994) Alien plants of British isles. Botanical Society of the British Isles, UK
Clergeau P, Croci S, Jokimäki J, Kaisanlahti-Jokimäki ML, Dinetti M (2006) Avifauna homogenisation: analysis at different European latitudes. Biol Conserv 127:336–344
Coblenz BE (1978) The effects of feral goats (*Capra hircus*) on island ecosystems. Biol Conserv 13:279–286
Colautti RI, MacIsaac HJ (2004) A neutral terminology to define invasive species. Divers Distrib 10:135–141
Colautti RI, Bailey SA, van Overdijk CDA, Admunsen K, MacIsaac HJ (2006) Characterised and projected costs of nonindigenous species in Canada. Biol Invasions 8:45–59
Collins MD, Vázquez DP, Sanders NJ (2002) Species–area curves, homogenization and the loss of global diversity. Evol Ecol Res 4:457–464
Colwell RK, Rangel TF (2009) Hutchinson's duality: the once and future niche. Proc Natl Acad Sci U S A 106:19651–19658
Common MS, Norton TW (1992) Biodiversity: its conservation in Australia. Ambio 21:258–265
Convey P, Aitken S, Di Prisco G, Gill MJ, Coulson SJ, Barry T, Jónsdóttir IS, Dang PT, Hik D, Kulkarni T, Lewis G (2012) The impacts of climate change on circumpolar biodiversity. Biodiversity 13:134–143
Corbet SA, Bee J, Dasmahaoatra K, Gale S, Gorringe E, La Ferla B, Moorhouse T, Trevail A, Van Bergen Y, Yorontsova M (2001) Native or exotic? Double or single? Evaluating plants for pollinator-friendly gardens. Ann Bot 87:219–232
Córdova-Stroobandt KL (2013) Caracterización de la biodiversidad urbana en la cuenca central de Cochabamba, Bolivia. Acta Nova 6:94–121
Cornwell WK, Grubb PJ (2003) Regional and local patterns in plant species richness with respect to resource availability. Oikos 100:417–428
Coronato A, Escobar J, Mallea C, Roig C, Lizarralde M (2003) Características geomorfológicas de ríos de montaña colonizados por *Castor canadensis* en Tierra del Fuego, Argentina. Ecol Austral 13:15–26

Correa C, Bravo AP, AP Hendry AP. (2012) Reciprocal trophic niche shifts in native and invasive fish: salmonids and galaxiids in Patagonian lakes. Freshw Biol 57:1769–1781
Cossíos ED (2004) La liebre europea, *Lepus europaeus* (Mammalia, Leporidae), especie invasora en el sur del Perú. Rev Peru Biol 11:209–212
Cosyns E, Delporte A, Lens L, Hoffmann M (2005a) Germination success of temperate grassland species after passage through ungulate and rabbit guts. J Ecol 9:353–361
Cosyns E, Claerbout S, Lamoot I, Hoffmann M (2005b) Endozoochorous seed dispersal by cattle and horse in a spatially heterogeneous landscape. Plant Ecol 178:149–162
Courchamp F, Chapuis JL, Pascal M (2003) Mammal invaders on islands: impact, control and control impact. Biol Rev 78:347–383
Crall AW, Meyerson LA, Stohlgren TJ, Jarnevich CS, Newman GJ, Graham J (2006) Show me the numbers: what data currently exist for non-native species in the USA? Front Ecol Environ 4:414–418
Crawley MJ (1987) What makes a community invasible? In: Gray AJ, Crawley MJ, Edwards PJ (eds) Colonization, succession and stability. Blackwell Scientific Publications, Oxford, pp 429–453
Crawley MJ (1997) Plant ecology. Blackwell Science, Oxford
Crawley MJ, Brown SL, Heard NS, Edwards GR (1999) Invasion-resistance in experimental grassland communities: species richness or species identity? Ecol Lett 2:140–148
Crooks JA (2002) Characterizing ecosystem-level consequences of biological invasion: the role of ecosystem engineers. Oikos 97:153–166
Crosby AW (2004) Ecological imperialism: the biological expansion of Europe, 900–1900. Cambridge University Press, Cambridge
Cuevas J, Marticorena A, Cavieres LA (2004) New additions to the introduced flora de of the Juan Fernandez Islands: Origin, distribution, life history traits and, potential of invasion. Rev Chil Hist Nat 77:523–538
D'Antonio CM, Vitousek PM (1992) Biological invasions by exotic grasses, the grass/fire cycle, and global change. Annu Rev Ecol Syst 23:63–87
Daehler CC (2001) Darwin's naturalization hypothesis revisited. Am Nat 158:324–330
Daehler CC (2003) Performance comparisons of co-occurring native and alien plants: implications for conservation and restoration. Annu Rev Ecol Syst 34:183–211
Daehler CC (2006) Invasibility of tropical islands by introduced plants: partitioning the influence of isolation and propagule pressure. Preslia 78:361–374
Daehler CC, Strong DR (1997) Reduced herbivore resistence in introduced smooth cordgrass (*Spartina alterniflora*) after a century of herbivore-free growth. Oecologia 110:99–108
Danton P (2004) Plantas silvestres de la Isla Robinson Crusoe guía de reconocimiento. In: Corporación Nacional Forestal, Región de Valparaíso. Chile, Viña del Mar
Darwin C (1859) The origin of species. John Murray, London
Davidson AM, Jennions M, Nicotra AM (2011) Do invasive species show higher phenotypic plasticity than native species and, if so, is it adaptive? A meta-analysis. Ecol Lett 14:419–431
Davis MA (2009) Invasion biology. Oxford University Press, Oxford
Davis MA, Pelsor M (2001) Experimental support for a resource-based mechanistic model of invasibility. Ecol Lett 4:421–428
Davis MA, Thompson K (2001) Invasion terminology: should ecologists define their terms differently than others? No, not if we want to be of any help. Ecol Soc Am Bull 82:206
Davis MA, Grime JP, Thompson K (2000) Fluctuating resources in plant communities: a general theory of invisibility. J Ecol 88:528–534
Davis MA, Thompson K, Grime JP (2001) Charles S. Elton and the dissociation of invasion ecology from the rest of ecology. Divers Distrib 7:97–102
Dawson W, Rohr RP, van Kleunen M, Fischer M (2012) Alien plant species with a wider global distribution are better able to capitalize on increased resource availability. New Phytol 194:859–867
De Candolle ALP (1855) Géographie botanique raissoné. V. Masson, Paris

de la Maza CL, Rodríguez R, Bown H, Hernández J, Escobedo F (2002) Vegetation diversity in the Santiago de Chile urban ecosystem. Arboricult J 26:347–357

Dehnen-Schmutz K, Touza J (2008) Plant invasions and ornamental horticulture: pathway, propagule pressure and the legal framework. In: Teixeira da Silva JA (ed) Floriculture, ornamental and plant biotechnology: advances and topical issues. Global Science Books, Isleworth, pp 15–21

Dennis B (2002) Allee effects in stochastic populations. Oikos 96:389–401

Devictor V, Julliard R, Couvet D, Lee A, Jiguet F (2007) Functional homogenization effect of urbanization on bird communities. Conserv Biol 21:741–751

Devictor V, Julliard R, Clavel J, Jiguet F, Lee A, Couvet D (2008) Functional biotic homogenization of bird communities in disturbed landscapes. Glob Ecol Biogeogr 17:252–261

Di Castri F (1989) History of biological invasions with special emphasis on the Old World. In: Drake JA, Mooney HA, di Castri F, Groves RH, Kruger FJ, Rejmánek M, Williamson M (eds) Biological invasions: a global perspective. Wiley, Chichester, pp 1–30

Di Tomaso JM (1998) Impact, biology, and ecology of saltcedar (*Tamarix* spp.) in the southwestern United States. Weed Technol 12:326–336

Di Tomaso JM (2000) Invasive weeds in rangelands: species, impacts, and management. Weed Sci 48:255–265

Didham RK, Tylianakis JM, Hutchinson MA, Ewers RM, Gemmel NJ (2005) Are invasive species the drivers of ecological change? Trends Ecol Evol 20:470–474

Diez JM, Sullivan JJ, Hulme PE, Edwards G, Duncan RP (2008) Darwin's naturalization conundrum: dissecting taxonomic patterns of species invasions. Ecol Lett 11:674–681

Dirnböck T, Geimler J, López SP, Stuessy TF (2003) Predicting future threats to the native vegetation of Robinson Crusoe Island, Juan Fernández archipelago, Chile. Conserv Biol 17:1650–1659

Drake JM, Lodge DM (2004) Global hot spots of biological invasions: evaluating options for ballast-water management. Proc R Soc Lond B 271:575–580

Drake JR, Mooney HA, di Castri F, Groves RH, Kruger FJ, Rejmánek M, Williamson M (1989) Biological invasions: a global perspective. In: Scientific committee on problems of the environment (SCOPE). Wiley, Chichester

Dukes JS, Mooney HA (1999) Does global change increase the success of biological invaders? Trends Ecol Evol 14:135–139

Duncan RP, Williams PA (2002) Darwin's naturalization hypothesis challenged. Nature 417:608–609

Dunham JB, Rieman BE, Peterson JT (2002) Patch-based models of species occurrence: lessons from salmonid fishes in streams. In: Scott JM, Heglund PJ, Samson F, Haufler J, Morrison M, Raphael M, Wall B (eds) Predicting species occurrences: issues of accuracy and scale. Island Press, Covelo, pp 327–334

Dyer B (2000) Systematic review and biogeography of the freshwater fishes of Chile. Estudios Oceanológicos 19:77–98

Ehrlich PR (1989) Attributes of invaders and the invading processes: vertebrates. In: Drake JA, Mooney HA, di Castri F, Groves RH, Kruger FJ, Rejmánek M, Williamson M (eds) Biological invasions: a global perspective. John Wiley & Sons, Chichester, pp 315–328

Elgueta A, González J, Ruzzante D, Walde S, Habit E (2012) Trophic interference by *Salmo trutta* on *Aplochiton zebra* and *Aplochiton taeniatus* in Southern Patagonian lakes. J Fish Biol 82:430–443

Elton CS (1958) The ecology of invasions by animals and plants. University of Chicago Press, Chicago

Eritja R, Escosa R, Lucientes J, Marques E, Roiz D, Ruiz S (2005) Worldwide invasion of vector mosquitoes: present European distribution and challenges for Spain. Biol Invasions 7:87–97

Escobedo VM, Aranda JE, Castro SA (2011) Hipótesis de Naturalización de Darwin evaluada en la flora exótica de Chile continental. Rev Chil Hist Nat 84:543–552

Essl F, Stefan D, Rabitsch W, Hulme PE, Hülber K, Jarošík V, Kleinbauer I, Krausmann F, Kühn I, Nentwig W, Vilà M, Genovesi P, Gherardi F, Desprez-Loustau ML, Roques A, Pyšek P (2011) Socioeconomic legacy yields an invasion debt. Proc Natl Acad Sci U S A 108:203–207

Fanelli G, Tescarollo P, Testi A (2006) Ecological indicators applied to urban and suburban flora. Ecol Indic 6:444–457

Fenner F, Ratcliffe FN (1965) Myxomatosis. Cambridge University Press, Cambridge

Fenner F, Ross J (1994) Myxomatosis. In: Thompson HV, King CM (eds) The European rabbit: the history and biology of a successful colonizer. Oxford University Press, Oxford, pp 205–240

Fernández IC, Simonetti JA (2013) Small mammal assemblages in fragmented shrublands of urban areas of Central Chile. Urban Ecosyst 16:377–387

Figueroa JA, Castro SA, Marquet PM, Jaksic FM (2004) Exotic plant invasions to the mediterranean region of Chile: causes, history and impacts. Rev Chil Hist Nat 77:465–483

Figueroa JA, Cavieres LA, Gómez-González S, Molina-Montenegro MA, Jaksic FM (2009) Do heat and smoke increase emergence of exotic and native plants in the matorral of central Chile? Acta Oecol 35:335–340

Figueroa JA, Teillier S, Castro SA (2011) Diversity patterns and composition of native and exotic floras in central Chile. Acta Oecol 37:103–109

Figueroa JA, Castro SA, Abarca L, Contreras TE, Jaksic FM (2014) El fuego facilita las invasiones biológicas de plantas en comunidades mediterráneas, Invasiones biológicas en Chile. In: Jaksic FM, Castro SA (eds) . Ediciones Universidad Católica de Chile, pp 343–355

Figueroa JA, Teillier S, Guerrero N, Ray C, Rivano S, Saavedra D, Castro SA (2016) Vascular flora in public space of Santiago, Chile. Gayana Botánica 73:44–69

Flux JEC, Fullagar PJ (1992) World distribution of the rabbit, *Oryctolagus cuniculus*, on islands. Mammal Rev 22:151–205

Fonseca DM, LaPointe DA, Fleischer RC (2000) Bottlenecks and multiple introductions: population genetics of the vector of avian malaria in Hawaii. Mol Ecol 9:1803–1814

Fox MD, Fox BJ (1986) The susceptibility of natural communities to invasion. In: Groves RH, Burdon JJ (eds) Ecology of biological invasions: an Australian perspective. Australian Academy of Science, Canberra, pp 57–66

Frenot Y, Chown SL, Whinam J, Selkirk PM, Convey P, Skotnicki M, Bergstrom DM (2005) Biological invasions in the Antarctic: extent, impacts and implications. Biol Rev 80:45–72

Friebe A, Schulz M, Kuck P, Schnabl H (1995) Phytotoxins from shoot extracts and root exudates of *Agropyron repens* seedlings. Phytochemistry 38:1157–1159

Fuentes ER, Otaiza RD, Alliende MC, Poiani AJ (1984) Shrub clumps of the Chilean matorral vegetation: structure and possible maintenance mechanisms. Oecologia 42:405–411

Fuentes ER, Espinoza GA, Gajardo G (1987) Allelopathic effects of the Chilean matorral shrub *Flourencia thurifera*. Rev Chil Hist Nat 60:57–62

Fuentes N, Ugarte E, Kühn I, Klotz S (2007) Alien plants in Chile: inferring invasion periods from herbarium records. Biol Invasions 10:649–657

Fuentes-Contreras E, Muñoz R, Niemeyer HM (1997) Diversidad de áfidos (Hemiptera: Aphidoidea) en Chile. Rev Chil Hist Nat 70:531–542

Funk JL (2008) Differences in plasticity between invasive and native plants from a low resource environment. J Ecol 96:1162–1173

Fussell M, Corbet SA (1991) Forage for bumble bees and honeybees in farmland: a case-study. J Apic Res 30:87–97

Galil BS, Zenetos A (2002) A sea change: exotics in the eastern Mediterranean Sea. In: Leppäkoski E, Gollasch S, Olenin S (eds) Invasive aquatic species of Europe: distribution, impacts and management. Springer, Dordrecht, pp 325–336

García RA, Pauchard A, Cavieres LA, Peña E, Rodríguez MF (2010) Fire promotes *Teline monspessulana* (Fabaceae) invasion by increasing its germination. Rev Chil Hist Nat 83:443–452

Gardner J (2001) Rehabilitating mines to meet land use objectives: bauxite mining in Jarrah forest of Western Australia. Unasylva 52:3–8

Gardner SM, Cabido MR, Valladares GR, Diaz S (1995) The influence of habitat structure on arthropod diversity in Argentine semiarid Chaco forest. J Veg Sci 6:349–356

Gaskin JF, Kazmer DJ (2008) Introgression between invasive saltcedars (*Tamarix chinensis* and *T. ramosissima*) in the USA. Biol Invasions 11:1121–1130

Gaston KJ, Warren PH, Thompson K, Smith RM (2005a) Urban domestic gardens (IV): the extent of the resource and its associated features. Biodivers Conserv 14:3327–3349

Gaston KJ, Smith RM, Thompson K, Warren PH (2005b) Urban domestic gardens (II): experimental test of methods for increasing biodiversity. Biodivers Conserv 14:395–413

Gaston KJ, Warren PH, Thompson K, Smith RM (2006) The extent of the resource and its associated features. Biodivers Conserv 14:3327–3349

Geller JB (1999) Decline of a native species masked by sibling species invasion: the case of Californian marine mussels. Conserv Biol 13:661–664

Genovesi P, Scalera R, Solarz W, Roy D (2009) Towards an early warning and information system for invasive alien species threatening biodiversity in Europe. Eur Environ Agency 5:1–52

Gentle CB, Duggin JA (1997) Allelopathy as a competitive strategy in persistent thickets of *Lantana camara* L. in three Australian forest communities. Plant Ecol 132:85–95

Gianoli E (2004) Plasticity traits and correlation in two population of *Convolvulus arvensis* (Convolvulaceae) differing in environmental heterogeneity. Int J Plant Sci 165:825–832

Godoy O, Valladares F, Castro-Díez P (2011) Multispecies comparison reveals that invasive and native plants differ in their traits but not in their plasticity. Funct Ecol 25:1248–1253

Golani D, Azzurro E, Corsini-Foka M, Falautano M, Andaloro F, Bernardi G (2007) Genetic bottlenecks and successful biological invasions: the case of a recent Lessepsian migrant. Biol Lett 3:541–545

Gómez S, Cavieres LA (2009) Litter burning does not equally affect seedling emergence of native and alien species of the Mediterranean-type Chilean matorral. Int J Wildland Fire 18:213–221

Gómez-González S, Lohengrin LA, Cavieres A, Torres P, Torres-Díaz C (2009) Competitive effects of the alien invasive *Centaurea solstitialis* on two Chilean *Baccharis* species at different life-cycle stages. Gayana Botánica 66:71–83

González AL, Kominoski JS, Danger M, Ishida S, Iwai N, Rubach A (2010) Can ecological stoichiometry help explain patterns of biological invasions? Oikos 119:779–790

Gotelli NJ (2001) A primer of ecology. Sinauer Press, Sunderland

Goulson D (2003) Effects of introduced bees on native ecosystems. Annu Rev Ecol Evol Syst 34:1–26

Graves HB (1984) Behavior and ecology of wild and feral swine (*Sus scrofa*). J Anim Sci 58:482–492

Greimler J, Stuessy TF, Swenson U, Baeza CM, Matthei O (2002) Plants invasions on an oceanic archipelago. Biol Invasions 4:73–85

Grigera DE, Rapoport EH (1983) Status and distribution of the European hare in South America. J Mammal 64:163–166

Grosholz ED (2005) Recent biological invasion may hasten invasional meltdown by accelerating historical introductions. Proc Natl Acad Sci U S A 102:1088–1091

Groves RH (1996) Recent incursions of weeds to Australia 1971–1995, Technical series 3. CRC Weed Management Systems

Groves RH (2002) The impacts of alien plants in Australia. In: Pimentel D (ed) Biological invasions: economic and environmental costs of alien plant, animal and microbe species. CRC Press LLC, Boca Raton, pp 11–23

Guichón ML, Bello M, Fasola L (2005) Expansión poblacional de una especie introducida en la Argentina: la ardilla de vientre rojo *Callosciurus erythraeus*. Mastozoología Neotropical 12:189–197

Gundale MJ, Almeida JP, Wallander H, Wardle DA, Kardol P, Nilsson MC, Fajardo A, Pauchard A, Peltzer DA, Ruotsalainen S, Mason B, Rosenstock N (2016) Differences in endophyte communities of introduced trees depend on the phylogenetic relatedness of the receiving forest. J Ecol 104:1219–1232

Gurevitch J, Padilla DK (2004) Are invasive species a major cause of extinctions? Trends Ecol Evol 19:470–474

Gutiérrez-Tapia P, Azócar MI, Ignacio M, Castro SA (2018) A citizen-based platform reveals the distribution of functional groups inside a large city from the Southern Hemisphere: e-Bird and the urban birds of Santiago (Central Chile). Rev Chil Hist Nat 91:3

Habit E, Cussac V (2016) Conservation of the freshwater fauna of Patagonia: An alert to the urgent need for integrative management and sustainable development. J Fish Biol 89:369–371

Habit E, Dyer B, Vila I (2006) Estado de conocimiento de los peces dulceacuícolas de Chile. Gayana 70:100–113

Haeuser E, Dawson W, Thuiller W, Dullinger S, Block S, Bossdorf O, Carboni M, Conti L, Dullinger I, Essl F, Klonner G, Moser D, Münkemüller T, Parepa M, Talluto MV, Kreft H, Pergl J, Pyšek P, Weigelt P, Winter M, Hermy M, van der Veken S, Roquet C, van Kleunen M (2018) European ornamental garden flora as an invasion debt under climate change. J Appl Ecol 55:2386–2395

Haider S, Kueffer C, Bruelheide H, Seipel T, Alexander JM, Rew LJ, Arévalo JR, Cavieres LA, McDougall KL, Milbau A, Naylor BJ, Speziale K, Pauchard A (2018) Mountain roads and non-native species modify elevational patterns of plant diversity. Glob Ecol Biogeogr 27:667–678

Hails R, Timms-Wilson T (2007) Genetically modified organisms as invasive species? In: Caldwell MM, Heldmaier G, Jackson RB, Lange OL, Mooney HA, Schulze ED, Sommer U (eds) Biological invasions: analysis and synthesis. Springer, Berlin/Heidelberg/New York, pp 293–310

Harris CJ, Murray BR, Hose GC, Hamilton MA (2007) Introduction history and invasion success in exotic vines introduced to Australia. Divers Distrib 13:467–475

Hechenleitner VP, Gardner MF, Thomas PI, Echeverría C, Escobar B, Brownless P, Martínez C (2005) Plantas amenazadas del centro-sur de Chile. Distribución, Conservación y Propagación. Ediciones de la Universidad Austral de Chile y Real Jardín Botánico de Edimburgo

Henderson L (2001) Alien weeds and invasive plants, Plant Protection Research Institute Handbook No. 12. Agricultural Research Council, Pretoria

Hendry AP, Nosil P, Rieseberg LH (2007) The speed of ecological speciation. Funct Ecol 21:455–464

Hengeveld R (1989) Dynamics of biological invasions. Chapman & Hall, London

Hernández JH, Villaseñor NR (2018) Twelve-year change in tree diversity and spatial segregation in the Mediterranean city of Santiago, Chile. Urban Forest Urban Green 29:10–18

Heywood VH (1989) Patterns, extents and modes of invasions by terrestrial plants. In: Drake JA, Mooney HA, di Castri F, Groves RH, Kruger FJ, Rejmánek M, Williamson M (eds) Biological invasions: a global perspective. Wiley, Chichester, pp 31–60

Hierro JL, Callaway RM (2003) Allelopathy and exotic plant invasion. Plant Soil 256:29–39

Hierro JL, Maron JL, Callaway RM (2005) A biogeographic approach to plant invasions: the importance of studying exotics in their introduced and native range. J Ecol 93:5–15

Higgins TM, McCutchan JH, Lewis WM (2008) Nitrogen ebullition in a Colorado plains river. Biogeochemistry 89:367–377

Hillebrand H, Watermann F, Karez R, Berninger UG (2001) Differences in species richness patterns between unicellular and multicellular organisms. Oecologia 126:114–124

Hodges KE (2008) Defining the problem: terminology and progress in ecology. Front Ecol Environ 6:35–42

Hodkinson DJ, Thompson K (1997) Plant dispersal: the role of man. J Appl Ecol 34:1484–1496

Hoffmann A (1998) El árbol urbano en Chile. Ediciones Claudio Gay, Santiago de Chile

Holmgren M (2002) Exotic herbivores as drivers of plant invasion and switch to ecosystem alternatives states. Biol Invasions 4:25–33

Holmgren M, Avilés R, Sierralta L, Segura A, Fuentes ER (2000) Why have European herbs so successfully invaded the Chilean matorral? Effects of herbivory, soil nutrients, and fire. J Arid Environ 44:197–211

Holt RD, Lawton JH (1994) The ecological consequences of shared natural enemies. Annu Rev Ecol Syst 25:495–520
Holway DA (1999) Competitive mechanisms underlying the displacement of native ants by the invasive Argentine ant. Ecology 80:238–251
Holway DA, Suarez AV (1999) Animal behavior: an essential component of invasion biology. Trends Ecol Evol 14:328–330
Holway DA, Suaarez AV, Case TJ (2002) Role of abiotic factors in governing susceptibility to invasion: a test with argentine ants. Ecology 83:1610–1619
Hone J (1994) Analysis of vertebrate pest control. Cambridge University Press, Cambridge
Hopkins CCE (2002) Introduced marine organisms in Norwegian waters, including Svalbard. In: Leppäkoski E, Gollasch S, Olenin S (eds) Invasive aquatic species of Europe: distribution, impacts and management. Springer, pp 240–252
Huffaker CB, Messenger PS (1976) Theory and practice of biological control. Academic, New York
Hughes KA, Pertierra LR, Molina-Montenegro MA, Convey P (2015) Biological invasions in terrestrial Antarctica: what is the current status and can we respond? Biodivers Conserv 24:1031–1055
Hulme PE (2007) Biological invasions in Europe: drivers, pressures, states, impacts and responses. In: Hester R, Harrison RM (eds) Biodiversity under threat. Cambridge University Press, Cambridge, pp 56–80
Hulme PE (2008) Phenotypic plasticity and plant invasions: is it all jack? Funct Ecol 22:3–7
Hulme PE (2009) Trade, transport and trouble: managing invasive species pathways in an era of globalization. J Appl Ecol 46:10–18
Hulme PE, Pauchard A, Pyšek P, Vilà M, Alba C, Blackburn TM, Bullock JM, Chytrý M, Dawson W, Dunn AM, Essl F, Genovesi P, Maskell LC, Meyerson LA, Nuñez MA, Pergl J, Pescott OL, Pocock MJO, Richardson DM, Roy HE, Smart SM, Štajerová K, Stohlgren T, van Kleunen M, Winter M (2015) Challenging the view that invasive non-native plants are not a significant threat to the floristic diversity of Great Britain. Proc Natl Acad Sci U S A 112:2988–2989
Hurka H, Neuffer B (1997) Evolution processes in the genus *Capsella* (Brassicaceae). Plant Syst Evol 206:295–316
Ibarra JT, Fasola L, MacDonald DW, Rozzi R, Bonacic C (2009) Invasive American mink (*Mustela vison*) in wetlands of the Cape Horn biosphere reserve, southern Chile: what are they eating? Oryx 43:87–90
Instituto Forestas, INFOR (2010) Estadísticas del sector forestal (en línea). Santiago, Chile. Disponible en http//www.infor.cl Consultado 20 julio. 2018
Iponga DM, Milton SJ, Richardson DM (2008) Superiority in competition for light: a crucial attribute of the invasive alien tree *Schinus molle* (Peruvian pepper tree) in shaping its impact in semi-arid South African savanna. J Arid Environ 72:612–623
Iriarte JA, Lobos GA, Jaksic FM (2005) Invasive vertebrate species in Chile and their control and monitoring by governmental agencies. Rev Chil Hist Nat 78:143–154
Irimia RE, Montesinos D, Eren O, Lortie C, Kristine F, Cavieres LA, Sotes GJ, Hierro JL, Jorge A, Loureiro J (2017) Extensive analysis of native and non-native *Centaurea solstitialis* L. populations across the world shows no traces of polyploidization. PeerJ 14:e3531
Jaksic FM (1996) Ecología de los vertebrados en Chile. Ediciones Universidad Católica de Chile, Santiago
Jaksic FM (1998) Vertebrate invaders and their ecological impacts in Chile. Biodivers Conserv 7:1427–1445
Jaksic FM, Fuentes ER (1980) Why are native herbs in the Chilean matorral more abundant beneath bushes: microclimate or grazing? J Ecol 68:665–669
Jaksic FM, Fuentes ER (1991) Ecology of a successful invader: the European rabbit in Central Chile. In: Grooves RH, di Castri F (eds) Biogeography of mediterranean invasions. University of Cambridge, Cambridge, pp 273–283
Jaksic FM, Ostfeld RS (1983) Numerical and behavioral estimates of predation upon rabbits in mediterranean-type shrublands: a paradoxical case. Rev Chil Hist Nat 56:39–49

Jaksic FM, Soriguer RC (1981) Predation upon the European rabbit (*Oryctolagus cuniculus*) in mediterranean habitats of Chile and Spain: a comparative analysis. J Anim Ecol 50:269–281

Jaksic FM, Yáñez JL (1980) ¿Quién controla las poblaciones de conejos introducidos? Medio Ambiente 4:41–44

Jaksic FM, Yáñez JL (1983) Rabbit and fox introductions in Tierra del Fuego: history and assessment of the attempts at biological control of the rabbit infestation. Biol Conserv 26:367–374

Jaksic FM, Iriarte JA, Jiménez JE, Martínez DR (2002) Invaders without frontiers: cross-border invasions of exotic mammals. Biol Invasions 4:157–173

James WF, Barko JW, Eakin HL (1997) Nutrient regeneration by the zebra mussel (*Dreissena polymorpha*). J Freshw Ecol 12:209–216

Jeschke J, Aparicio LG, Haider S, Heger T, Lortie C, Pyšek P, Strayer D (2012) Support for major hypotheses in invasion biology is uneven and declining. NeoBiota 14:1–20

Jiang L, Tan J, Pu Z (2010) An experimental test of Darwin's naturalization hypothesis. Am Nat 175:415–423

Johow F (1948) Flora de las plantas vasculares de Zapallar. Rev Chil Hist Nat 49:8–566

Jones CG, Lawton JH, Shachak M (1994) Organisms as ecosystem engineers. Oikos 69:373–386

Jones CG, Lawton JH, Shachak M (1997) Positive and negative effects of organisms as physical ecosystem engineers. Ecology 78:1946–1957

Jones EI, Nuismer SL, Gomulkiewicz R (2013) Revisiting Darwin's conundrum reveals a twist on the relationship between phylogenetic distance and invasibility. Proc Natl Acad Sci U S A 110:20627–20632

Kareiva P (1983) Local movement in herbivorous insects: applying a passive diffusion model to mark-recapture field experiments. Oecologia 57:322–327

Kark S, Iwaniuk A, Schalimtzek A, E Banker (2007) Living in the city: can anyone become an "urban exploiter"? J Biogeogr 34:638–651

Kark S, Solarz W, Chiron F, Clergeau P, Shirley S (2009) Alien birds, amphibians and reptiles of Europe. In: DAISIE (ed) Handbook of alien species in Europe. Springer, Dordrecht, pp 105–118

Kauussaari M, Bommarco R, Heikkinen RK, Helm A, Krauss J, Lindborg R, Ockinger E, Pärtel M, Pino J, Rodà F, Stefanescu C, Teder T, Zobel M, Steffan-Dewenter I (2009) Extinction debt: a challenge for biodiversity conservation. Trends Ecol Evol 24:564–571

Keane RM, Crawley MJ (2002) Exotic plant invasions and the enemy release hypothesis. Trends Ecol Evol 17:164–170

Kennedy TA, Naeem S, Howe KM, Knops JMH, Tilman D, Reich P (2002) Biodiversity as a barrier to ecological invasion. Nature 417:636–638

Khuroo AA, Rashid I, Reshi Z, Dar GH, Wafai BA (2007) The alien flora of Kashmir Himalaya. Biol Invasions 9:269–292

Kikvidze Z, Brooker RW, Butterfield BJ, Callaway RM, Cavieres LA, Cook BJ, Lortie CJ, Michalet R, Pugnaire FI, Xiao S, Anthelme F, Björk RG, Cranston BH, Gavilán RG, Kanka R, Lingua E, Maalouf JP, Noroozi J, Parajuli R, Phoenix GK, Reid A, Ridenour WM, Rixen C, Schöb C (2015) The effects of foundation species on community assembly: a global study on alpine cushion plant communities. Ecology 96:2064–2069

King CM (1990) The handbook of New Zealand mammals. Oxford University Press, Oxford

Kitano S (2004) Ecological impacts of rainbow, brown, and brook trout in Japanese inland waters. Glob Environ Res 8:41–50

Klein AM, Vaissière BE, Cane JH, Steffan-Dewenter I, Cunningham SA, Kremen C, Tscharntke T (2007) Importance of pollinators in changing landscapes for world crops. Proc R Soc Lond B 274:303–313

Knapp S, Kühn I, Schweiger O, Klotz S (2008a) Challenging urban species diversity: contrasting phylogenetic patterns across plant functional groups in Germany. Ecol Lett 11:1054–1064

Knapp S, Kühn I, Wittig R, Ozinga WA, Poschlod P, Klotz S (2008b) Urbanization causes shifts in species' trait state frequencies. Preslia 80:375–388

Knop E (2016) Biotic homogenization of three insect groups due to urbanization. Glob Chang Biol 22:228–236

Kobelt M, Nentwig W (2008) Alien spider introductions to Europe supported by global trade. Divers Distrib 14:273–280
Kolar CS, Lodge DM (2001) Progress in invasion biology: predicting invaders. Trends Ecol Evol 16:199–204
Kollmann J, Bañuelos MJ (2004) Latitudinal trends in growth and phenology of the invasive alien plant *Impatiens glandulifera* (Balsaminaceae). Divers Distrib 10:377–385
Kot M, Lewis MA, van den Driessche P (1996) Dispersal data and the spread of invading organisms. Ecology 77:2027–2042
Kowarik I (1995) On the role of alien species in urban flora and vegetation. In: Pyšek P, Prach K, Rejmánek M, Wade M (eds) Plant invasions: general aspects and special problems. PB Academic Publishing, Amsterdam, pp 85–103
Kowarik I (2003) Human agency in biological invasions: secondary releases foster naturalisation and population expansion of alien plant species. Biol Invasions 5:293–312
Kowarik I (2005) Urban ornamentals escaped from cultivation. In: Gressel J (ed) Crop ferality and volunteerism. CRC Press, Boca Raton, pp 97–121
Kowarik I (2011) Novel urban ecosystems, biodiversity, and conservation. Environ Pollut 159:1974–1983
Kowarik I, von der Lippe M, Cierjacks A (2013) Prevalence of alien versus native species of woody plants in Berlin differs between habitats and at different scales. Preslia 85:113–132
Kowark I (2008) On the role of alien species in urban flora and vegetation. Urban Ecol 4:321–338
Krauss H, Weber A, Appel M, Enders B, Graevenitz A, Isenberg HD, Schiefer HG, Slenczka W, Zahner H (2003) Zoonoses: infectious diseases transmissible from animals to humans. ASM Press, American Society for Microbiology, Washington, DC
Krebs CJ (2001) Ecology: the experimental analysis of distribution and abundance. Benjamin Cummings, Menlo Park
Kruger FJ, Breytenbach GJ, MacDonald IAW, Richardson DM (1989) The characteristics of invaded Mediterranean-climate regions. In: Drake JA, Mooney HA, di Castri F, Groves RH, Kruger FJ, Rejmánek M, Williamson M (eds) Biological invasions: a global perspective. Wiley, Chichester, UK, pp 181–213
Kruse M, Strandberg M, Strandberg B (2010) Ecological effects of allelopathic plants: a review, Technical report no. 315. National Environmental Research Institute, Silkeborg
Kühn I, Klotz S (2002) Floristischer status und gebietsfremde Arten. In: Klotz S, Kühn I, Durka W (eds) BIOLFLOR: Eine Datenbank zu biologisch okologischen Merkmalen der Gefaßpflanzen in Deutschland. Bundesamt fur Naturschutz, Bonn, pp 47–56
Kühn I, Klotz S (2006) Urbanization and homogenization: comparing the floras of urban and rural areas in Germany. Biol Conserv 127:292–300
Kühn I, Brandl R, Klotz S (2004) The flora of German cities is naturally species rich. Evol Ecol Res 6:749–764
Kuri-Morales PA, Guzmán-Morales E, de la Paz-Nicolau E, Salas-Fernández A (2015) Enfermedades emergentes y reemergentes. Gaceta de Médica Mexico 151:674–680
La Sorte FA, McKinney ML (2006) Compositional similarity and the distribution of geographical range size for native and non-native species in urban floras. Divers Distrib 12:679–686
La Sorte FA, Pyšek P (2009) Extra-regional residence time as a correlate of plant invasiveness: European archaeophytes in North America. Ecology 90:2589–2597
La Sorte FA, Aronson MF, Williams NS, Celesti-Grapow L, Cilliers S, Clarkson BD, Pyšek P (2014) Beta diversity of urban floras among European and non-European cities. Glob Ecol Biogeogr 23:769–779
Lambdon P, Pyšek P, Basnou C, Delipetrou P, Essl F, Hejda M, Jarošík V, Pergl J, Winter M, Andriopoulos P, Arianoutsou M, Bazos I, Brundu G, Celesti-Grapow L, Chassot P, Jogan N, Josefsson M, Kark S, Klotz S, Kokkoris Y, Kühn I, Marchante H, Perglová I, Vilà M, Zikos A, Hulme PE (2008) Alien flora of Europe: species diversity, geographical pattern and state of the art of research. Preslia 80:101–149

Lander TA, Boshier DH, Harris SA (2009) Flower and fruit production and insect pollination of the endangered Chilean tree, *Gomortega keule* in native forest, exotic pine plantation and agricultural environments. Rev Chil Hist Nat 82:403–412

Largiadèr CR (2008) Hybridization and introgression between native and alien species. In: Nentwig W (ed) Biological invasions. ecological studies (Analysis and synthesis). Springer, Berlin/Heidelberg, pp 275–292

Latcham RE (1936) La agricultura precolombina en Chile y los países vecinos. Ediciones de la Universidad de Chile Santiago, Chile

Le Maitre DC, van Wilgen BW, Gelderblom CM, Bailey C, Chapman RA, Nel JA (2002) Invasive alien trees and water resources in South Africa: Case studies of the costs and benefits of management. For Ecol Manag 160:143–159

Lee CE (2002) Evolutionary genetics of invasive species. Trends Ecol Evol 17:386–391

Legendre P (2014) Interpreting the replacement and richness difference components of beta diversity. Glob Ecol Biogeogr 23:1324–1334

Levine JM (2000) Species diversity and biological invasions: relating local process to community pattern. Science 288:852–854

Levine JM, D'Antonio CM (1999) Elton revisited: a review of evidence linking diversity and invasibility. Oikos 87:15–26

Levine JM, Alder PB, Yelenik SG (2004) A meta-analysis of biotic resistance to exotic plant invasions. Ecol Lett 7:975–989

Liebhold AM, Halverson JA, Elmes GA (1992) Gypsy moth invasion in North America: a quantitative analysis. J Biogeogr 19:513–520

Lim HC, Sodhi NS (2004) Responses of avian guilds to urbanisation in a tropical city. Landsc Urban Plan 66:199–215

Lizarralde M (1993) Current status of the introduced beaver (*Castor canadensis*) population in Tierra del Fuego, Argentina. Ambio 22:351–358

Lizarralde MS, Bailliet G, Poljak S, Fasanella M, Giulivi C (2007) Assessing genetic variation and population structure of invasive North American beaver (*Castor canadensis* Kuhl, 1820) in Tierra Del Fuego (Argentina). Biol Invasions 10:673–683

Lobo D, Leao T, Melo FPL, Santos AMM, Tabarelli M (2011) Forest fragmentation drives Atlantic forest of northeastern Brazil to biotic homogenization. Divers Distrib 17:287–296

Lockwood JL, McKinney ML (2001) Biotic homogenization. Kluwer/Academic Press, New York

Lockwood JL, Cassey P, Blackburn T (2005) The role of propagule pressure in explaining species invasions. Trends Ecol Evol 20:223–228

Lockwood JL, Hoopes M, Marchetti M (2007) Invasion ecology. Blackwell Publishing, Massachusetts

Lonsdale WM (1999) Global patterns of plant invasions and the concept of invasibility. Ecology 80:1522–1536

Lorenzo P, González L, Reigosa MJ (2010) The genus *Acacia* as invader: the characteristic case of *Acacia dealbata* link in Europe. Ann For Sci 67:1–11

Lososová Z, Chytrý M, Tichý L, Danihelka J, Fajmon K, Hájek O, Kintrová K, Kühn I, Láníková D, Otýpková Z, Řehořek V (2012) Native and alien floras in urban habitats: a comparison across 32 cities of Central Europe. Glob Ecol Biogeogr 21:545–555

Lovell SJ, Stone SF, Fernandez L (2006) The economic impacts of aquatic invasive species: a review of the literature. Agricult Resour Econom Rev 35:178–194

Lowe S, Browne M, Boudjelas S, De Poorter M (2000) 100 of the world's worst invasive alien species: a selection from the global invasive species database. Published by The Invasive Species Specialist Group (ISSG) a specialist group of the Species Survival Commission (SSC) of the World Conservation Union (IUCN)

Lubina JA, Levin SA (1988) The spread of a reinvading species: range expansion in the California Sea otter. Am Nat 131:526–543

Ma C, Li SP, Pu Z, Tan J, Liu M, Zhou J, Li H, Jiang L (2016) Different effects of invader – native phylogenetic relatedness on invasion success and impact: a meta-analysis of Darwin's naturalization hypothesis. Proc R Soc Lond B 283:20160663

MacArthur R (1970) Species packing and competitive equilibrium for many species. Theor Popul Biol 1:1–11

MacDougall AS, Turkington R (2005) Are invasive species the drivers or passengers of change in degraded ecosystems? Ecology 86:42–52

Macgregor-Fors I (2008) Relation between habitat attributes and bird richness in a western Mexico suburb. Landsc Urban Plan 84:92–98

Machordom A, García-Marín JL, Sanz N, Almodóvar A, Pla C (1999) Allozyme diversity in brown trout (*Salmo trutta*) from Central Spain: genetic consequences of restocking. Freshw Biol 41:707–717

Mack RN (1996) Predicting the identity and fate of plant invaders: emergent and emerging approaches. Biol Conserv 78:107–121

Mack RN (2003) Global plant dispersal, naturalization and invasion: pathways, modes and circumstances. In: Ruiz G, Carlton J (eds) Global pathways of biotic invasions. Island Press, Washington, DC, pp 3–30

Mack RN, Erneberg M (2002) The United States naturalized flora: largely the product of deliberate introductions. Ann Mo Bot Gard 89:176–189

Mack RN, Lonsdale WM (2001) Humans as global plant dispersers: getting more than we bargained for. Bioscience 51:95–102

Mack RN, Simberloff D, Lonsdale WM, Evans H, Clout M, Bazzaz FA (2000) Biotic invasions, causes, epidemiology, global consequences, and control. Ecol Appl 10:689–710

Mackie GL, Schloesser DW (1996) Comparative biology of zebra mussels in Europe and North America: an overview. Am Zool 36:244–258

Malhi Y (2017) The concept of the Anthropocene. Annu Rev Environ Resour 42:25.1–25.28

Marchetti MP, Light T, Feliciano J, Armstrong T, Hogan Z, Viers J, Moyle PB (2001) Homogenization of California's fish fauna through abiotic change. In: Lockwood JL, McKinney ML (eds) Biotic homogenization. Kluwer Academic/Plenum, New York, pp 259–278

Marchetti MP, Lockwood JL, Light T (2006) Effects of urbanization on California's fish diversity: differentiation, homogenization and the influence of spatial scale. Biol Conserv 127:310–318

Marco A, Barthelemy C, Dutoit T, Bertaudière-Montes V (2010) Bridging human and natural sciences for a better understanding of urban floral patterns: the role of planting practices in Mediterranean gardens. Ecol Soc 15:2

Maron JL, Vilà M (2001) When do herbivores affect plant invasion? Evidence for the natural enemies and biotic resistance hypotheses. Oikos 95:361–373

Marr SM, Marchetti MP, Olden JD, García-Berthou E, Morgan DL, Arismendi I, Day JA, Griffiths CL, Skelton PH (2010) Freshwater fish introductions in mediterranean-climate regions: are there commonalities in the conservation problem? Divers Distrib 16:606–619

Martí R (1993) The spreading of the Ruddy duck *Oxyura jamaicensis* in Europe and its effect on conservation of the White-headed duck *Oxyura leucocephala* in Spain. SEO/BirdLife Spain, Madrid

Martí R, Del Moral JC (2003) La invernada de aves acuáticas en España. Dirección General de Conservación de la Naturaleza-SEO/BirdLife. Organismo Autónomo Parques Nacionales, Ministerio de Medio Ambiente, Madrid, España

Martínez-Carretero EEM (2010) The synanthropic flora in the Mendoza (Argentina) urban area. Urban Ecosyst 13:237–242

Martín-Forés I, Casado MA, Castro I, Ovalle C, del Pozo A, Acosta-Gallo B, Sánchez-Jardón L, de Miguel JM (2012) Flora of the Mediterranean Basin in the Chilean *espinales*: evidence of colonisation. Pastos 42:137–160

Marx HE, Giblin DE, Dunwiddie PW, Tank DC (2016) Deconstructing Darwin's naturalization conundrum in the San Juan Islands using community phylogenetics and functional traits. Divers Distrib 22:318–331

Marzluff JM (2016) A decadal review of urban ornithology and a prospectus for the future. Ibis 159:1–13

Massoia E, Chébez JC (1993) Mamíferos silvestres del archipiélago fueguino. Ediciones LOLA, Buenos Aires

Matthei O (1995) Manual de las malezas que crecen en Chile. Alfabeta Impresores, Santiago, Chile
Mattila HR, Otis GW (2003) A comparison of the host preference of monarch butterflies (*Danaus plexippus*) for milkweed (*Asclepias syriaca*) over dog-strangler vine (*Vincetoxicum rossicum*). Entomol Exp Appl 107:193–199
McDougall K, Alexander J, Haider S, Pauchard A, Walsh N, Kueffer C (2011) Alien flora of mountains: global comparisons for the development of local preventive measures against plant invasions. Divers Distrib 17:103–111
McKinney ML (2004a) Measuring floristic homogenization by non-native plants in North America. Glob Ecol Biogeogr 13:4–53
McKinney ML (2004b) Do exotics homogenize or differentiate communities? Roles of sampling and exotic species richness. Biol Invasions 6:495–504
McKinney ML (2006) Urbanization as a major cause of biotic homogenization. Biol Conserv 127:247–260
McKinney ML (2008) Effects of urbanization on species richness: a review of plants and animals. Urban Ecosyst 11:161–176
McKinney ML, Lockwood JL (1999) Biotic homogenization: a few winners replacing many losers in the next mass extinction. Trends Ecol Evol 14:450–453
McKinney ML, Lockwood JL (2005a) Biotic homogenization and the breakdown of biogeographic barriers. In: Sax DF, Gaines SD, Stachowicz JJ (eds) Exotic species: a source of insight into ecology, evolution and biogeography. Sinauer Associated, Sunderland, pp 365–380
McKinney ML, Lockwood JL (2005b) Community composition and homogenization: evenness and abundance of native and exotic plant species. In: Sax DF, Stachowicz JJ, Gaines SD (eds) Species invasions: insights into ecology, evolution and biogeography. Sinauer Associates, Sunderland, pp 365–380
McNeely JA (2001) The great reshuffling: human dimensions of invasive alien species. IUCN, Gland/Cambridge
Meijer K, Schilthuizen M, Beukeboom L, Smit C (2016) A review and meta-analysis of the enemy release hypothesis in plant-herbivorous insect systems. PeerJ 4:e2778
Méndez E (2005) Flora y vegetación del centro urbano de Luján de Cuyo, Mendoza (Argentina). Revista de la Facultad de Ciencias Agrarias 37:67–74
Menge BA, Chan F, Lubchenco J (2008) Response of a rocky intertidal ecosystem engineer and community dominant to climate change. Ecol Lett 11:151–162
Meyerson LA, Mooney HA (2007) Invasive alien species in an era of globalization. Front Ecol Environ 5:199–208
Milbau A, Stout JC (2007) Database of alien plants in Ireland. Irish Biochange Project, Dublin. http://www.biochange.ie/alienplants
Minchin D (2007) A checklist of alien and cryptogenic aquatic species in Ireland. Aquat Invasions 2:341–366
Miura O (2007) Molecular genetic approaches to elucidate ecological and evolutionary issues of the biological invasions. Ecol Res 22:876–883
Molina-Montenegro MA, Naya DE (2012) Latitudinal patterns in phenotypic plasticity and fitness-related traits: assessing the climatic variability hypothesis (CVH) with an invasive plant species. PLoS One 7:e47620
Molina-Montenegro MA, del Pozo A, Gianoli E (2018) Ecophysiological basis of the Jack-and-Master strategy: *Taraxacum officinale* (dandelion) as an example of a successful invader. J Plant Ecol 11:147–157
Montenegro G, Rivera O, Bas F (1978) Herbaceous vegetation in the Chilean matorral: dynamic of growth and evaluation of allelopathic effects of some dominant shrubs. Oecologia 36:237–244
Montenegro G, Teillier S, Arce P, Poblete V (1991) Introductions of plants and animals into the mediterrean-type climate area of Chile. In: Groves RH, di Castri F (eds) Biogeography of mediterranean invasions. Cambridge University Press, Cambridge, pp 103–109
Mooney HA, Cleland EE (2001) The evolutionary impact of invasive species. Proc Natl Acad Sci U S A 98:5446–5451

Mooney HA, Hobbs HA (2000) Invasive species in a changing world. Island Press, California
Mooney HA, Vitousek PM, Matson PA (1987) Exchange of materials between terrestrial ecosystems and the atmosphere. Science 238:926–932
Moors PJ, Atkinson IAE (1984) Predation on seabirds by introduced animals, and factors affecting its severity. In: Croxall JP, Evans PGH, Schreiber RW (eds) Status and conservation of the world's seabirds. ICBP Technical Publication, Cambridge, pp 667–690
Mora JP, Smith-Ramírez C (2016) Are birds, wind and gravity legitimate dispersers of fleshy-fruited invasive plants on Robinson Crusoe Island, Chile? Flora 224:167–171
Morales CL (2007) Introducción de abejorros (*Bombus*) no nativos: causas, consecuencias ecológicas y perspectivas. Ecol Austral 17:51–65
Morales MC, Verdejo V, Orlando J, Carú-Marambio M (2016) Fungal communities as an experimental approach to Darwin's naturalization hypothesis. Res Microbiol 167:126–132
Moran VC, Zimmerman HG (1991) Biological control of jointed cactus, *Opuntia aurantiaca* (Cactaceae), in South Africa. Agric Ecosyst Environ 37:5–27
Moro MF, Farías-Castro ASF (2015) A check list of plant species in the urban forestry of Fortaleza, Brazil: where are the native species in the country of megadiversity? Urban Ecosyst 18:47–71
Morueta-Holme N, Engemann K, Sandoval-Acuña P, Jonas J, Segnitz M, Svenning JC (2015) Strong upslope shifts in Chimborazo's vegetation over two centuries since Humboldt. Proc Natl Acad Sci U S A 112:12741–12745
Moyle PB (2002) The decline of anadromous fishes in California. Conserv Biol 8:869–870
Müller CH (1966) The role of chemical inhibition (allelopathy) in vegetational composition. Bull Torrey Pines Botanical Club 93:332–351
Muñoz AA, Cavieres LA (2008) The presence of a showy invasive plant disrupts pollinator service and reproductive output in native alpine species only at high densities. J Ecol 96:459–467
Muñoz-Fuentes V, Vilà C, Green AJ, Negro JJ, Sorenson MD (2007) Hybridization between white-headed ducks and introduced ruddy ducks in Spain. Mol Ecol 16:629–638
Muñoz-Pedreros A, Yáñez JL (2000) Mamíferos de Chile. Centro de Estudios Agrarios y Ambientales (CEA) Ediciones, Valdivia, Chile
Naeem S, Knops JMH, Tilman D, Howe KM, Kennedy T, Gale S (2000) Plant diversity increases resistance to invasion in the absence of covarying extrinsic factors. Oikos 91:97–108
Nagy L, Grabherr G (2009) The biology of alpine habitats. Oxford University Press, Oxford
National Institutes of Health U.S. (2007) Understanding emerging and re-emerging infectious disease. Biological Sciences Curriculum Study
Nehring S, Hesse KJ (2008) Invasive alien plants in marine protected areas: the *Spartina anglica* affair in the European Wadden Sea. Biol Invasions 10:937–950
Newsome AE, Noble IR (1986) Ecological and physiological characters of invading species. In: Groves RH, Burdon JJ (eds) Biological invasions. Cambridge University Press, Cambridge, pp 1–20
Niemeyer HN (1992) Aspectos ecológicos y moleculares de la interacción entre áfidos y sus plantas hospederas. Rev Chil Hist Nat 65:103–114
Nobel JV (1974) Geographic and temporal development of plagues. Nature 250:276–279
Noble IR (1989) Attributes of invaders and the invading process: terrestrial and vascular plants. In: Drake JA, Mooney HA, di Castri F, Groves RH, Kruger FJ, Rejmánek M, Williamson M (eds) Biological invasions: a global perspective. Wiley, Chichester, pp 1–30
Novillo A, Ojeda RA (2008) The exotic mammals of Argentina. Biol Invasions 10:1333–1344
O'Dowd DJ, Green PT, Lake PS (2003) Invasional 'meltdown' on an oceanic island. Ecol Lett 6:812–817
Office of Technology Assessment, OTA (1993) Harmful non-indigenous species in the United States, Report OTA-F-565. US Government Printing Office, Washington, DC
OIKONOS (2009) Juan Fernández islands conservancy. In: www.oikonos.org/projects/jfic.htm
Okubo A (1980) Diffusion and ecological problems: mathematical models. Springer, Berlin
Olden JD (2006) Biotic homogenization: a new research agenda for conservation biogeography. J Biogeogr 33:2027–2039

Olden JD, Poff NL (2003) Toward a mechanistic understanding and prediction of biotic homogenization. Am Nat 163:442–460
Olden JD, Poff NL (2004) Ecological processes driving homogenization: testing a mechanistic model using fish faunas. Ecology 85:1867–1875
Olden JD, Rooney TP (2006) On defining and quantifying biotic homogenization. Glob Ecol Biogeogr 15:113–120
Olden JD, Poff NL, Douglas M, Douglas ME, Fausch KD (2004) Ecological and evolutionary consequences of biotic homogenization. Trends Ecol Evol 19:18–24
Olden JD, Poff NL, McKinney M (2006) Forecasting faunal and floral homogenization associated with human population geography in North America. Biol Conserv 127:261–271
Ortega-Álvarez R, MacGregor-Fors (2009) Living in the big city: effects of urban land-use on bird community structure, diversity, and composition. Landsc Urban Plan 90:189–195
Panzacchi M, Bertolino S, Cocchi R, Genovesi P (2007) Population control of coypu *Myocastor coypus* in Italy compared to eradication in UK: a cost-benefit analysis. Wildl Biol 13:159–171
Pardo R, Vila I, Capella JJ (2009) Competitive interaction between introduced rainbow trout and native silverside in a Chilean stream. Environ Biol Fish 86:353–359
Parker IM, Simberloff D, Lonsdale WM, Goodell K, Wonham M, Kareiva PM, Williamson MH, von Holle B, Moyle PB, Byers JE, Goldwasser L (1999) Impact: toward a framework for understanding the ecological effects of invaders. Biol Invasions 1:3–19
Parker JD, Burkepile DE, Hay ME (2006) Opposing effects of native and exotic herbivores on plant invasions. Science 311:1459–1461
Parkes JP, Robley A, Forsyth DM, Choquenot D (2006) Adaptive management experiments in vertebrate pest control in New Zealand and Australia. Wildl Soc Bull 34:229–236
Parsons J (1972) Spread of African pasture grasses to the American tropics. J Range Manag 25:12–17
Pauchard A, Kueffer C, Dietz H, Daehler CC, Alexander J, Edwards PJ, Arévalo JR, Cavieres LA, Guisan A, Haider S, Jakobs G, McDougall K, Millar CI, Naylor BJ, Parks CG, Rew LJ, Seipel T (2009) Ain't no mountain high enough: plant invasions reaching new elevations. Front Ecol Environ 7:479–486
Pauchard A, García RA, Langdon B, Fuentes N (2011) Invasiones de plantas introducidas en Chile y su impacto en la biodiversidad: historia, estado actual y desafíos para su gestión. In: Figueroa E (ed) Conservación de la biodiversidad en las Américas: lecciones y recomendaciones de política. Editorial de la Universidad de Chile, Santiago, pp 147–182
Pauchard A, Langdon B, Jiménez A, Cavieres L, Peña E, Núñez MA (2014) Pináceas invasoras en el sur de Sudamérica: patrones, mecanismos e impactos potenciales. In: Jaksic FM, Castro SA (eds) Invasiones biológicas en Chile. Ediciones Universidad Católica de Chile, pp 283–308
Pausas JG, Keeley JE (2009) A burning story: the role of fire in the history of life. Bioscience 59:593–601
Pavez EF, González CA, Jiménez JE (1992) Diet shifts of black-chested eagles (*Geranoaetus melanoleucus*) from native prey to European rabbits. J Raptor Res 26:27–32
Pavez EF, Lobos GA, Jaksic FM (2010) Cambios de largo plazo en el paisaje y los ensambles de micromamíferos y rapaces en Chile central. Rev Chil Hist Nat 83:99–111
Pavoine S, Ricotta C (2014) Functional and phylogenetic similarity among communities. Methods Ecol Evol 5:666–675
Peay KG, Belisle M, Fukami T (2012) Phylogenetic relatedness predicts priority effects in nectar yeast communities. Proc R Soc Lond B 279:749–758
Pederson P, Bullock R, Carlton J, Dijkstra J, Dobroski N, Dyrynda P, Fisher R, Harris L, Hobbs N, Lambert G, Lazo-Wasem E, Mathieson A, Miglietta M-P, Smith J, Smith J III, Tyrrell M (2003) Marine invaders in the Northeast. Rapid assessment survey of non-native and native marine species of floating dock communities. Massachusetts Institute of Technology Sea Grant College Program, Massachusetts
Peña-Gómez FT, Guerrero PC, Bizama G, Duarte M, Bustamante RO (2014) Climatic niche conservatism and biogeographical non-equilibrium in *Eschscholzia californica* (Papaveraceae), an invasive plant in the Chilean Mediterranean region. PLoS One 9:e105025

Penaluna B, Arismendi I (2009) Peces nativos en aguas continentales del Sur de Chile. Imprenta América Ltda
Pérez JE, Alfonsi C, Salazar SK, Mascotay O (2007) Especies marinas exóticas y criptogénicas en las costas de Venezuela. Boletín del Instituto Oceanográfico 46:79–96
Perrings C, Dehnen-Schmutz K, Touza J, Williamson M (2005) How to manage biological invasions under globalization. Trends Ecol Evol 20:212–215
Petit RJ (2004) Biological invasions at the gene level. Divers Distrib 10:159–165
Petren K, Case TJ (1996) An experimental demonstration of exploitation competition in an ongoing invasion. Ecology 77:118–132
Pimentel D, Lach L, Zuniga R, Morrison D (2000) Environmental and economic cost of nonindigenous species in the United States. Bioscience 50:53–65
Pimentel D, Zúñiga R, Morrison D (2005) Update on the environmental and economic costs associated with alien-invasive species in the United States. Ecol Econ 52:273–288
Pimm SL (1991) The balance of nature? Ecological issues in the conservation of species and communities. University of Chicago Press, Chicago
Pool TK, Olden JD (2012) Taxonomic and functional homogenization of an endemic desert fish fauna diversity and distributions. 18:366–376
Porten M, Huber L (2003) An assessment method for the quantification of *Daktulosphaira vitifoliae* (Fitch) (hem. Phylloxeridae) populations in the field. J Appl Entomol 127:157–162
Prenter J, MacNeil C, Dick JTA, Dunn AM (2004) Roles of parasites in animal invasions. Trends Ecol Evol 19:385–390
Prentice HC, Malm JU, Hathaway L (2008) Chloroplast DNA variation in the European herb *Silene dioica* (red campion): postglacial migration and interspecific introgression. Plant Syst Evol 272:23–37
Preston CD, Pearman DA, Hall AR (2004) Archaeophytes in Britain. Bot J Linn Soc 145:257–294
Prigioni C, Balestrieri A, Remonti L (2005) Food habits of the coypu, *Myocastor coypus*, and its impact on aquatic vegetation in a freshwater habitat of NW Italy. Folia Zool 54:269–277
Pringle RM (2005) The origins of the Nile perch in Lake Victoria. Bioscience 55:780–787
Programa de las Naciones Unidas, PNUD (2017) Catálogo de las especies exóticas asilvestradas/naturalizadas en Chile. Laboratorio de Invasiones Biológicas (LIB). Universidad de Concepción, Proyecto GEF/MMA/PNUD. Fortalecimiento de los Marcos Nacionales para la Gobernabilidad de las Especies Exóticas Invasoras: Proyecto Piloto en el Archipiélago de Juan Fernández, Santiago de Chile, 61 pp
Pyšek P (1998a) Alien and native species in Central European urban floras: a quantitative comparison. J Biogeogr 25:155–163
Pyšek P (1998b) Is there a taxonomic pattern to plant invasions? Oikos 82:282–294
Pyšek P (2003) How reliable are data on alien species in Flora Europaea? Flora 198:499–507
Pyšek P, Prach K (2003) Research into plant invasions in a cross-roads region: history and focus. Biol Invasions 5:349–356
Pyšek P, Richardson DM (2007) Traits associated with invasiveness in alien plants: where do we stand? In: Nentwig W (ed) Biological invasions. Ecological studies 193. Springer, Berlin/Heidelberg, pp 97–126
Pyšek P, Richardson DM (2010) Invasive species, environmental change and management, and health. Annu Rev Environ Resour 35:25–55
Pyšek P, Sádlo J, Mandák B (2002) Catalogue of alien plants of the Czech Republic. Preslia 74:97–186
Pyšek P, Sádlo J, Mandák B, Jarosik V (2003) Czech alien flora and the historical pattern of its formation: what came first to Central Europe? Oecologia 135:122–130
Pyšek P, Cock MJW, Nentwig W, Ravn HP (2007) Ecology and management of giant hogweed (*Heracleum mantegazzianum*). CABI, Wallingford
Pyšek P, Richardson DM, Pergl J, Jarošik V, Sixtová Z, Weber E (2008a) Geographical and taxonomic biases in invasion ecology. Trends Ecol Evol 23:237–244

Pyšek P, Lambdon PW, Arianoutsou M, Kühn I, Pino J, Winter M (2008b) Alien vascular plants of Europe. In: Drake JA (ed) Handbook of alien species in Europe. Invading nature, Springer Series in Invasion Ecology. Springer, New York, pp 43–61

Pyšek P, Jarošik V, Pergl J, Randall R, Chytry M, Kühn I, Tichý L, Danihelka J, Chrtek J, Sádlo J (2009) The global invasion success of Central European plants is related to distribution characteristics in their native range and species traits. Divers Distrib 15:891–903

Quiroz C, Pauchard A, Cavieres LA, Anderson CB (2009) Análisis cuantitativo de la investigación en invasiones biológicas en Chile: tendencias y desafíos. Rev Chil Hist Nat 82:497–505

Rahel FJ (2000) Homogenization of fish faunas across the United States. Science 288:854–856

Rahel FJ (2002) Homogenization of freshwater faunas. Annu Rev Ecol Syst 33:291–315

Raikow DF, Sarnelle O, Wilson AE, Hamilton SK (2004) Dominance of the noxious cyanobacterium *Microcystis aeruginosa* in low-nutrient lakes is associated with exotic zebra mussels. Limnol Oceanogr 49:482–487

Randall RP (2002) A global compendium of weeds. RG & FJ Richardson, Melbourne

Randall JM, Benton N, Morse LE (2001) Categorizing invasive plants: the challenge of rating the weeds already in California. In: Groves RH, Panetta FD, Virtue JG (eds) Weed risk assessment. CSIRO Publishing, Collingwood, pp 203–216

Ratcliffe FN, Gay FJ, Greaves T (1952) Australian termites. Commonwealth Scientific and Industrial Research Organization, Melbourne

Raven PH (1988) The Californian flora. In: Barbour MG, Mayor J (eds) Terrestrial vegetation of California. Sacramento. California Native Plant Society, California, pp 109–137

Reichard SH (1997) Prevention of invasive plant introductions on national and local levels. In: Luken JA, Thieret JA (eds) Assessment and management of plant invasions. Springer, New York, pp 215–227

Reichard SH, White P (2001a) Horticultural introductions of invasive plant species: a north American perspective. In: McNeely JA (ed) The great reshuffling: human dimensions of invasive alien species. IUCN, Gland, pp 161–170

Reichard SH, White P (2001b) Horticultural as a pathway of invasive plant introductions in the United States. Bioscience 51:103–113

Reiche K (1899) Estudios críticos sobre la flora de Chile. Anales de la Universidad de Chile 104:767–847

Rejmánek M (1989) Invasibility of plant communities. In: Drake JA, Mooney HA, di Castri F, Groves RH, Kruger FJ, Rejmánek M, Williamson M (eds) Biological invasions: a global perspective. Wiley, Chichester, pp 369–388

Rejmánek M (1995) What makes a species invasive? In: Pyšek P, Prach K, Rejmánek M, Wade M (eds) Plant invasions: general aspects and special problems. SPB Academic Publishing, Amsterdam, pp 3–13

Rejmánek M (1996) A theory of seed plant invasiveness: the first sketch. Biol Conserv 78:171–180

Rejmánek M (1998) Invasive plant species and invadible ecosystems. In: Sandlund OT, Schei PJ, Vilken A (eds) Invasive species and biodiversity management. Kluwer, Dordrecht, pp 79–102

Rejmánek M (2000) Invasive plants: approaches and predictions. Austral Ecol 25:497–506

Rejmánek M, Richardson DM (1996) What attributes make some plant species more invasive? Ecology 77:1655–1661

Rejmánek M, Thomsen CD, Peters ID (1991) Invasive vascular plants of California. In: Groves RH, di Castri F (eds) Biogeography of mediterranean invasions. Cambridge University Press, Cambridge, pp 81–101

Rejmánek M, Richardson DM, Barbour MG, Crawley MJ, Hrusa GF, Moyle PB, Randall JM, Simberloff D, Williamson M (2002) Biological invasions: politics and the discontinuity of ecological terminology. Ecol Soc Am Bull 83:131–133

Rejmánek M, Richardson DM, Higgins SI, Pitcairn MJ, Grotkopp E (2005) Ecology of invasive plants: state of the art. In: Mooney HA, Mack RN, McNeely JA, Neville L, Schei PJ, Waage J (eds) Invasive alien species: a new synthesis. Island Press, Washington, DC, pp 104–161

Revenga C, Campbell I, Abell R, de Villiers P, Bryer M (2005) Prospects for monitoring freshwater ecosystems towards the 2010 targets. Philos Trans R Soc B 360:397–413

Ricciardi A (2004) Assessing species invasions as a cause of extinction. Trends Ecol Evol 19:619
Ricciardi A, Mottiar M (2006) Does Darwin's naturalization hypothesis explain fish invasions? Biol Invasions 8:1403–1407
Richardson DM (1997) Forestry trees as invasive aliens. Conserv Biol 12:18–26
Richardson DM (1998) Ecology and biogeography of *Pinus*. Cambridge University Press, Cambridge
Richardson DM, Allsopp N, D'Antonio CM, Milton SJ, Rejmánek M (2000a) Plant invasions: the role of mutualisms. Biol Rev 75:65–93
Richardson DM, Pyšek P, Rejmánek M, Barbour MG, Panetta FD, West CJ (2000b) Naturalization and invasion of alien plants: concepts and definitions. Divers Distrib 6:93–107
Richardson DM, van Wilgen BW, Nuñez MA (2008) Alien conifer invasions in South America: short fuse burning? Biol Invasions 10:1387–3547
Richardson DM, Hui C, Núñez MA, Pauchard A (2014) Tree invasions: patterns, processes, challenges and opportunities. Biol Invasions 16:473–481
Ricklefs RE (1987) Community diversity: relative roles of local and regional processes. Science 235:167–171
Ricotta C, La Sorte FA, Pyšek P, Rapson GL, Celesti-Grapow L, Thompson K (2009) Phyloecology of urban alien floras. J Ecol 97:1243–1251
Rodda G, Fritts T, Chiszar D (1997) The disappearance of Guam's wildlife. Bioscience 47:565–575
Rodríguez JP (2001) Exotic species introductions into South America: an underestimated threat? Biodivers Conserv 10:1981–1994
Rodríguez R, Ruiz E, Elissetche JP (2005) Árboles en Chile. Editorial Universidad de Concepción, Chile
Rodriguez R, Marticorena C, Alarcón D, Baeza C, Cavieres L, Finot VL, Fuentes N, Kiessling A, Mihoc M, Pauchard A, Ruiz E, Sánchez P, Marticorena A (2018) Catálogo de las plantas vasculares de Chile. Gayana Botánica 75:1–430
Rogers PM, Arthur CP, Soriger RC (1994) The rabbit in continental Europe. In: Thompson HV, King CM (eds) The European rabbit: the history and biology of a successful colonizer. Oxford Science Publications, Oxford, pp 22–63
Rojas G, Roure J (2001) Atmospheric pollen in Santiago, Chile. Grana 40:126–132
Roman J (2006) Diluting the founder effect: cryptic invasions expand a marine invader's range. Proc R Soc Lond B 273:2453–2459
Rooney TP, Wiegmann SM, Rogers DA, Waller DM (2004) Biotic impoverishment and homogenization in unfragmented forest understory communities. Conserv Biol 18:787–798
Rooney TP, Olden JD, Leach MK, Rogers DA (2007) Biotic homogenization and conservation priorization. Biol Conserv 134:447–450
Roques A, Rabitsch W, Lopez-Vaamonde C, Nentwig W, Kenis M (2008) Alien terrestrial invertebrates of Europe. In: Drake JA (ed) Handbook of alien species in Europe. Invading nature. Springer Series in Invasion Ecology, Amsterdam, pp 63–79
Rosenzweig ML (2001) The four questions: what does the introduction of exotic species do to diversity? Evol Ecol Res 3:361–367
Roura-Pascual N, Suarez AV, McNyset K, Gomez C, Pons P, Touyama Y, Wild AL, Gascon F, Peterson AT (2006) Niche differentiation and fine-scale projections for Argentine ants based on remotely sensed data. Ecol Appl 16:1832–1841
Rozebelds ACF, Cave L, Morris DI, Buchanan AM (1999) The weed invasions in Tasmania since 1970. Aust J Bot 47:23–48
Rozzi R, Sherriffs M (2003) El visón (*Mustela vison*, Schereber), un nuevo mamífero exótico para la Isla Navarino. Anales del Instituto de la Patagonia, Serie Ciencias Naturales 31:97–104
Ruiz GM, Fofnoff PW, Carlton JT, Wonham MJ, Hines AH (2000) Invasion of coastal marine communities in North America: apparent patterns, processes, and biases. Annu Rev Ecol Syst 31:481–531
Sakai AK, Allendorf FW, Holt JS, Lodge DM, Molofsky J, With KA, Baughman S, Cabin RJ, Cohen JE, Ellstrand NC, McCauley DE, O'Neil P, Parker MI, Thompson JN, Weller SG (2001) The population biology of invasive species. Annu Rev Ecol Syst 32:305–332

Sala OE, Chapin FS, Armesto JJ, Berlow E, Bloomfield J, Dirzo R, Huber-Sanwald E, Huenneke LF, Jackson RB, Kinzig A, Leemans R, Lodge DM, Mooney HA, Oesterheld M, Poff NL, Sykes MT, Walker BH, Walker M, Wall DH (2000) Global biodiversity scenarios for the year 2100. Science 287:1770–1774

Sandoya V, Pauchard A, Cavieres LA (2017) Natives and non-natives plants show different responses to elevation and disturbance on the tropical high Andes of Ecuador. Ecol Evol 2017:1–11

Santilli L, Castro SA, Figueroa JA, Guerrero N, Ray C, Romero M, Rojas G, Lavanderos N (2018) Composition and diversity patterns of the ornamental flora of Central Chile: implication for urban planning. Gayana Botánica 72:568–588

Sanz V, Caula S (2015) Assessing bird assemblages along an urban gradient in a Caribbean island (Margarita, Venezuela). Urban Ecosyst 18:729–746

Sanz-Elorza M, Dana ED, Sobrino E (2004) Atlas de las plantas alóctonas invasoras en España. Dirección General para la Biodiversidad. Ministerio de Medio Ambiente, Madrid

Savidge J (1987) Extinction of and island forest avifauna by an introduced snake. Ecology 68:660–667

Sax DF (2001) Latitudinal gradients and geographic ranges of exotic species: implications for biogeography. J Biogeogr 28:139–150

Sax DF (2002) Native and naturalized plant diversity are positively correlated in scrub communites of California and Chile. Divers Distrib 8:193–210

Sax DF, Gaines SD, Brown JH (2002) Species invasions exceed extinctions on islands worldwide: a comparative study of plants and birds. Am Nat 160:766–783

Sax DF, Stachowicz JJ, Gaines SD (2005) Species invasions: insights into ecology, evolution and biogeography. Sinauer Associates Incorporated, Sunderland

Scalera R, Zaghi D (2004) Alien species and nature conservation in the EU: the role of the LIFE program. European Commission, Brussels

Scherer-Lorenzen M, Venterink HO, Buschmann H (2007) Nitrogen enrichment and plant invasions: the importance of nitrogen-fixing plants and anthropogenic eutrophication. Biol Invasions 193:163–180

Schloesser DW, Nalepa TF, Mackie GL (1996) Infestation of unionid bivalves (Unionidae) in North America. Am Zool 36:300–310

Schüttler E, Klenke R, McGehee S, Rozzi R, Jax K (2009) Vulnerability of ground-nesting waterbirds to predation by invasive American mink in the Cape Horn Biosphere Reserve, Chile. Biol Conserv 142:1450–1460

Schwindt E (2008) Especies exóticas en el Mar Patagónico y sectores aledaños. In: Campagna C (ed) Síntesis del estado de conservación del Mar Patagónico y áreas de influencia. Foro para la Conservación del Mar Patagónico y Áreas de Influencia, Buenos Aires, pp 57–60

Scott MC, Helfman GS (2001) Native invasions, homogenization, and the mismeasure of integrity of fish assemblages. Fisheries 26:6–15

Seipel TC, Kueffer C, Rew LJ, Daehler CC, Pauchard A, Naylor BJ, Alexander JM, Edwards PJ, Parks CG, Arevalo JR, Cavieres LA, Dietz H, Jakobs G, Mcdougall K, Otto R, Walsh N (2012) Processes at multiple scales affect richness and similarity of non-native plant species in mountains around the world. Glob Ecol Biogeogr 21:236–246

Shea K, Chesson P (2002) Community ecology theory as a framework for biological invasions. Trends Ecol Evol 17:170–176

Shigesada N, Kawasaki K (1997) Biological invasions: theory and practice. Oxford University Press, Oxford

Sick H (1984) Ornitologia Brasileira: uma introdução. Editora Universidade de Brasília, Brasil

Sielfeld KW, Venegas C (1980) Poblamiento e impacto ambiental de *Castor canadensis* Kuhl, en Isla Navarino, Chile. Anales del Instituto de la Patagonia (Chile) 11:247–257

Sierra C (2001) El cerdo cimarrón (*Sus scrofa*, Suidae) en la Isla del Coco, Costa Rica: escarbaduras, alteraciones al suelo y erosión. Revista de Biología Tropical (Costa Rica) 49:1158–1170

Silverman J, Brightwell RJ (2008) The argentine ant: challenges in managing an invasive unicolonial pest. Annu Rev Entomol 53:231–252

Simberloff D (1989) Which insect introductions succeed and which fail? In: Drake JA, Mooney HA, di Castri F, Groves RH, Kruger FJ, Rejmánek M, Williamson M (eds) Biological invasions: a global perspective. Wiley, Chichester, pp 61–76
Simberloff D (2005) Non-native species do threaten the natural environment. J Agric Environ Ethics 18:595–607
Simberloff D (2006) Invasional meltdown 6 years later: important phenomenon, unfortunate metaphor, or both? Ecol Lett 9:912–919
Simberloff D (2009) The role of propagule pressure in biological invasions. Annu Rev Ecol Evol Syst 40:81–102
Simberloff D, Stiling P (1996) How risky is biological control? Ecol Soc Am 77:1965–1974
Simberloff D, von Holle B (1999) Positive interactions of nonindigenous species: Invasional meltdown? Biol Invasions 1:21–32
Simões M, Baruch Z (1991) Responses to simulated herbivory and water stress in two tropical C4 grasses. Oecologia 88:173–180
Simonetti JA (1986) Human-induced dietary shift in *Dusicyon culpaeus*. Mammalia 50:406–408
Skellam JG (1951) Random dispersal in theoretical populations. Biometrika 38:196–218
Skewes RO, Gonzalez F, Rubilar L, Quezada OM, Olave IT, Vargas V, Avila AC (1999) Investigación, aprovechamiento y control del castor, islas Tierra del Fuego y Navarino. Servicio de Gobierno Regional XII Región. Magallanes y Antártica Chilena, Punta Arenas
Skewes O, González F, Olave R, Ávila A, Vargas V, Paulsen P, König HE (2006) Abundance and distribution of American beaver, *Castor canadensis* (Kuhl 1820), in Tierra del Fuego and Navarino islands, Chile. Eur J Wildl Res 52:292–296
Slotte T, Ceplitis A, Neuffer B, Hurka H, Lascoux M (2006) Intrageneric phylogeny of *Capsella* (Brassicaceae) and the origin of the tetraploid *C. bursa-pastoris* based on chloroplast and nuclear DNA sequences. Am J Bot 93:1714–1724
Smart SM, Thompson K, Marrs RH, Le Duc MG, Maskell LC, Firbank LG (2006) Biotic homogenization and changes in species diversity across human-modified ecosystems. Proc R Soc Lond B 22:2659–2665
Smith KG (2006) Patterns of nonindigenous herpetofaunal richness and biotic homogenization among Florida counties. Biol Conserv 127:327–335
Smith RM, Thompson K, Hodgson JG, Warren PH, Gaton KJ (2006) Urban domestic gardens (IX): composition and richness of the vascular plant flora, and implications for native biodiversity. Biol Conserv 129:312–322
Soto D, Arismendi I, González J, Sanzana J, Jara F, Jara C, Guzmán E, Lara A (2007) Southern Chile, trout and salmon country: invasion patterns and threats for native species. Rev Chil Hist Nat 79.97–117
Spear D, Chown SL (2008) Taxonomic homogenization in ungulates: patterns and mechanisms at local and global scales. J Biogeogr 35:1962–1975
Stanton ML (1987) Reproductive biology of petal color variants in wild populations of *Raphnus sativus*: I. Pollinator response to color morphs. Am J Bot 74:178–187
Stearns S (1992) The evolution of life histories. Oxford University Press, Oxford
Stebbins GL (1959) The role of hybridization in evolution. Proc Am Philos Soc 103:231–251
Stenquist S (2000) Saltcedar integrated weed management and the endangered species act. In: Spencer N (ed) Proceedings of the X international symposium on biological control of weeds. Montana State University, Bozeman, pp 487–504
Stohlgren TJ, Binkley D, Chong GW, Kalkhan MA (1999) Exotic plant species invade hot spots of native plant diversity. Ecol Monogr 69:25–46
Storey AA, Ramírez JM, Quiroz D, Burley D, Addison DJ, Walter R, Anderson J, Hunt TL, Athens JS, Huynen L, Matisoo-Smith EA (2007) Radiocarbon and DNA evidence for a pre-Columbian introduction of Polynesian chickens to Chile. Proc Natl Acad Sci U S A 104:10335–10339
Stotz GC, Gianoli E, Jr James JF (2016) Spatial pattern of invasion and the evolutionary responses of native plant species. Evol Appl 9:939–951
Strauss SY, Webb CO, Salamin N (2006) Exotic taxa less related to native species are more invasive. Proc Natl Acad Sci U S A 103:5841–5845

Suárez CL, Berdasquera D (2000) Enfermedades emergentes y reemergentes: factores causales y vigilancia. Revista Cubana de Medicina General Integral 16:593–597

Suarez AV, Holway DA, Case TJ (2001) Patterns of spread in biological invasions dominated by long-distance jump dispersal: insights from argentine ants. Proc Natl Acad Sci U S A 98:1095–1100

Taylor CM, Hastings A (2005) Allee effects in biological invasions. Ecol Lett 8:895–908

Taylor KT, Maxwell BD, Pauchard A, Nuñez MA, Peltzer DA, Terwei A, Rew LJ (2015) Drivers of plant invasion vary globally: evidence from pine invasions within six ecoregions. Glob Ecol Biogeogr 25:96–106

Te Beest M, Le Roux JJ, Richardson DM, Brysting AK, Suda J, Kubešová M, Pyšek P (2011) The more the better? The role of polyploidy in facilitating plant invasions. Ann Bot 109:19–45

Thomas CD, Palmer G (2015) Non native plants add to the British flora without negative consequences for native diversity. Proc Natl Acad Sci U S A 112:4387–4392

Thompson JN (1988) Evolutionary ecology of the relationship between oviposition preference and performance of offspring in phytophagous insects. Entomol Exp Appl 47:3–14

Thompson JD (1991) The biology of an invasive plant: what makes *Spartina anglica* so successful? Bioscience 41:393–401

Thuiller W, Gallien L, Boulangeat I, Bello F, Munkemüller T, Roquet C, Lavergne S (2010) Resolving Darwin's naturalization conundrum: a quest for evidence. Divers Distrib 16:461–475

Tilman D (1988) Plant strategies and the dynamics and structure of plant communities. Princeton University Press, Princeton

Tilman D (1997) Distinguishing between the effects of species diversity and species composition. Oikos 80:185

Tilman D (2004) Niche tradeoffs, neutrality, and community structure: a stochastic theory of resource competition, invasion, and community assembly. Proc Natl Acad Sci U S A 101:10854–10861

Torres JA, Moreno B, Alcalá-Zamora A (1994) La Malvasía Canela (*Oxyura jamaicensis*) en España y su relación con la Malvasía Cabeciblanca (*Oxyura leucocephala*). Oxyura 7:5–27

Traveset A, Richardson DM (2014) Mutualistic interactions and biological invasions. Annu Rev Ecol Evol Syst 45:89–113

Tsutsui ND, Suarez AV, Holway DA (2000) Reduced genetic variation and the success of an invasive species. Proc Natl Acad Sci U S A 97:5948–5953

Ulbrich J (1930) Die bisamratte. Heinrich, Dresde

Usher MB (1986) Invasibility and wildlife conservation: invasive species on nature reserves. Philos Trans R Soc Lond B 314:695–709

van der Gaag DJ (2007) Report workshop management of *Anoplophora*, Wageningen. www.minlnv.nl/cdlpub/servlet/CDLServlet?p_file_id=22662

van Wilgen BW, Richardson DM, Le Maitre DC, Marais C, Magadlela D (2001) The economic consequences of alien plant invasions: examples of impacts and approaches to sustainable management in South Africa. Environ Dev Sustain 3:145–168

Vargas PV, Arismendi I, Gomez-Uchida D (2015) Evaluating taxonomic homogenization of freshwater fish assemblages in Chile. Rev Chil Hist Nat 88:16–26

Vellend M (2001) Do commonly used indices of β-diversity measure species turnover? J Veg Sci 12:545–552

Velthuis HHW, van Doorn A (2006) A century of advances in bumblebee domestication and the economic and environmental aspects of its commercialization for pollination. Apidologie 37:421–451

Verloove F (2006) Catalogue of neophytes in Belgium (1800–2005). In: Branquart E, Baus E, Pieret N, Vanderhoeven S, Desmet P (eds) . SOS invasions. Conference Abstract Book, Bruselas, Bélgica, pp 10–11

Verspoor E (1988) Widespread hybridization between native Atlantic salmon, *Salmo salar*, and introduced brown trout, *S. trutta*, in eastern Newfoundland. J Fish Biol 32:327–334

Vidal-Figueroa T, Delibes M (1987) Primeros datos sobre el visón Americano (*Mustela vison*) en el suroeste de Galicia y noroeste de Portugal. Ecología 1:145–152

Vila I, Habit E (2015) Current situation of the fish fauna in the Mediterranean region of Andean river systems in Chile. Fishes Mediterranean Environ 2015:002, 19p

Vilà M, Basnou C, Pyšek P, Josefsson M, Genovesi P, Gollasch S, Nentwig W, Olenin S, Roques A, Roy D, Hulme PE, DAISIE Partners (2010) How well do we understand the impacts of alien species on ecological services? A pan-European cross-taxa assessment. Front Ecol Environ 8:135–144

Villagrán C, Le-Quesne C, Aravena JC, Jiménez H, Hinojosa F (1998) El rol de los cambios de clima del Cuaternario en la distribución actual de la vegetación de Chile central-sur. Bamberger Geographische Schriften 15:227–242

Villaseñor JL, Espinosa-García FJ (2004) The alien flowering plants of Mexico. Divers Distrib 10:113–123

Villéger S, Blancheta S, Beauchard O, Oberdorff T, Brosse S (2011) Homogenization patterns of the world's freshwater fish faunas. Proc Natl Acad Sci U S A 108:18003–18008

Vitousek PM (1990) Biological invasions and ecosystem processes: towards an integration of population biology and ecosystem studies. Oikos 57:7–13

Vitousek PM, Walder LR, Whiteaker LD, Mueller-Dombois D, Matson PA (1987) Biological invasion by *Myrica faya*, alters ecosystem development in Hawaii. Science 238:802–804

Vitousek PM, D'Antonio CM, Loope LL, Westbrooks R (1996) Biological invasions as global environmental change. Am Sci 84:468–478

Vitousek PM, D'Antonio CM, Loope LL, Rejmánek M, Westbrooks R (1997) Introduced species: a significant component of human-caused global change. N Z J Ecol 21:1–16

Wallem PK, Anderson CB, Martínez-Pastur G, Lencinas MV (2010) Community re-assembly by an exotic herbivore, *Castor canadensis*, in subantarctic forests, Chile and Argentina. Biol Invasions 12:325–335

Wallem PK, Soto N, Cabello JL, Castro SA, Jaksic FM (2014) La invasion del castor americano (*Castor canadensis*) en Tierra del Fuego como oportunidad para analizar la valoración del impacto y manejo de vertebrados exóticos invasivos en Chile. In: Jaksic FM, Castro SA (eds) Invasiones biológicas en Chile. Ediciones Universidad Católica de Chile, pp 209–233

Wan S, Yuan T, Bowdish S, Wallace L, Russell SD, Luo Y (2002) Response of an allergenic species, *Ambrosia psilostachya* (Asteraceae), to experimental warming and clipping: implications for public health. Am J Bot 89:1843–1846

Wayne P, Foster S, Connolly J, Bazzaz F, Epstein P (2002) Production of allergic pollen by ragweed (*Ambrosia artemisiifolia* L.) is increase in CO_2-enriched atmospheres. Anna Allerg Asthma Immunol 88:279–282

Webb SD (1991) Ecogeography and the Great American interchange. Paleobiology 17:266–280

Whitcraft CR (2007) Wetland plant influence on sediment ecosystem structure and trophic function. Ph. D. Thesis, University of California, San Diego

White PCL, Harris S (2002) Economic and environmental costs of alien vertebrate species in Britain. In: Pimentel D (ed) Biological invasions: economic and environmental costs of alien plant, animal and microbe species. CRC Press, Boca Raton, pp 113–149

Wiklund C (1975) The evolutionary relationship between adult oviposition preferences and larval host plant range in *Papilio machaon* L. Oecologia 18:156–197

Wilcove DS, Rothstein D, Dubow J, Phillips A, Losos E (1998) Quantifying threats to imperiled species in the United States. Bioscience 48:607–615

Williams MC, Wardle GM (2007) Pinus radiata invasion in Australia: identifying key knowledge gaps and research directions. Austral Ecol 32:721–739

Williamson M (1996) Biological invasions. Chapman & Hall, London

Willis J, Thomas MB, Lawton JH (1999) Is the increased vigour of invasive weeds explained by a trade-off between growth and herbivore resistance? Oecologia 120:632–640

Willmer PG, Bataw AAM, Highes JP (1994) The superiority of bumblebees to honeybees as pollinators: insect visits to raspberry flowers. Ecol Entomol 9:271–284

Winter K, Adams L, Thorp R, Inouye D, Day L, Ascher J, Buchmann S (2006) Importation of non-native bumble bees into North America: potential consequences of using *Bombus terrestris* and other non-native bumblebees for greenhouse crop pollination in Canada, Mexico, and the United States. White Paper of the North American Pollinator Protection Campaign

Winter M, Schweigera O, Klotz S, Nentwig W, Andriopoulos P, Arianoutsou M, Basnou C, Delipetrou P, Didžiulis V, Hejda M, Hulme PE, Lambdon PW, Pergl J, Pyšek P, Royl DB, Kühn I (2009) Plant extinctions and introductions lead to phylogenetic and taxonomic homogenization of the European flora. Proc Natl Acad Sci U S A 106:21721–21725

Wittenberg R, Cock MJW (2005) Best practices for the prevention and management of invasive alien species. In: Mooney HA, Mack RN, McNeely JA, Neville LE, Schei PJ, Waage JK (eds) Invasive alien species. Scope, Island Press, Washington/Covelo/London, pp 209–284

Wright JP, Jones CG (2006) The concept of organisms as ecosystem engineers ten years on: Progress, limitations, and challenges. Bioscience 56:203–209

Wu SH, Hsieh CF, Chaw SM, Rejmánek M (2004) Plant invasions in Taiwan: insights from the flora of casual and naturalized alien species. Divers Distrib 10:349–336

Yachi S, Kawasaki K, Shigesada N, Teremoto E (1989) Spatial patterns in propagating waves of fox rabies. Forma 4:3–12

Zavaleta ES, Hulvey KB (2004) Realistic species losses disproportionately reduce grassland resistance to biological invaders. Science 306:1175–1177

Zavaleta ES, Hobbs R, Mooney H (2001) Viewing invasive species removal in a whole-ecosystem context. Trends Ecol Evol 16:454–459

Ziska LH, Caulfield FA (2000) Rising CO_2 and pollen production of common ragweed (*Ambrosia artemisiifolia*), a known allergy-inducing species: implications for public health. Aust J Plant Physiol 27:893–898

The manufacturer's authorised representative in the EU is Springer
Nature Customer Service Centre GmbH, Europaplatz 3, 69115 Heidelberg,
Germany. If you have any concerns regarding our products, please
contact ProductSafety@springernature.com

Printed and bound by CPI Group (UK) Ltd, Croydon, CR0 4YY
22/09/2026
02220981-0015

Fabián M. Jaksic · Sergio A. Castro

Biological Invasions in the South American Anthropocene

Global Causes and Local Impacts

This book provides a conceptually organized framework to understand the phenomenon of biological invasions at the Anthropocene global scale. Most advances toward that aim have been provided from North American and European researchers, with fewer contributions from Australia and South Africa. Here we fill the void from the Neotropics, focusing on the research experience in South American countries, with a strong emphasis on Argentina and Chile.

The text is divided into two parts: the first half comprises self-contained chapters, providing a conceptual and bibliographic foundation in the field of invasion biology, from an Anthropocene perspective. The second half reviews the ecology, biogeography, and local impacts in South America of exotic species groups (European rabbit, Eurasian wild boar, Canadian beaver, North American mink, and Holarctic freshwater fishes), which are shown to be useful models for case studies of global relevance.

ISBN 978-3-030-56381-3

9 783030 563813